T0235844

**Statistics and Computing**

Statistics and Computing (SC) includes monographs and advanced texts on statistical computing and statistical packages.

Changquan Huang • Alla Petukhina

# Applied Time Series Analysis and Forecasting with Python

Changquan Huang
School of Economics
Xiamen University
Xiamen, China

Alla Petukhina
School of Computing, Communication and Business
HTW Berlin
Berlin, Germany

ISSN 1431-8784 ISSN 2197-1706 (electronic)
Statistics and Computing
ISBN 978-3-031-13586-6 ISBN 978-3-031-13584-2 (eBook)
https://doi.org/10.1007/978-3-031-13584-2

Mathematics Subject Classification: 62-01, 62M10, 62-04, 62-07, 62M20, 62M45, 62F03, 62F10



This Springer imprint is published by the registered company Springer Nature Switzerland AG
The registered company address is: Gewerbestrasse 11, 6330 Cham, Switzerland

# Preface

It is a data-rich era, and data exist almost everywhere. One of the greatest challenges nowadays is how to deal with various kinds of data. It is well known that time series data are the most common data type. And thus methods and techniques for analyzing and forecasting time series have become one of the indispensable tools to handle real-world data problems. Part of this book is rightly concerned with these methods and techniques. It will introduce a wide range of time series models and approaches to building adequate models. Another part is about the general-purpose programming language Python. Python's history is relatively short, but its popularity has been rising steadily. In recent years, Python has been continually taking leading position in solving data-scientific problems and artificial intelligence challenges. The book will show you how to use Python and its extension packages to implement time series analysis and forecasting. Therefore, it is an organic combination of the principle of time series analysis and Python programming.

This book has grown out of a course in time series analysis that Changquan Huang has been teaching at Xiamen University since 2003. More than 18 years of experience in teaching the time series analysis course has made him realize and understand the difficulties of students taking this course. For this reason, during the course of writing the book, he has always been doing his best to let the book be reader friendly and interesting in the hope that the reader can grasp the essence of time series analysis thoroughly and quickly.

The book is intended for an undergraduate and graduate audience as well as for everyone interested in time series analysis and forecasting with Python. To understand the book, only a prerequisite knowledge in probability theory and statistics is needed, which is equivalent to an undergraduate's probability and statistics course for two semesters. Besides, a knowledge of linear (matrix) algebra is helpful in better understanding Chaps. 7–9 of the book.

Changquan Huang wrote every chapter of this book as well as the Python code, and is responsible for the whole book. Alla Petukhina validated the Python code and created Quantlets, and the code of numerical examples has been indicated with a small sign Q. We believe that these publicly available Quantlets on www.quantlet.

com and https://github.com/QuantLet/pyTSA/ create a valuable contribution to the distribution of knowledge in statistical science. We welcome all readers of this book to propose changes to our existing codes or add codes in other programming languages. A free online companion course to the book developed together with Professor Wolfgang Karl Härdle is available through https://quantinar.com.

Our thanks go to Guido van Rossum, the Python Software Foundation, and all the open-source Python package developers for making Python applications in various fields possible. In particular, our thanks go to Kevin Sheppard from University of Oxford for answering our consultation about his excellent Python package `arch` and to the anonymous referee for suggestions. We also thank Veronika Rosteck, Daniel Ignatius Jagadisan, and the Springer team for their support and patience.

Alla Petukhina would like to thank Prof. Wolfgang Karl Härdle and all colleagues from IRTG 1792 at Humboldt University of Berlin for their help and valuable advice, as well as her parents, sister, and son for unconditional support and inspiration.

Changquan Huang would like to thank Prof. Yang Xiangqun, Prof. Wang Zikun, and Prof. GU Ming-gao for mentoring him during the course of his postgraduate study. He particularly thanks Prof. Wolfgang Karl Härdle for inviting him to visit Humboldt University of Berlin. He is grateful to Jiancheng Jiang from University of North Carolina at Charlotte, Hongtu Zhu from University of North Carolina at Chapel Hill, and Xin Yuan Song from The Chinese University of Hong Kong for their encouragement. Finally, he is indebted to his family for their love and understanding.

Xiamen, China
Berlin, Germany
May 2022

Changquan Huang
Alla Petukhina

# Contents


| | | |
|---|---|---|
| **1** | **Time Series Concepts and Python** | 1 |
| 1.1 | The Concept of Time Series | 1 |
| 1.1.1 | What Is Time Series | 1 |
| 1.1.2 | Brief History of Time Series Analysis | 4 |
| 1.1.3 | Objectives of Time Series Analysis | 6 |
| 1.2 | The Programming Language Python | 6 |
| 1.2.1 | Introduction and Installing | 7 |
| 1.2.2 | Demonstrations | 7 |
| 1.2.3 | Python Extension Packages and Some Usages | 11 |
| 1.3 | Time Series Moment Functions and Stationarity | 15 |
| 1.3.1 | Moment Functions | 15 |
| 1.3.2 | Stationarity and Ergodicity | 16 |
| 1.3.3 | Sample Autocorrelation Function | 18 |
| 1.3.4 | White Noise and Random Walk | 21 |
| 1.4 | Time Series Data Visualization | 29 |
| | Problems | 34 |
| **2** | **Exploratory Time Series Data Analysis** | 37 |
| 2.1 | Partial Autocorrelation Functions | 37 |
| 2.1.1 | Definition of PACF | 37 |
| 2.1.2 | Sample PACF and PACF Plot | 39 |
| 2.2 | White Noise Test | 42 |
| 2.3 | Simple Time Series Compositions | 47 |
| 2.4 | Time Series Decomposition and Smoothing | 53 |
| 2.4.1 | Deterministic Components and Decomposition Models | 53 |
| 2.4.2 | Decomposition and Smoothing Methods | 58 |
| 2.4.3 | Example | 61 |
| | Problems | 68 |

**3 Stationary Time Series Models** .... 71
3.1 Backshift Operator, Differencing, and Stationarity Test .... 71
3.1.1 Backshift Operator .... 71
3.1.2 Differencing and Stationarity .... 72
3.1.3 KPSS Stationarity Test .... 73
3.2 Moving Average Models .... 80
3.2.1 Definition of Moving Average Models .... 80
3.2.2 Properties of MA Models .... 84
3.2.3 Invertibility .... 85
3.3 Autoregressive Models .... 88
3.3.1 Definition of Autoregressive Models .... 88
3.3.2 Durbin-Levinson Recursion Algorithm .... 90
3.3.3 Properties of Autoregressive Models .... 92
3.3.4 Stationarity and Causality of AR Models .... 94
3.4 Autoregressive Moving Average Models .... 98
3.4.1 Definitions .... 98
3.4.2 Properties of ARMA Models .... 100
Problems .... 105

**4 ARMA and ARIMA Modeling and Forecasting** .... 107
4.1 Model Building Problems .... 107
4.2 Estimation Methods .... 108
4.2.1 The Innovations Algorithm .... 109
4.2.2 Method of Moments .... 110
4.2.3 Method of Conditional Least Squares .... 111
4.2.4 Method of Maximum Likelihood .... 113
4.3 Order Determination .... 115
4.4 Diagnosis of Models .... 116
4.5 Forecasting .... 118
4.6 Examples .... 119
Problems .... 142

**5 Nonstationary Time Series Models** .... 143
5.1 The Box-Jenkins Method .... 143
5.1.1 Seasonal Differencing .... 143
5.1.2 SARIMA Models .... 147
5.2 SARIMA Model Building .... 155
5.2.1 General Idea .... 155
5.2.2 Case Studies .... 156
5.3 REGARMA Models .... 165
Problems .... 174

**6 Financial Time Series and Related Models** .... 177
6.1 Stylized Facts of Financial Time Series .... 177
6.1.1 Examples of Return Series .... 177
6.1.2 Stylized Facts of Financial Time Series .... 182

6.2 GARCH Models ... 183
6.2.1 ARCH Models ... 183
6.2.2 GARCH Models ... 185
6.2.3 Estimation and Testing ... 188
6.2.4 Examples ... 190
6.3 Other Extensions ... 204
6.3.1 EGARCH Models ... 204
6.3.2 TGARCH Models ... 205
6.3.3 An Example ... 205
Problems ... 212

**7 Multivariate Time Series Analysis** ... 215
7.1 Basic Concepts ... 215
7.1.1 Covariance and Correlation Matrix Functions ... 215
7.1.2 Stationarity and Vector White Noise ... 217
7.1.3 Sample Covariance and Correlation Matrices ... 219
7.1.4 Multivariate Portmanteau Test ... 220
7.2 VARMA Models ... 226
7.2.1 Definitions ... 227
7.2.2 Properties ... 229
7.3 VAR Model Building and Analysis ... 233
7.3.1 VAR(1) Representation of VARMA Processes ... 233
7.3.2 VAR Model Building Steps ... 233
7.3.3 Granger Causality ... 235
7.3.4 Impulse Response Analysis ... 236
7.4 Examples ... 237
Problems ... 255

**8 State Space Models and Markov Switching Models** ... 257
8.1 State Space Models and Representations ... 257
8.1.1 State Space Models ... 258
8.1.2 State Space Representations of Time Series ... 259
8.2 Kalman Recursions ... 261
8.3 Local-Level Model and SARIMAX Models ... 263
8.3.1 Local-Level Model ... 263
8.3.2 SARIMAX Models ... 265
8.4 Markov Switching Models ... 271
8.4.1 Definitions ... 271
8.4.2 Examples ... 273
Problems ... 285

**9 Nonstationarity and Cointegrations** ... 287
9.1 Stochastic Trend and Stochastic Seasonality ... 287
9.1.1 Deterministic Trend and Stochastic Trend ... 287
9.1.2 Deterministic Seasonality and Stochastic Seasonality ... 293

9.2 Brownian Motions and Simulation ... 302
9.2.1 Probability Space ... 302
9.2.2 Brownian Motions ... 304
9.3 Stationarity, Nonstationarity, and Unit Root Tests ... 306
9.3.1 Trend Stationarity and Difference Stationarity ... 306
9.3.2 Unit Root Tests ... 308
9.3.3 Stationarity Tests ... 316
9.4 Cointegrations and Granger's Representation Theorem ... 318
9.4.1 Spurious Regressions and $I(d)$ Processes ... 318
9.4.2 Cointegrations ... 322
9.4.3 Granger's Representation Theorem ... 324
9.4.4 Estimation of Vector Error Correction Models ... 328
9.4.5 Real Case of Spurious Regression and Noncointegration ... 334
Problems ... 339

**10 Modern Machine Learning Methods for Time Series Analysis** ... 341
10.1 Introduction ... 341
10.1.1 Brief History of Artificial Intelligence ... 341
10.1.2 AI in Time Series Analysis ... 343
10.2 Artificial Neural Networks ... 344
10.2.1 Artificial Neural Network Developments ... 344
10.2.2 Neural Network Models ... 347
10.3 Deep Learning and Backpropagation Algorithms ... 350
10.3.1 What Is Deep Learning? ... 350
10.3.2 Gradient Descent and Backpropagation Algorithms ... 350
10.4 Time Series Forecasting and TensorFlow ... 351
10.4.1 Time Series Forecasting ... 351
10.4.2 TensorFlow and Keras ... 351
10.5 Implementation and Example ... 352
10.5.1 Implementation Steps ... 352
10.5.2 An Example ... 353
10.6 Concluding Remarks ... 360
Problems ... 360

**References** ... 363

**Index** ... 369

# Chapter 1
# Time Series Concepts and Python

Well begun is half done. In this chapter, we will understand the concept of time series by observing some real-life examples of time series and learn about brief history and objectives of time series analysis. We introduce the programming language Python and its extension packages and demonstrate some useful usages in the field of time series. We also introduce the concept of stationarity and two important time series models: white noise and random walk. At last, we discuss different ways for visualization of time series data with Python so as to further check time series.

## 1.1 The Concept of Time Series

Properly understanding the concept of time series is prerequisite for any time series analysis. In this section, in order to better understand the concept of time series, we observe three real examples of time series by showing their time series plots and then give the definition of time series. We also present brief history and objectives of time series analysis.

### *1.1.1 What Is Time Series*

It is a big data era nowadays. The world is full of various kinds of data. And time series are one of the most common data type. There is a great deal of time series data in both everyday life and fields of sciences and technology. However, what is a time series? Roughly speaking, the time series is a sequence of data from a natural or social process (or a trial) observed over time. So it is "time" ordered, and we must not interchange the positions of any two values of the time series. There are

 
C. Huang, A. Petukhina, *Applied Time Series Analysis and Forecasting with Python*, Statistics and Computing,
https://doi.org/10.1007/978-3-031-13584-2_1

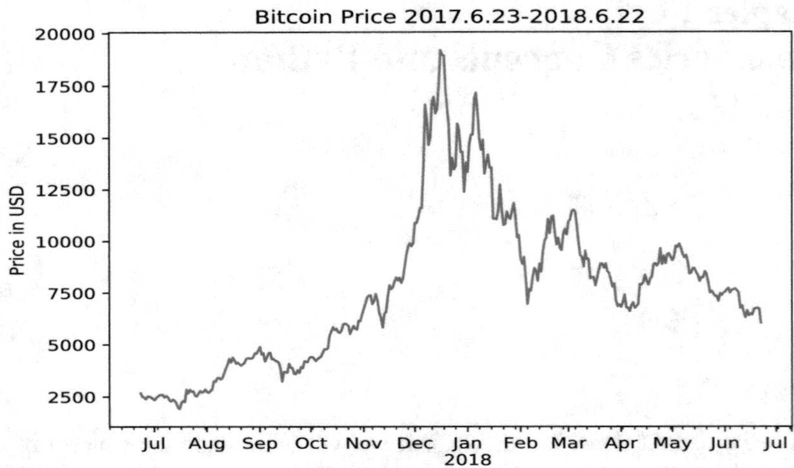


**Fig. 1.1** Bitcoin price time series plot from June 23, 2017, to June 22, 2018

pyTSA_BTC

a lot of time series examples in our everyday life. For instance, your body weight measured on every Sunday forms a time series; the daily closing price of a listed company stock also shapes a time series. The first step in any time series analysis is always to plot the recorded time series data and carefully examine this graph. In a coordinate system, let the index of time $t$ be on the horizontal axis and the observed value on the vertical axis. Thus the time series data are plotted against time. Such a graph is known as a *time series plot* or *time plot*. It is easy to plot but very useful in analyzing time series. It displays the evolution pattern of a time series. The pattern includes the trend and/or seasonality (period) of the time series. Below are a few examples of time series.

*Example 1.1 (Bitcoin Price Time Series)* In recent years, digital currencies are hot and Bitcoin is the most famous. From June 23, 2017, to June 22, 2018, the Bitcoin price was recorded at the end of every day, and the recorded data forms the Bitcoin price time series. Now we can use Python to plot it (in the next sections, we will learn how to plot it with Python). Figure 1.1 is the Bitcoin price time series plot. From the plot, we observe that the Bitcoin price dramatically increases from October 2017 and reaches the peak at the end of 2017 (actually, the peak is 19187.0 USD reached in December 16, 2017). And since then, the price fluctuantly decreases. This time series plot clearly reflects some people's craziness about Bitcoin.[1]

[1] On November 9, 2021, the highest price of Bitcoin had been beyond 67331 USD.

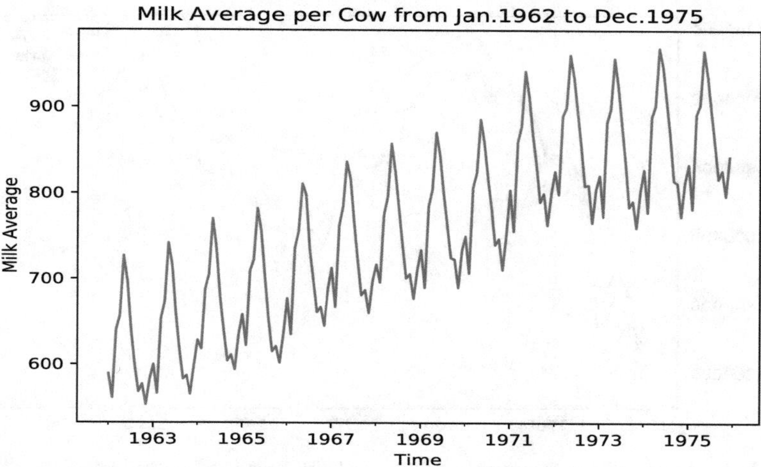


**Fig. 1.2** Time series plot of monthly milk average per cow

pyTSA_MilkAv

*Example 1.2 (Milk Average per Cow)* A dairy cattle ranch monthly recorded milk average per cow from January 1962 to December 1975. Figure 1.2 is the time series plot of the data. From this graph, we clearly see that the time series possesses a linear ascending trend, and so we say that this time series has a deterministic increasing trend (component). At the same time, we also see that the time series has a period of 12 months, and thus we say that this time series possesses deterministic seasonality (component). Besides, the milk production is high at the middle of year and low at the beginning and end of year.

*Example 1.3 (Numbers of Boys and Girls Born in the United States)* Figure 1.3 is the time series plot of the numbers of boys and girls born in the United States per year from 1940 to 2002. From the plot, we observe that the numbers of boys and girls born in the United States reached the highest in the early 1960s and decreased fast in the 1970s and then rapidly increased again in the 1980s. Interesting is that the yearly numbers of boys are always larger than the yearly numbers of girls. Maybe it is because the female lifetime is averagely longer than the male and, in order to balance the male and female in the world, nature arranges such a kind of mechanism and intentionally does so.

In the examples above, we see that a time series is a sequence of data in time order. However, we must realize that we cannot obtain the time series value at time $t$ before time $t$ arrives. At the same time, the time series value is influenced by a lot of random factors. As a result, the time series value at time $t$ is random. For example, we cannot know the number of boys born in 2001 in the United States until the year of 2001 ends. In addition, we often use notation such as $X_t$ to denote

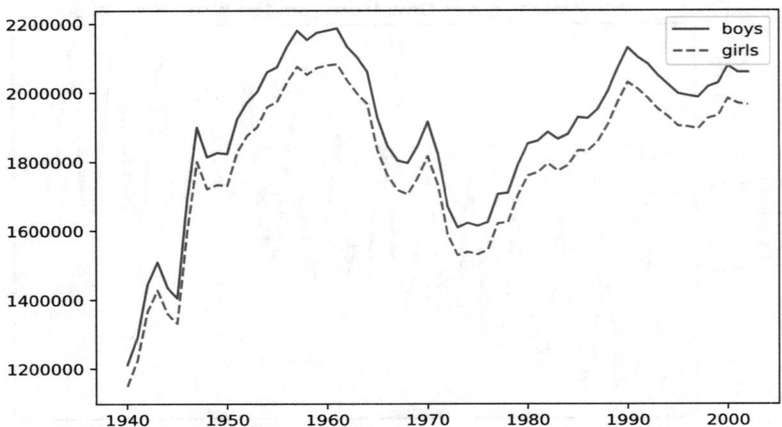


**Fig. 1.3** Time series plot of the numbers of boys and girls born in the United States per year
pyTSA_USBoysGirls

the time series variable at time $t$. That is to say, the value we obtain when time $t$ arrives is a realization (or sample or observation) of the random variable $X_t$, and it can be denoted by $x_t$. Now we formally give the definition below.

**Definition 1.1** An ordered sequence of random variables $\{X_t\}$ is called a time series where $t$ is time index and $t \in T = \{\cdots, -n, \cdots, -1, 0, 1, \cdots, n, \cdots\}$. The time series values which have been recorded are called a realization (or sample or observation) of the time series.

Obviously the time series is in order, and sometimes it is also referred as to a *(discrete) stochastic process* (see Sect. 9.2). In clear context, a realization (or sample or observation) of the time series can also be simply called a time series and often denoted by $\{x_{1:n}\} = \{x_t; t = 1, 2, \cdots, n\}$ since the number of observations is always finite. Please note that we sometimes use $\{a : b\}$ to represent $\{a, a+1, \cdots, b\}$ for any integers $a \leq b$ and $x_{1:n} = \{x_{1:n}\}$ to represent $\{x_1, x_2, \cdots, x_n\} = \{x_t; t = 1, 2, \cdots, n\}$.

### *1.1.2 Brief History of Time Series Analysis*

As early as 28 BC during the Western Han Dynasty in China, there exists the recording of sunspots (e.g., see Needham (1959) and Tong (1990)). Nowadays we can obtain the annual mean total sunspot number from 1700 to 2017 from the Royal Observatory of Belgium, Brussels. It forms a time series of length 318 years. Figure 1.4 is its time series plot, and we see that there is a some kind of period but no trend in it. In other words, the evolution pattern of the sunspot number time series has hardly changed since 1700.

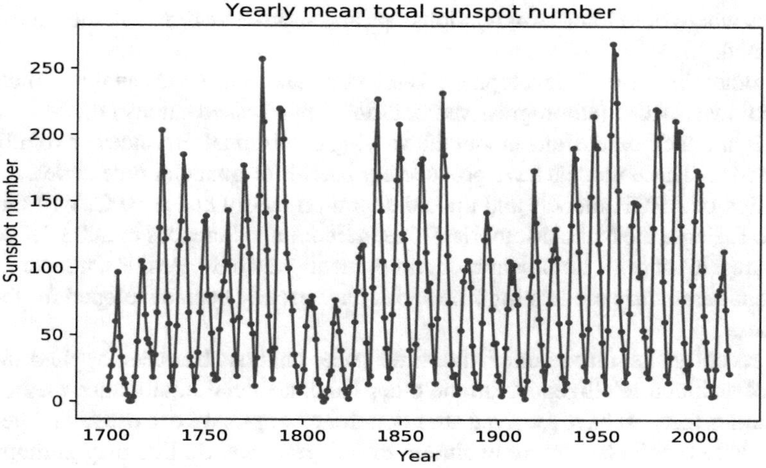


**Fig. 1.4** Time plot of the yearly mean total sunspot number from 1700 to 2017

pyTSA_Sunspot

The theoretical developments in time series analysis started in the 1920s and 1930s when the strict basics of probability theory was being constructed by the Russian mathematician A. N. Kolmogorov. During that time, the Russian statistician E. Slutsky and the British statisticians G. U. Yule and J. Walker introduced MA (moving average) models and AR (autoregressive) models to represent time series and used the moving average method to remove seasonal effect in a time series. See, for example, Slutsky (1927) and Yule (1927). And then H. Wold developed ARMA (autoregressive moving average) models for stationary time series. In 1970, the classic book *Time Series Analysis: Forecasting and Control* by G. E. P. Box and G. M. Jenkins was published, which contains the full modeling procedure for a (maybe nonstationary) time series: identification, estimation, diagnostics, and forecasting.[2] Today, the Box-Jenkins models (or procedure) are perhaps the most commonly used, and many methods for forecasting and seasonal adjustment can be traced back to these models.

One direction of continuing development in time series analysis is multivariate ARMA models, among which VAR (vector autoregressive) models have especially become popular. The pity is that these techniques are only applicable for stationary time series. However, real-life time series often possess an ascending trend which suggests nonstationarity (viz., a unit root). In the 1980s and 1990s, a few of important tests for unit roots developed. On the other hand, it was found that nonstationary multivariate time series could have a common unit root. These time series are known as cointegrated time series and can be used in the error correction

[2] We notice that the fifth edition of this book also came out in 2016.

models where both long-term relationships and short-term dynamic mechanism are estimated.

Another direction of development is the nonlinear time series analysis, including well-known ARCH (autoregressive conditional heteroskedasticity) models. ARCH models and their extensions model the varying conditional variances or volatility of time series. These models have proved very useful for financial time series. It is the invention of ARCH models and the cointegration theory that gave C. W. J Granger and R. F. Engle the Nobel Memorial Prize in Economic Sciences in 2003. Of course, there are a lot of other nonlinear models not mentioned here. Besides, nonparametric and semiparametric models for time series have greatly been developed for the past decades.

In recent years, on the one hand, time series analysis has been applied into the field of artificial intelligence. On the other hand, artificial intelligence methods for time series forecast have received attention, for example, we can use neural network methods to predict the future evolution of a time series. So this may perhaps be a new direction for investigating time series.

### *1.1.3 Objectives of Time Series Analysis*

The objectives of time series analysis rely on the background of applications. However, a common objective is to use descriptive statistics methods to analyze time series data, where the time series plot is especially useful. From a time series plot, we should examine whether the time series has (1) a deterministic trend; (2) deterministic seasonality; (3) dramatic change in its behavior pattern; and (4) outliers and/or anomalies. On the other hand, in the fields of econometrics, quantitative finance, seismology, meteorology, and geophysics, the most important objective of time series analysis is forecasting time series future evolution; in the fields of signal processing and communication engineering, time series analysis is used especially for signal detection and estimation; in the fields of data mining, pattern recognition, and machine learning, time series analysis can be used for clustering, classification, anomaly detection, as well as forecasting.

## 1.2 The Programming Language Python

In this section, first, we briefly introduce the programming language Python and how to install it. Second, we demonstrate some important Python usage. Third, we discuss how to install Python extension packages. At last, we illustrate several Python extension packages frequently used in data analysis and scientific computing.

### 1.2.1 Introduction and Installing

Python is an interpreted high-level language for general-purpose programming. It was first created by Mr. Guido van Rossum from the Netherlands (now he lives in the United States) and released for people to free use in 1991. Presently it is run by the Python Software Foundation, which is a non-profit organization created specifically to own Python-related Intellectual Property. Python has a simple but effective approach to object-oriented programming and design philosophy that emphasizes code readability, notably using significant whitespace. It provides architecture that enables clearly programming on both small and large scales. It has a large and comprehensive standard library that supports many common programming tasks such as connecting to web servers, searching text with regular expressions, and reading and modifying files. It is easily extended by adding new modules implemented in a compiled language such as C or C++. Python has a variety of basic data types available: numbers (floating point, complex, and unlimited-length long integers), strings (both ASCII and Unicode), lists, and dictionaries. It also supports raising and catching exceptions, resulting in cleaner error handling. Python's automatic memory management frees you from having to manually allocate and free memory in your code. In recent years, Python has been continuing to take leading positions in solving big data science tasks and artificial intelligence challenges.

The Python interpreter and the standard library are freely available in source or binary form for all major operational systems (OS) from the Python website: https://www.python.org/. According to the OS of your computer, you can download the corresponding Python interpreter to your computer and easily install it. Since the standard library is distributed with the Python interpreter, it is installed together with Python. In this book, we install Python-3.6.5-amd64 for Windows 10, and all the codes are run on this version of Python. We do not intend to totally introduce the Python syntax and semantics about which you can refer to the documentations in the Python website and related books. We only demonstrate some usages of Python, which are often used in the book. At present, there are many so-called enhanced or integrated versions of Python. If you are a new learner, we do not recommend any of them because using the plain Python, we can better understand it. If you have some experience with Python, you can use any of Python distributions that you are used to. And Anaconda is perhaps your favorite platform.

### 1.2.2 Demonstrations

After installing Python well, we can start up it by pressing the IDLE button from the Start menu and get a prompt similar to the following:

```
Python 3.6.5 (v3.6.5:f59c0932b4, Mar 28 2018, 17:00:18)
[MSC v.1900 64 bit (AMD64)] on win32
Type "copyright", "credits" or "license()" for more information.
>>>
```

The exact appearance of the Python interpreter will depend on which version you are using. This window is the console of Python where we type Python commands. The interpreter reads and executes the commands interactively. In order to see if it works, let us try the following command:

```
>>> print("I love Python!")
```

When we press the Enter key to run the command, the following output appears:

```
I love Python!
```

From Python 3.0 and later, the command `print` is used as a function, and the syntax becomes

```
print("string" or variable_name)
```

Let us see the following examples:

```
>>> str_variable = "so far so good"
>>> print(str_variable)
so far so good
>>> int_variable = 9876
>>> print(int_variable)
9876
>>> print(str_variable , "and", int_variable)
so far so good and 9876
```

It is worth noting that as a rule of Python, if you use the equal sign (=) to assign a variable to another, then the two variables actually become the same variable, that is, the change in the one makes the same change in the other. Below are two examples:

```
>>> x=[5,6,8]   #example No.1
>>> y=x
>>> y
[5, 6, 8]
>>> y[0]=9
>>> y
[9, 6, 8]
>>> x
[9, 6, 8]
>>> x[2]=3
>>> y
[9, 6, 3]
>>> a = ['Mary', 'Susan', 'Lamb']  #example No.2
>>> b=a
>>> b[2]='Candy'
>>> a
['Mary', 'Susan', 'Candy']
>>> a[1]='Tom'
>>> b
['Mary', 'Tom', 'Candy']
```

where we see that the two variables $y$ and $x$ are exactly the same and so are the variables $b$ and $a$. Besides, by Python default, a sequence is indexed from 0, not from 1. For example, the variable $x$ is indexed from 0, and if you try to run `x[3]`, there appears an error:

```
>>> x=[5,6,8]
>>> x[0];x[1];x[2]
5
6
8
>>> x[3]
Traceback (most recent call last):
  File "<pyshell#1>", line 1, in <module>
    x[3]
IndexError: list index out of range
```

On the other hand, it accepts negative indices for indexing from the end of sequence:

```
>>> x=[5,6,8]
>>> x[-1];x[-2];x[-3]
8
6
5
>>> x[-4]
Traceback (most recent call last):
  File "<pyshell#1>", line 1, in <module>
    x[-4]
IndexError: list index out of range
```

There is a very useful function `range(a,b,s)` in Python, which generates a sequence of integers. Please note that this sequence belongs to an independent class "range" and in order to turn it into a list we must use the function `list()`. The function `range(a,b,s)` is often used in `for` loops. All the three arguments $a$, $b$, and $s$ are integers and can be positive or negative. If the argument $s$ (called *step* and not equal to zero) is omitted, it defaults to 1. If the start argument $a$ is omitted, it defaults to 0, at which time the only argument is $b$. In addition, the sequence starts with $a$ and is generated by step $s$. If $s$ is positive, the last element is the largest $a + i * s$ less than $b$. if $s$ is negative, the last element is the smallest $a + i * s$ greater than $b$. Let us look at the following examples:

```
>>> x=range(10); y=range(2,10); z=range(1,10, 2)
>>> x; y; z
range(0, 10)
range(2, 10)
range(1, 10, 2)
# the three sequences do not appear.
>>> type(x); type(y); type(z)
<class 'range'>
<class 'range'>
<class 'range'>
>>> list(x); list(y); list(z)
[0, 1, 2, 3, 4, 5, 6, 7, 8, 9]
[2, 3, 4, 5, 6, 7, 8, 9]
[1, 3, 5, 7, 9]
# the three sequences appear and belong to class 'list'.
```

```
>>> for i in range(5,-5,-2):
        print(i)


5
3
1
-1
-3
>>> a = ['Mary', 'Susan', 'Lamb']
>>> for i in range(3):
        print(i, a[i])


0 Mary
1 Susan
2 Lamb
```

A module is a file that contains Python definitions and statements intended for use in other Python programs. There are a lot of built-in modules in the Python standard library. Although the modules are built-in, they must be imported to the Console when being used. Now we introduce several Python built-in modules, which are often used in this book.

- `math` The module provides mathematical functions such as `math.gcd(x,y)`, `math.exp(x)`, `math.sqrt(x)`, `math.sin(x)`, and many others. It also gives important constants such as `math.pi` and `math.e` and so on.

```
>>> import math
>>> math.gcd(186,4); math.exp(3.2); math.sqrt(3)
2
24.532530197109352
1.7320508075688772
>>> math.sin(2); math.pi; math.e
0.909297426825681
3.141592653589793
2.718281828459045
```

- `statistics` The module provides functions for calculating statistics of data. For example, the `statistics.mean()`, `statistics.variance()`, and `statistics.median()`, respectively, calculate the average, variance, and median of data.

```
>>> import statistics as stat
>>> x=[2.75, 1.75, 1.25, 0.25, 0.5, 1.25, 3.5]
>>> stat.mean(x);stat.variance(x);stat.median(x)
1.6071428571428572
1.3720238095238095
1.25
```

- `random` The module implements pseudo-random number generators for various distributions. These distributions are uniform, normal (Gaussian), lognormal, gamma, and beta among others.

```
>>> import random
>>> random.random()
0.9397756353438375
# generating a random float uniformly in [0.0,1.0)
>>> random.uniform(0.1, 2)
0.532676850467863
# generating a random float uniformly in [0.1,2.0)
>>> random.gauss(1, 2.3)
1.1516985733451488
# generating a random float from the Gauss distribution
# with mean 1 and standard deviation 2.3
>>> random.sample(range(100), 10)
[44, 80, 23, 96, 59, 31, 15, 6, 41, 66]
# sampling ten integers from range(100) without replacement
```

- `datetime` The module supplies basic classes for manipulating dates and times. Below are some examples.

```
>>> from datetime import date
>>> nowday = date.today()
>>> nowday
datetime.date(2018, 8, 3)
>>> mybirthday = date(1999, 7, 31)
>>> age = nowday - mybirthday
>>> age.days
6943
>>> import datetime
>>> nowdatetime=datetime.datetime.now()
>>> print(nowdatetime)
2018-08-03 08:48:35.057793
```

### *1.2.3 Python Extension Packages and Some Usages*

There are a large number of the third-party Python extension packages, which are applied in various fields. In this subsection, we introduce a few of them. They are very useful in scientific computing, statistical analysis, data visualization, and so on. First of all, we recommend that you use `pip` to install and manage Python extension packages. Starting with Python 3.4, it is included with the Python installer by default and so has already been installed. Clicking the start menu with the right mouse button, we will see the Windows PowerShell (Administrator) menu. Opening it, we get a prompt like this:

```
PS C:\WINDOWS\system32>
```

If you want to install an extension package from the network, type and run the following command:

```
PS C:\WINDOWS\system32> pip install packagename
```

Here `pip` is a Python package installing and managing tool. As long as the network that you are using is well unobstructed, in general, the installation will be successful. If you have downloaded and stored a package (e.g., the package `pandas`) in your computer, then you can run the command below to install it:

```
PS C:\WINDOWS\system32> pip install J:\extension\pandas-0.23.4
                                        -cp36-cp36m-win_amd64.whl
# "J:\extension\pandas-0.23.4-cp36-cp36m-win_amd64.whl"
# is the path and name for pandas
```

The following is the five Python extension packages that we have installed:

- `Numpy` (version 1.15.4) is a fundamental and general-purpose array-processing package designed to manipulate large multi-dimensional arrays of arbitrary records (such as numbers, strings, objects, and others). It also includes tools for integrating C/C++ and Fortran code; random number generators; discrete Fourier transform; and so forth. Besides, `Numpy` can also be used as an efficient multi-dimensional container of generic data. It possesses the ability to define arbitrary data types. This allows it to seamlessly and speedily integrate with a wide variety of databases. Its homepage is https://www.numpy.org/.
- `Pandas` (version 0.23.4) is a Python extension package providing fast, flexible, and expressive data structures. It is designed to make working with structured (tabular, multi-dimensional, potentially heterogeneous) and time series data. Additionally, it is built on top of `Numpy` and is intended to integrate well within a scientific computing environment with many other third-party libraries. The two primary data structures of `Pandas` are one-dimensional Series and two-dimensional DataFrame. They handle the vast majority of typical use cases in finance, statistics, social science, and many areas of engineering. The homepage of Pandas is https://pandas.pydata.org/.
- `Scipy` (version 1.1.0, pronounced Sigh Pie) is an open-source software for mathematics, science, and engineering. The `Scipy` library depends on `Numpy` and works with `Numpy` arrays. It provides a number of user-friendly and efficient numerical routines for scientific and engineering computing. In the library, for example, there are modules for fast Fourier transform, numerical integration, optimization, signal processing, and others. Its homepage is https://www.scipy.org/.
- `Matplotlib` (version 3.3.3) is a Python 2D plotting library which produces publication quality figures in a variety of hardcopy formats and interactive environments across platforms. Using it we can generate time series plots, histograms, boxplots, bar charts, pie charts, scatter plots, etc., with just a few lines of code. For more details, see Hunter (2007).
- `Statsmodels` (version 0.12.1) is a Python package that provides classes and functions for the estimation of many different statistical models, as well as

for conducting statistical tests, and statistical data exploration. It covers linear regression models, generalized linear models, time series analysis, Markov switching models, survival analysis, multivariate statistics, and nonparametric statistics among others. For more details, refer to Seabold and Perktold (2010).

These five Python extension packages are very useful for big data analysis and artificial intelligence as well as time series analysis. Many other extension packages for big data or artificial intelligence are also dependent on them.

*Example 1.4 (Showing Some Usages of the Above Five Packages)* Draw 1000 random numbers (called simulated samples) from the normal (Gaussian) distribution with mean 4.1 and standard deviation 2.2. Then (1) plot the histogram of the simulated samples; (2) turn the simulated array into a pd.Series (time series), and plot the time series plot.

Below are the Python commands that we execute. Figure 1.5 is the histogram and Fig. 1.6 is the time series plot.

```
>>> import numpy as np
>>> import pandas as pd
>>> import matplotlib.pyplot as plt
>>> np.random.seed(135) # for repeat
>>> x=np.random.normal(loc=4.1, scale=2.2, size=1000)
# where loc=4.1= mean; scale=2.2= standard deviation
>>> type(x)
<class 'numpy.ndarray'>
# x belongs to class 'numpy.ndarray'
>>> h_fig=plt.hist(x, bins=25) #bins=25 is the number of the bins
>>> plt.xlabel('simulated sample'); plt.ylabel('frequency')
>>> plt.show()
>>> xts=pd.Series(x)
>>> type(xts)
<class 'pandas.core.series.Series'>
>>> xts.plot(); plt.xlabel('Time'); plt.ylabel('Simulated sample')
>>> plt.show()
```

The generated time series in Example 1.4 is actually a special white noise series. We will formally define and discuss it in Sect. 1.3.4.

There are a great deal of syntax and usages in the Python ecosystem (viz., Python and all its extension packages). Please note that the same function can sometimes be implemented with different modules or scripts, and we should select a simple and explicit one just as what the zen of Python says (running the command `import this`, you can see the zen ). We will study more and more syntax and usages in this ecosystem throughout the book.

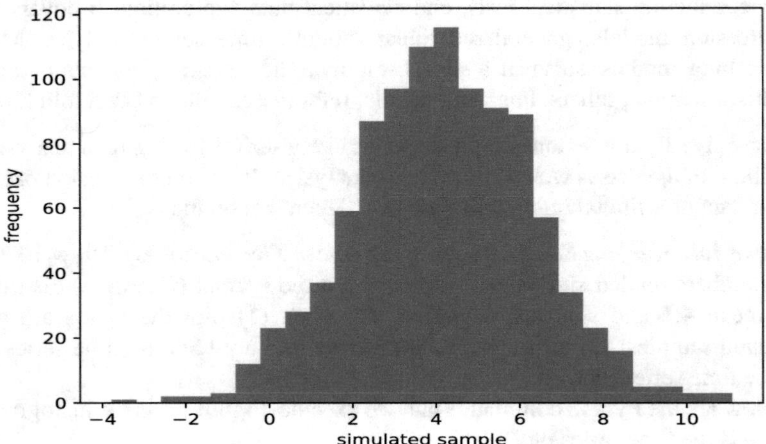


**Fig. 1.5** Histogram of the simulated samples in Example 1.4

pyTSA_SimGauss

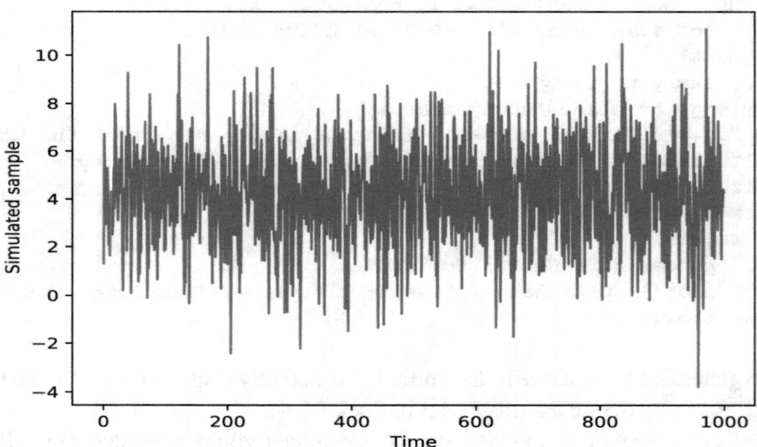


**Fig. 1.6** Time series plot of the simulated samples in Example 1.4

pyTSA_SimGauss

## 1.3 Time Series Moment Functions and Stationarity

In this section, we define and discuss several important concepts about time series analysis such as stationarity, autocorrelation function, and so forth. We also introduce two basic time series models: white noise and random walk.

### *1.3.1 Moment Functions*

In order to completely describe the probabilistic structure of a time series $\{X_t\}$, its finite dimensional distribution family has to be given, that is, for any positive integer $n$, and at arbitrary time points $t_1, t_2, \cdots, t_n$, the following distribution functions are given:

$$F_{t_{1:n}}(x_{1:n}) = F_{t_1,t_2,\cdots,t_n}(x_1, x_2, \cdots, x_n) = \Pr(X_{t_1} \le x_1, X_{t_2} \le x_2, \cdots, X_{t_n} \le x_n) \tag{1.1}$$

where $x_i \in R, i = 1, 2, \cdots, n$. At this point, we say that the time series $\{X_t\}$ follows the given distribution (family). In general, however, it is very difficult to calculate these distribution functions. Fortunately, much of the information in these multivariate distributions can be described in terms of their moments (viz., mean, deviation, and so on).

**Definition 1.2** (1) The mean function (first moment) of a time series is defined as

$$\mu_t = \mathsf{E}(X_t) \tag{1.2}$$

That is, $\mu_t$ is the expected value of the time series at time point $t$. (2) The variance (deviation) function (second central moment) of a time series is defined as

$$\sigma_t^2 = \text{Var}(X_t) = \mathsf{E}[(X_t - \mu_t)^2] \tag{1.3}$$

Furthermore, $\sigma_t = \sqrt{\sigma_t^2}$ is called the standard deviation function.

In general, $\mu_t$ (as well as $\sigma_t^2$) can be different at different time points. In addition, clearly they are natural extensions of the related concepts in classical statistics and have the similar properties.

**Definition 1.3**

(1) The autocovariance function of a time series is defined as

$$\gamma(s, t) = \text{Cov}(X_s, X_t) = \mathsf{E}[(X_s - \mu_s)(X_t - \mu_t)] \tag{1.4}$$

(2) The autocorrelation function of a time series is defined as

$$\rho(s,t) = \text{Corr}(X_s, X_t) = \frac{\gamma(s,t)}{\sigma_s \sigma_t} = \frac{\text{Cov}(X_s, X_t)}{\sqrt{\text{Var}(X_s)\text{Var}(X_t)}} \tag{1.5}$$

Both autocovariance and autocorrelation measure the (linear) correlation between two points $X_s$ and $X_t$ on the same time series, but the latter is dimensionless and easier to use and interpret. The frequently used properties of them are as follows: for any integers $s$ and $t$,

$$\gamma(t,t) = \sigma_t^2;\ \gamma(s,t) = \gamma(t,s);\ |\gamma(s,t)| \le \sigma_s \sigma_t \tag{1.6}$$

$$\rho(t,t) \equiv 1;\ \rho(s,t) = \rho(t,s);\ |\rho(s,t)| \le 1 \tag{1.7}$$

### *1.3.2 Stationarity and Ergodicity*

Now we define the stationarity for time series. If a time series possesses this property, then it is in some type of statistical equilibrium. There are two frequently used kinds of stationarity and the definitions are as follows.

**Definition 1.4** A time series $\{X_t\}$ is strictly stationary or has strict stationarity if $\{X_1, \cdots, X_n\}$ and $\{X_{1+k}, \cdots, X_{n+k}\}$ possess the same joint distribution for any integer $n \ge 1$ and any integer $k$.

Evidently the strict stationarity is a very restrictive condition and is often difficult to verify. Fortunately, we mainly use a weaker form of stationarity in this book and it is relatively easy to test.

**Definition 1.5** A time series $\{X_t\}$ is weakly stationary or has weak stationarity if (1) $\mathsf{E}(X_t) = \mu$ is a constant and (2) for any time $t$, $\mathsf{E}(X_t^2) < \infty$ and $\text{Cov}(X_t, X_{t+k}) = \gamma(k)$ is independent of $t$ for each integer $k$.

Note the following remarks:

- A time series $\{X_t\}$ is said to be *mean stationary* if $\mathsf{E}(X_t) = \mu$ is a constant.
- A time series $\{X_t\}$ is said to be *variance stationary* if $\text{Var}(X_t) = \mathsf{E}[(X_t - \mu_t)^2] = \sigma^2$ is a constant.
- If a time series is both mean stationary and variance stationary, then its time series plot fluctuates around the mean line $y = \mu$.
- Obviously the weak stationarity includes both the mean stationarity and the variance stationarity.
- If a time series has the weak stationarity, then $\gamma_k = \gamma(k) = \text{Cov}(X_t, X_{t+k})$ is called the lag $k$ *autocovariance function* of $\{X_t\}$, and $\rho_k = \rho(k) = \gamma(k)/\gamma(0)$ is called the lag $k$ *autocorrelation function* of $\{X_t\}$. In addition, for any integer $k$, $\gamma(-k) = \gamma(k)$ and $\rho(-k) = \rho(k)$.

- If $\mathsf{E}(X_t^2) < \infty$, then the strict stationarity implies the weak stationarity but not vice versa. In other words, generally speaking, the strict stationarity is more difficult to apply than the weak stationarity.
- In this book, like most textbooks we let the stationarity represent the weak stationarity, and thus we mostly drop the word "weak" or "weakly."

*Example 1.5 (Two Special Stationary Time Series)*

(1) Assume that $\{X_t\}$ is independent and identically distributed and the first and second moments exist. Then it is a stationary time series and its autocorrelation function

$$\rho(k) = \begin{cases} 1, \ k = 0 \\ 0, \ \text{otherwise} \end{cases} \tag{1.8}$$

Such a time series is called a white noise (see Sect. 1.3.4). In addition, according to the law of large numbers, as $n \to \infty$

$$\frac{1}{n}(X_1 + \cdots + X_n) \to \mu \tag{1.9}$$

where $\mu$ is the expected value (mean) of $\{X_t\}$. In other words, we can use the sample mean $\sum_{t=1}^{n} x_t/n$ of $\{X_t\}$ to estimate the mean $\mu$ of $\{X_t\}$.

(2) Suppose that $\{Y_t\}$ is a time series where for all $t$, $Y_t = Y$ and $Y$ is a random variable with $\mathsf{E}(Y) = \nu$ and $\text{Var}(Y) = \sigma^2$. By Definition 1.5, $\{Y_t\}$ is obviously stationary. However, for any integer $n \geq 1$

$$\frac{1}{n}(Y_1 + \cdots + Y_n) = Y \tag{1.10}$$

This implies that the sample mean $\sum_{t=1}^{n} y_t/n$ of $\{Y_t\}$ cannot be used to estimate the mean $\nu$ of $\{Y_t\}$ even though $\{Y_t\}$ is stationary.

To prevent situations like Example 1.5 (2), we introduce the following definition:

**Definition 1.6** Suppose that $\{X_t\}$ is a stationary time series with the mean $\mu$ and the variance $\sigma^2$.

(1) $\{X_t\}$ is said to be mean ergodic if

$$\lim_{n\to\infty} \mathsf{E}[(\frac{1}{n}\sum_{t=1}^{n} X_t - \mu)^2] = 0 \tag{1.11}$$

(2) $\{X_t\}$ is said to be variance ergodic if

$$\lim_{n\to\infty} \mathsf{E}[\frac{1}{n}\sum_{t=1}^{n}(X_t-\mu)^2-\sigma^2]^2=0 \tag{1.12}$$

(3) $\{X_t\}$ is said to be ergodic if it is both mean ergodic and variance ergodic.

It is evident that time series can often be observed only one time during a certain period and a time series is usually correlated at different time points. So we have to put some conditions on time series in order to estimate its underlying parameters. From now on, we assume that all the stationary time series studied in this book are ergodic. Therefore for any stationary time series with the mean $\mu$ and the variance $\sigma^2$, we can well estimate $\mu$ and $\sigma^2$ by using sufficiently long sample $\{x_t;\ 1\le t\le n\}$.

### *1.3.3 Sample Autocorrelation Function*

In practice, $\mu$, $\gamma(k)$, and $\rho(k)$ are all unknown and have to be estimated from the time series data. This leads to the following definition:

**Definition 1.7** Let $\{x_t;\ 1\le t\le n\}$ be a time series sample of size $n$ from $\{X_t\}$.

(1) $\bar{x}=\sum_{t=1}^{n}x_t/n$ is called the sample mean of $\{X_t\}$.
(2) $c_k=\sum_{t=1}^{n-k}(x_{t+k}-\bar{x})(x_t-\bar{x})/n$ is known as the sample autocovariance function of $\{X_t\}$.
(3) $r_k=c_k/c_0$ is said to be the sample autocorrelation function of $\{X_t\}$.

Note the following remarks about this definition:

- Like most literature, this book uses ACF to denote the sample autocorrelation function as well as the autocorrelation function. What is denoted by ACF can easily be identified in context.
- Clearly $c_0$ is the sample variance of $\{X_t\}$. Besides, $r_0=c_0/c_0=1$ and for any integer $k$, $|r_k|\le 1$.
- When we compute the ACF of any sample series with a fixed length $n$, we cannot put too much confidence in the values of $r_k$ for large $k$'s, since fewer pairs of $(x_{t+k}, x_t)$ are available for calculating $r_k$ as $k$ is large. One rule of thumb is not to estimate $r_k$ for $k>n/3$ (see Chan 2010, pp.19–20), and another is $n\ge 50$, $k\le n/4$ (see Box et al. 2016, p31). In any case, it is always a good idea to be careful.
- We also compute the ACF of a nonstationary time series sample by Definition 1.7. In this case, however, the ACF or $r_k$ very slowly or hardly tapers off as $k$ increases.
- Plotting the ACF ($r_k$) against lag $k$ is easy but very helpful in analyzing time series sample. Such an ACF plot is known as a *correlogram*.

- If $\{X_t\}$ is stationary with $\mathsf{E}(X_t) = 0$ and $\rho_k = 0$ for all $k \neq 0$, that is, it is a white noise series (see next subsection), then the sampling distribution of $r_k$ is asymptotically normal with the mean 0 and the variance of $1/n$. Hence, there is about 95% chance that $r_k$ falls in the interval $[-1.96/\sqrt{n}, 1.96/\sqrt{n}]$.

Now we can give a summary that (1) if the time series plot of a time series clearly shows a trend or/and seasonality, it is surely nonstationary; (2) if the ACF $r_k$ very slowly or hardly tapers off as lag $k$ increases, the time series should also be nonstationary.

Note that from now on we often use the Python package `PythonTsa`, which is specially prepared for this book and contains several important Python functions for analyzing time series and most datasets analyzed in the book. The reader can install it using `pip` like other Python packages. The data folder Ptsadata is also in the `PythonTsa` package and from the folder we can get the dataset that we want to analyze. An example of the related Python code can be seen in Example 1.6. Besides, the function `acf_pacf_fig` in `PythonTsa` is a command for plotting ACF and PACF. The argument `both` in the `acf_pacf_fig` is Boolean: when `both=True`, plot both ACF and PACF; when `both=False`, plot ACF only.

*Example 1.6 (Chinese Quarterly GDP)* The dataset "gdpquarterlychina1992.1-2017.4" in the folder Ptsadata is the Chinese quarterly GDP from 1992 to 2017. Now we are producing its time series plot and correlogram with Python. The Python commands are as follows:

```
>>> import pandas as pd
>>> import numpy as np
>>> from PythonTsa.datadir import getdtapath
>>> dtapath=getdtapath()
>>> x=pd.read_csv(dtapath + 'gdpquarterlychina1992.1-2017.4.
                  csv',header=0)
>>> dates = pd.date_range(start='1992',periods=len(x),freq='Q')
>>> x.index=dates
>>> import matplotlib.pyplot as plt
>>> x.plot(); plt.title('Chinese Quarterly GDP 1992-2017')
>>> plt.ylabel('billions of RMB')
>>> plt.show()
>>> from PythonTsa.plot_acf_pacf import acf_pacf_fig
>>> acf_pacf_fig(x, both=False, lag=60)  #plotting ACF
>>> plt.show()
```

Figure 1.7 is the time series plot of the Chinese quarterly GDP. It clearly shows an increasing trend and seasonal effects, and so this time series is nonstationary. Figure 1.8 is the correlogram (ACF plot), and we see that the ACF values very slowly fall into the 95% confidence (interval) band and then continue to fall out of the confidence band.

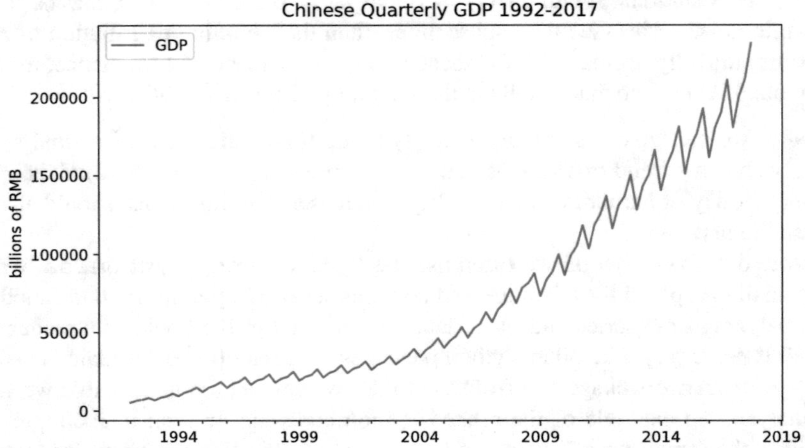


**Fig. 1.7** Time series plot of Chinese quarterly GDP from 1992 to 2017

pyTSA_GDPChina

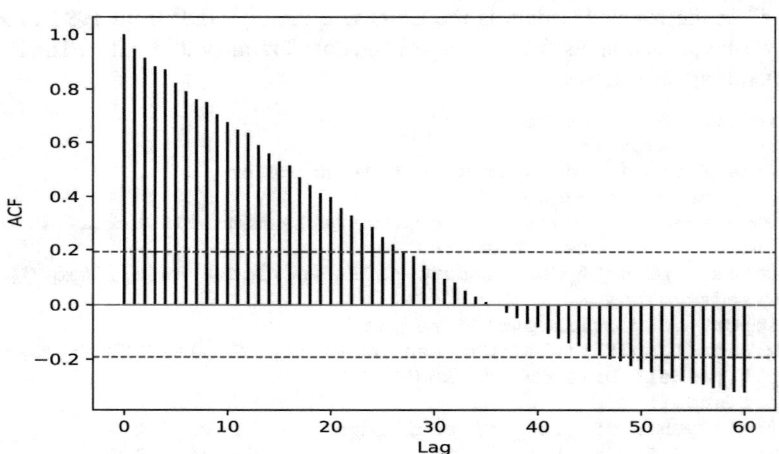


**Fig. 1.8** Correlogram (ACF plot) of Chinese quarterly GDP from 1992 to 2017

pyTSA_GDPChina

### 1.3.4 White Noise and Random Walk

The white noise acts as a building block of time series and is vitally important in time series analysis. We formally define it below.

**Definition 1.8** A time series $\{W_t\}$ is called a white noise series or purely random series if it satisfies the following conditions: (1) for all $t$, $\mathsf{E}(W_t) = \mu$ is a constant; (2) for all $t$, $\text{Var}(W_t) = \sigma_w^2$ is a constant; (3) it is uncorrelated at different time points, that is, when $t \neq s$, $\text{Cov}(W_t, W_s) = 0$.

In time series models, the white noise reflects information that is not observable and is sometimes called an *innovation term (series)*. In various engineering applications, the white noise is used as a model for noise and a random signal having equal intensity at different frequencies. We denote it as $W_t \sim \text{WN}(\mu, \sigma_w^2)$. Sometimes we further require the white noise series to be independent and identically distributed and distinguish this by writing $W_t \sim \text{iid}(\mu, \sigma_w^2)$. What is more, if the distribution is normal (Gaussian), we denote the white noise as $W_t \sim \text{iidN}(\mu, \sigma_w^2)$. Without loss of generality, we usually assume that the mean $\mathsf{E}(W_t) = \mu = 0$. Explicitly $\{W_t\}$ is stationary, and the autocorrelation of $\{W_t\}$ possesses the following important property:

$$\rho(k) = \begin{cases} 1, & k = 0 \\ 0, & \text{otherwise} \end{cases} \tag{1.13}$$

*Example 1.7 (Simulating a Gaussian White Noise)* Simulate a Gaussian white noise with the mean 0 and the variance 1, and then produce its time series plot and correlogram. The Python code is below:

```
>>> from numpy import random
>>> import pandas as pd
>>> random.seed(135) # for repeat
>>> x=random.normal(loc=0, scale=1, size=1000)
>>> xts=pd.Series(x)
>>> import matplotlib.pyplot as plt
>>> xts.plot(); plt.xlabel('Time')
>>> plt.ylabel('Simulated white noise'); plt.show()
>>> from PythonTsa.plot_acf_pacf import acf_pacf_fig
>>> acf_pacf_fig(xts, both=False, lag=30)  # plotting ACF
>>> plt.show()
```

Figure 1.9 is the time series plot of the simulated Gaussian white noise, and we see that it always fluctuates around the horizontal line $y = 0$. Figure 1.10 is the correlogram of the simulated Gaussian white noise. It displays the characteristics of a white noise's autocorrelation function, namely, $\rho(k) = 0$ for all $k \neq 0$. Thus, there is an intuitive method for testing that a stationary time series is a white noise or not: to examine its ACF plot and if the ACF plot is similar to Fig. 1.10, then we are apt to think that the time series is a white noise.

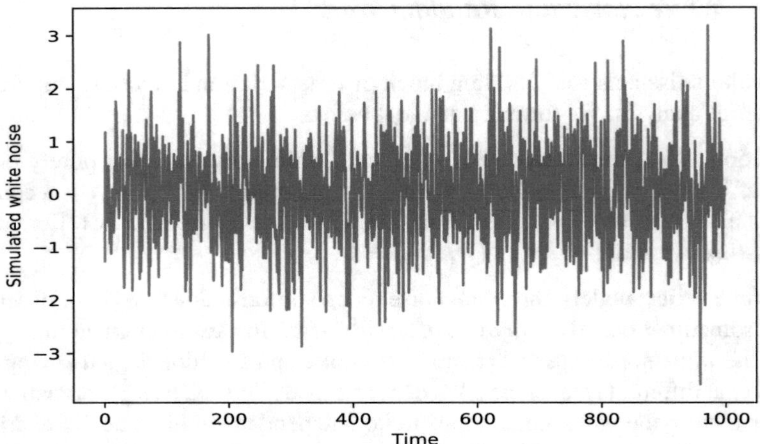


**Fig. 1.9** Time series plot of the simulated Gaussian white noise iidN(0,1)

pyTSA_GaussWN

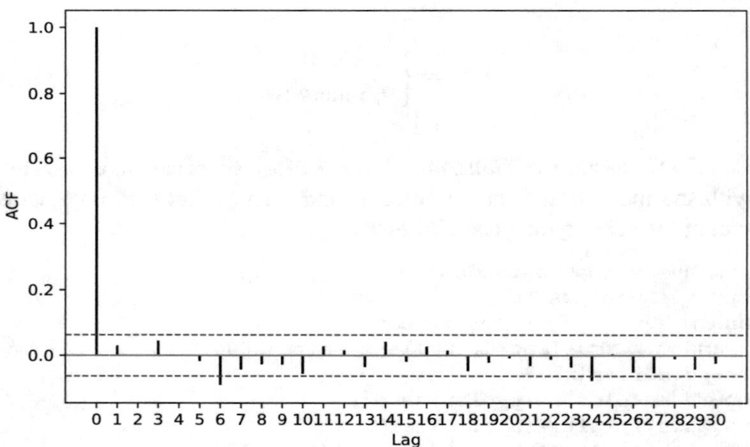


**Fig. 1.10** Correlogram of the simulated Gaussian white noise iidN(0,1)

pyTSA_GaussWN

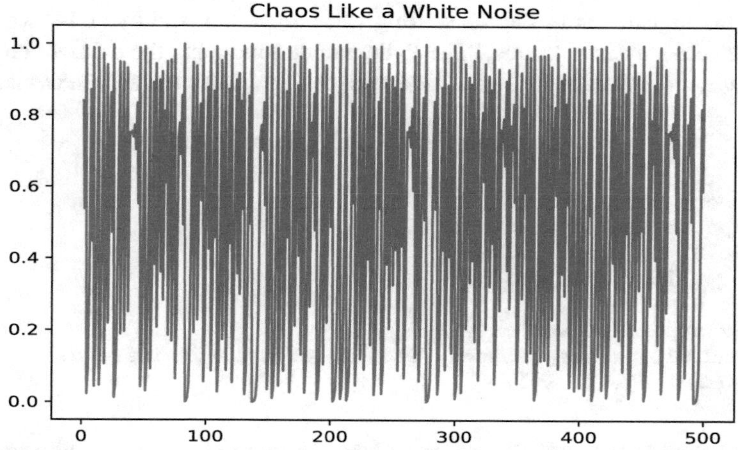


**Fig. 1.11** Time series plot of the chaos

pyTSA_Chaos

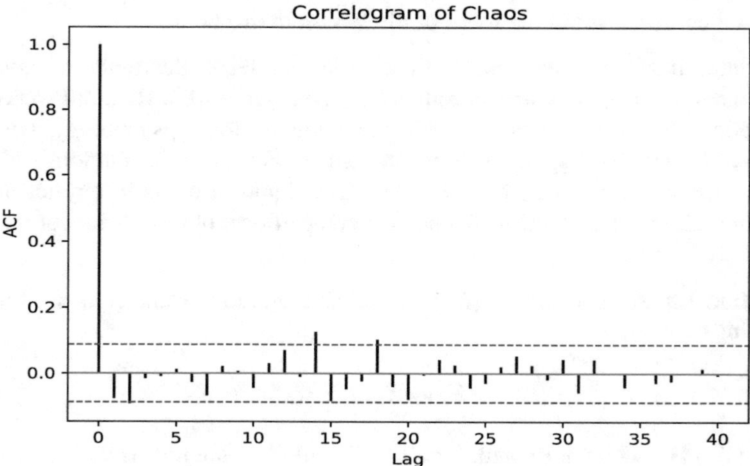


**Fig. 1.12** ACF plot of the chaos

pyTSA_Chaos

*Example 1.8 (Chaos Like a White Noise)* There is a time series dataset "chaos" in the folder "Ptsadata." Produce its time series plot and correlogram.

Figure 1.11 is the time series plot and Fig. 1.12 is the correlogram. Comparing Figs. 1.12 and 1.10 (the correlogram of the Gaussian white noise in Example 1.7), we can find that the two figures are so similar that we guess that Fig. 1.12 should be a correlogram of a certain white noise. However, we are wrong. The time series data "chaos" used to plot Fig. 1.12 actually come from the deterministic function (1.14) (called the logistic map). Let the starting value $x_0 = 0.3$, and by (1.14) we obtain the time series dataset "chaos." The Python code producing the dataset "chaos" is as follows, and the code for Fig. 1.11 as well as Fig. 1.12 is left as an exercise.

$$x_t = 4.0x_{t-1}(1 - x_{t-1}), x_0 \in (0, 1) \tag{1.14}$$

```
>>> import pandas as pd
>>> x=pd.Series(dtype=float)
>>> y=0.3   # start value
>>> for t in range(1,501):
    y=4.0*y*(1-y)
    x=x.append(pd.Series(y))
# append will be deprecated. Use "x.loc[t-1]=y" instead.
>>> index=range(1, 501)
>>> x.index=index
```

This example suggests that we must learn about the background of time series data before analyzing the data. In addition, note that by Definition 1.1 all time series are random, and so the time series discussed in this example should be seen as a special case. In general, a deterministic time series may be seen as a special time series. Of course, it is not the focus of discussion in this book.

Another interesting and useful time series model is the random walk. The term "random walk" was first introduced by Pearson (1905). Random walks have applications to many scientific fields including ecology, psychology, computer science, physics, biology, as well as economics. For example, random walks are used to depict a molecule's behavior path in a liquid or a gas in physics and the price movement of a stock in financial market in economics. A random walk is defined as follows:

**Definition 1.9** A time series $\{X_t\}$ is called a random walk if it satisfies the following equation

$$X_t = X_{t-1} + W_t \tag{1.15}$$

where $\{W_t\}$ is a white noise and, for all $t$, $W_t$ and $X_{t-1}$ are uncorrelated.

If $W_t \sim \text{iidN}(0, \sigma_w^2)$, then $\{X_t\}$ is a Gaussian (normal) random walk. By Eq. (1.15), the value of the time series $\{X_t\}$ at time $t$ is the value of the series at time $t-1$ plus a completely random value generated by the white noise $\{W_t\}$. Using Eq. (1.15), we can easily obtain

$$X_t = X_{t-1} + W_t = X_{t-2} + W_{t-1} + W_t = \cdots = X_0 + W_1 + W_2 + \cdots + W_{t-1} + W_t$$

Therefore, for all $t$, $\text{E}(X_t) = \text{E}(X_0)$ is a constant. That is, the random walk is mean stationary. On the other hand, however, we have

$$\text{Var}(X_t) = \text{Var}(X_{t-1}) + \sigma_w^2 > \text{Var}(X_{t-1})$$

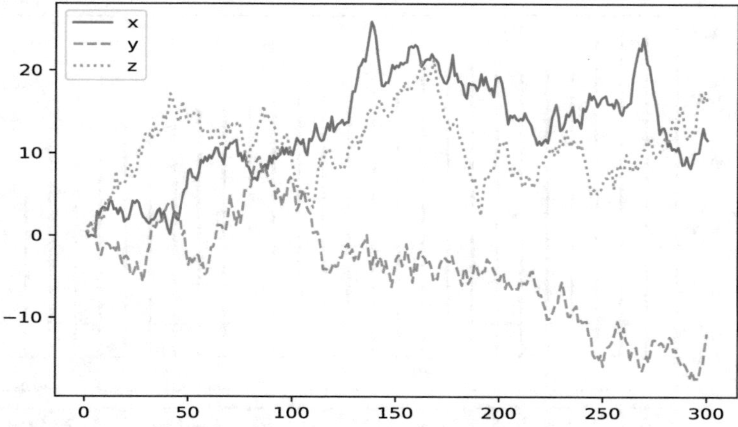


**Fig. 1.13** Three simulated paths (time plots) of the standard normal random walk

pyTSA_SimNormRW

Thus it can be seen that the random walk is not variance stationary. Naturally, it is more unlikely to be stationary. Therefore the random walk is a nonstationary time series that has neither deterministic trend nor deterministic seasonality. The below Python code generates three simulated paths of a standard ($\sigma_w^2 = 1$) Gaussian random walk. They are shown in Fig. 1.13, and from this figure we see that every path appears very different, but actually they are all generated by the same random walk.

```
>>> import numpy as np
>>> import pandas as pd
>>> from numpy.random import normal
>>> import matplotlib.pyplot as plt
>>> np.random.seed(1357)
>>> a=normal(size=300); b=normal(size=300); c=normal(size=300)
>>> x=np.cumsum(a); y=np.cumsum(b); z=np.cumsum(c)
>>> xyz=pd.DataFrame({'x': x, 'y': y, 'z': z})
>>> xyz.index=range(1,301)
>>> xyz.plot(style=['-', '--', ':']); plt.show()
#style means matplotlib line style per column
```

*Example 1.9 (Example 1.1 Continued)* The time series data in Example 1.1 is the Bitcoin closing price from June 23, 2017, to June 22, 2018. It is stored in Microsoft Excel xlsx format. In order to import it to the console, we must first install two packages: `xlrd` and `xlwt` in the same way that we install Python extension packages. From Figs. 1.1 (the time series plot) and 1.14 (the ACF plot), it is clear to see that the Bitcoin price time series is nonstationary. Now we use the notation $\{X_t\}$ to denote this time series and take natural logarithms on $\{X_t\}$. Thus we obtain the logarithm time series of the Bitcoin price $\{\log(X_t)\}$. Figures 1.15 and 1.16 are, respectively, the time series plot and ACF plot of $\{\log(X_t)\}$. From them, we see that

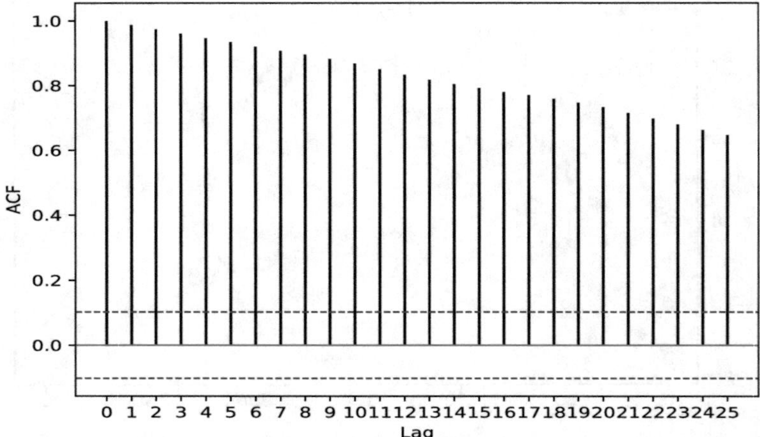


**Fig. 1.14** ACF plot of the Bitcoin price time series in Example 1.9

pyTSA_BTC

**Fig. 1.15** Time series plot of logarithm of the Bitcoin price series in Example 1.9

pyTSA_BTC

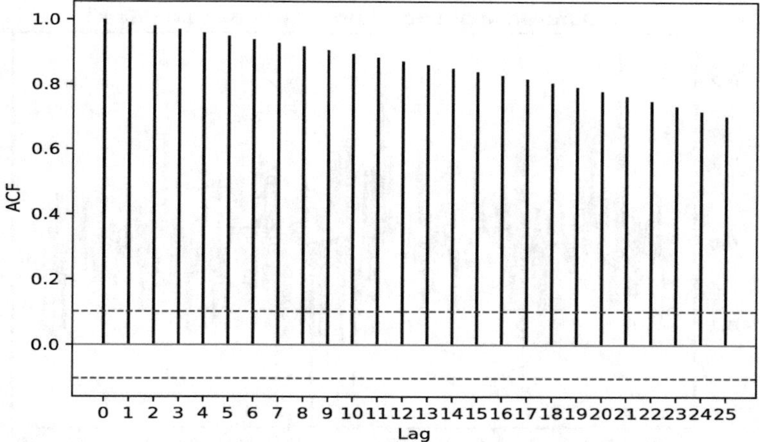


**Fig. 1.16** ACF plot of logarithm of the Bitcoin price time series in Example 1.9

pyTSA_BTC

the logarithm time series is also nonstationary. We continue to take a new action. Let

$$Y_t = \log(X_t) - \log(X_{t-1})$$

This operation is called the first difference operation. We will discuss it in detail in the next chapter. We are about to produce the time series plot and ACF plot of $\{Y_t\}$ and look at what happens. The two plots are, respectively, shown in Figs. 1.17 and 1.18. At this point, we see the very different situation that $\{Y_t\}$ seems stationary and its ACF appears very similar to the ACF of white noise. Actually this is a stylized fact from financial time series. Below is the Python code for this example.

```
>>> import pandas as pd
>>> import numpy as np
>>> from PythonTsa.datadir import getdtapath
>>> dtapath=getdtapath()
>>> bitcoin=pd.read_excel(dtapath +'BitcoinPrice17-6-23
                        -18-6-22.xlsx', header=0)
>>> dat=pd.date_range('2017-06-23',periods=len(bitcoin),freq='D')
>>> bitcoin.index=dat
>>> price=bitcoin['ClosingP']
>>> import matplotlib.pyplot as plt
>>> price.plot(); plt.title('Bitcoin Price 2017.6.23-2018.6.22')
>>> plt.ylabel('Price in USD'); plt.show()
>>> from PythonTsa.plot_acf_pacf import acf_pacf_fig
>>> acf_pacf_fig(price, lag=25)
>>> plt.show()
>>> logp=np.log(price)
>>> logp.plot(); plt.title('Logarithm of the Bitcoin Price')
```

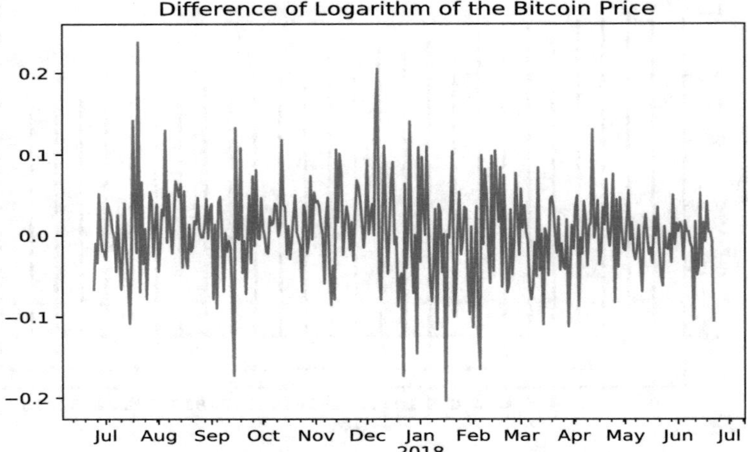


**Fig. 1.17** Time plot of differences of the Bitcoin log(price) series in Example 1.9

pyTSA_BTC

```
>>> plt.ylabel('log(rice)'); plt.show()
>>> acf_pacf_fig(logp, lag=25)
>>> plt.show()
>>> dlogp=logp.diff(1)
>>> dlogp=dlogp.dropna() #delete "NaN"
>>> dlogp.plot()
>>> plt.title('Difference of Logarithm of the Bitcoin Price')
>>> plt.show()
>>> acf_pacf_fig(dlogp, lag=25)
>>> plt.show()
```

Now suppose that $\{P_t\}$ is the closing price at time $t$ for a certain financial product (e.g., stock, bond, futures, exchange rate, and so on). Evidently the return on this product is $R_t = (P_t - P_{t-1})/P_{t-1}$. Notice that for the log function, we have $\log(1 + x) \approx x$ if $x$ is small. Furthermore, the yield rate of a financial product is usually small. Hence

$$R_t = \frac{P_t - P_{t-1}}{P_{t-1}} \approx \log(1 + \frac{P_t - P_{t-1}}{P_{t-1}}) = \log(P_t) - \log(P_{t-1})$$

That is, if we let $Z_t = \log(P_t)$, then

$$Z_t \approx Z_{t-1} + R_t \tag{1.16}$$

In finance, for mathematical convenience, we tend to use $Z_t - Z_{t-1}$ as the return on the financial product at time $t$ (usually called the *log return*). As we have seen from Example 1.9, in most cases, the return $R_t$ behaves like a white noise and by

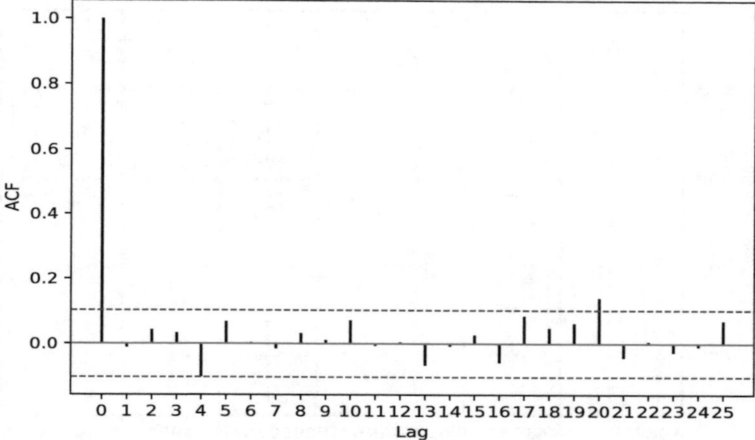


**Fig. 1.18** ACF plot of differences of the Bitcoin log(price) series in Example 1.9

pyTSA_BTC

(1.16) the log price $Z_t$ like a random walk. It is why many financial economists use the random walk as a key model for the log price of a financial asset.

## 1.4 Time Series Data Visualization

In this section, we discuss how to use Python and its extension packages to import and visualize time series data. We will see that the Python ecosystem is powerful and easy to use for those.

Visualization of time series is very useful for us to examine them. Actually, we have used this method to observe a few time series in the previous sections. Now we further illustrate how to import time series data to the console and visualize them with Python and its extensions. Time series data are stored in different format, and thus we have to use different approaches to import them to the console. Let us look at the two examples below.

*Example 1.10 (Example 1.2 Continued)* The time series in Example 1.2 is monthly recorded milk average per cow at a dairy cattle ranch, and its order is from left to right. We use the function `pd.concat()` in the extension package `pandas` to connect one row to the end of another in the DataFrame `milk` and so get the Series `mseries` (i.e., a time series). In addition to Fig. 1.2 illustrated in Example 1.2, here we, respectively, plot the yearly and monthly boxplots shown in Figs. 1.19 and 1.20. From Fig. 1.19, we clearly see that the milk average per cow linearly increases yearly, but since 1973 the growth almost stops. From Fig. 1.20, we observe that the milk average is the highest in May and June and the lowest in February and

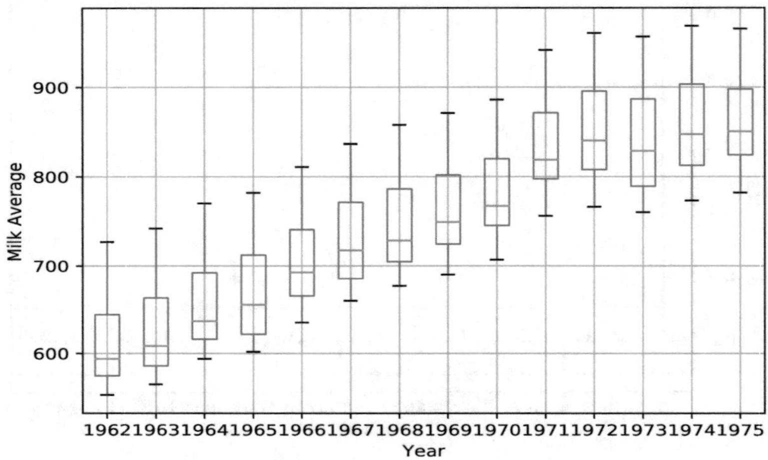


**Fig. 1.19** Yearly boxplot of the time series in Example 1.10

pyTSA_MilkAv

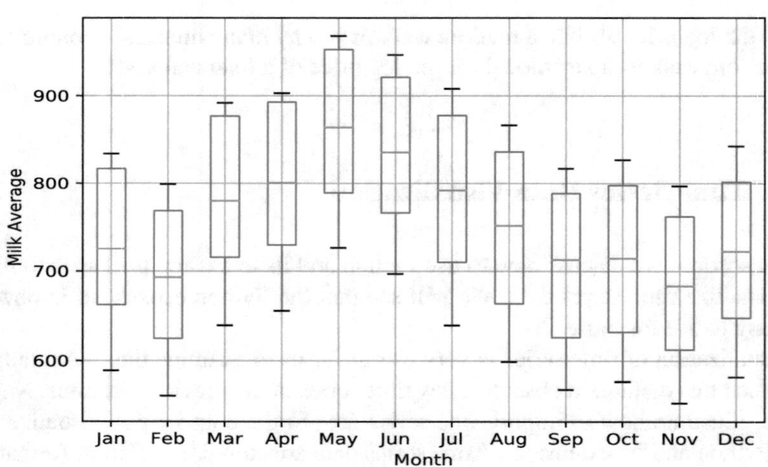


**Fig. 1.20** Monthly boxplot of the time series in Example 1.10

pyTSA_MilkAv

November. In other words, both of the boxplots have different functions and are all needed for examining the data. Below is the Python code for importing and handling the time series data.

```
>>> import numpy as np
>>> import pandas as pd
>>> import matplotlib.pyplot as plt
>>> from PythonTsa.datadir import getdtapath
```

```
>>> dtapath=getdtapath()
>>> milk=pd.read_excel(dtapath +'milk.xlsx',header=None)
>>> type(milk)
<class 'pandas.core.frame.DataFrame'>
>>> milk
      0    1    2    3    4    5    6    7      8      9
0   589  561  640  656  727  697  640  599  568.0  577.0
1   553  582  600  566  653  673  742  716  660.0  617.0
2   583  587  565  598  628  618  688  705  770.0  736.0
3   678  639  604  611  594  634  658  622  709.0  722.0
4   782  756  702  653  615  621  602  635  677.0  635.0
5   736  755  811  798  735  697  661  667  645.0  688.0
6   713  667  762  784  837  817  767  722  681.0  687.0
7   660  698  717  696  775  796  858  826  783.0  740.0
8   701  706  677  711  734  690  785  805  871.0  845.0
9   801  764  725  723  690  734  750  707  807.0  824.0
10  886  859  819  783  740  747  711  751  804.0  756.0
11  860  878  942  913  869  834  790  800  763.0  800.0
12  826  799  890  900  961  935  894  855  809.0  810.0
13  766  805  821  773  883  898  957  924  881.0  837.0
14  784  791  760  802  828  778  889  902  969.0  947.0
15  908  867  815  812  773  813  834  782  892.0  903.0
16  966  937  896  858  817  827  797  843    NaN    NaN
>>> mseries=pd.concat([milk.loc[0],milk.loc[1],milk.loc[2],
      milk.loc[3],milk.loc[4],milk.loc[5],milk.loc[6],
      milk.loc[7],milk.loc[8],milk.loc[9],milk.loc[10],
      milk.loc[11],milk.loc[12],milk.loc[13],milk.loc[14],
      milk.loc[15],milk.loc[16]],ignore_index='true')
# milk.loc[i] is the (i+1)th row in the DataFrame milk
>>> type(mseries)
<class 'pandas.core.series.Series'>
>>> mts=mseries.drop([168,169])
# delete the last two NaN's in mseries
>>> mts
0       589.0
1       561.0
2       640.0
        ...
165     827.0
166     797.0
167     843.0
Length: 168, dtype: float64
>>> timeindex=pd.date_range('1962-01',periods=168,freq='M')
# setting time index for the Series mts
>>> mts.index=timeindex
>>> mts
1962-01-31    589.0
1962-02-28    561.0
1962-03-31    640.0
              ...
1975-10-31    827.0
1975-11-30    797.0
1975-12-31    843.0
Freq: M, Length: 168, dtype: float64
```

```
#plotting the time series plot shown in Fig. 1.2
>>> mts.plot()
>>> plt.title('Milk Average per Cow from Jan.1962 to Dec.1975')
>>> plt.xlabel('Time'); plt.ylabel('Milk Average'); plt.show()
#plotting the yearly boxplot shown in Fig. 1.19
>>> my=np.array(mts).reshape(14,12)
>>> myt=np.transpose(my)
>>> year=[1962, 1963, 1964, 1965, 1966, 1967, 1968, 1969,
         1970, 1971,1972, 1973, 1974, 1975]
>>> myt=pd.DataFrame(myt, columns=year)
>>> bp=myt.boxplot()
>>> plt.xlabel('Year'); plt.ylabel('Milk Average'); plt.show()
#plotting the monthly boxplot shown in Fig. 1.20
>>> month=['Jan', 'Feb', 'Mar', 'Apr', 'May', 'Jun', 'Jul',
           'Aug', 'Sep', 'Oct', 'Nov','Dec']
>>> myd=pd.DataFrame(my, columns=month)
>>> bpm=myd.boxplot()
>>> plt.xlabel('Month'); plt.ylabel('Milk Average'); plt.show()
```

*Example 1.11 (Example 1.3 Continued)* The dataset in Example 1.3 is from the Internet, and so you must have access to the Internet. Figure 1.3 shows the time series plots and below are the Python commands for plotting them. Here we also plot the bar charts and lag plots for the time series data. In addition to the fact that the boys are always more than the girls every year, Fig. 1.21 also shows that the evolution pattern of the boy time series is almost the same as the one of the girl. A lag plot is used to help evaluate whether the time series are autocorrelated. If it is autocorrelated, the lag plot will exhibit an identifiable pattern. Otherwise the lag plot will demonstrate no identifiable pattern. Figure 1.22 is the lag plots of the boy and girl time series. From them we can clearly see that the both boy and girl time series are not purely random. That is, they both possess the autocorrelation property.

```
>>> import pandas as pd
>>> import matplotlib.pyplot as plt
>>> url= 'http://s3.amazonaws.com/assets.datacamp.com/course
        /dasi/present.txt'
>>> birth= pd.read_csv(url, sep=' ')
>>> birth.to_csv('G:\\datasets\\Noboyngirl.csv')
#storing the data birth into the file Noboyngirl.csv
>>> birth
    year     boys    girls
1   1940  1211684  1148715
2   1941  1289734  1223693
3   1942  1444365  1364631
..   ...      ...      ...
61  2000  2076969  1981845
62  2001  2057922  1968011
63  2002  2057979  1963747

[63 rows x 3 columns]
>>> birth_year = birth.set_index('year')
# the column 'year' of the data is used as index for the data
>>> birth_year
```

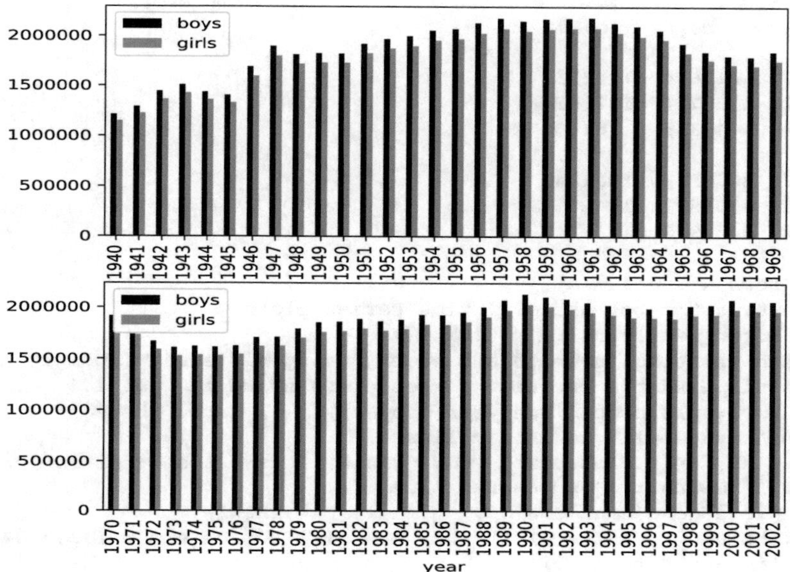


**Fig. 1.21** Bar charts of the boy and girl numbers time series in Example 1.1

pyTSA_USBoysGirls

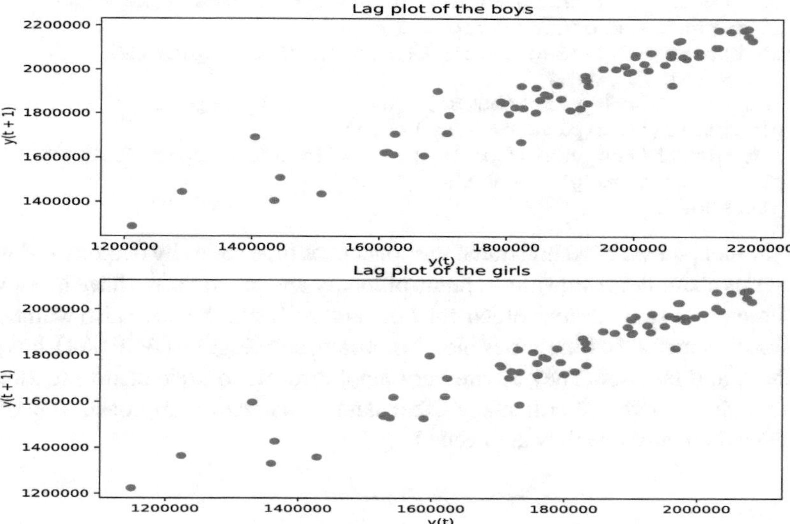


**Fig. 1.22** Lag plots of the boy and girl numbers time series in Example 1.1

pyTSA_USBoysGirls

```
         boys    girls
year
1940  1211684  1148715
1941  1289734  1223693
1942  1444365  1364631
...       ...      ...
2000  2076969  1981845
2001  2057922  1968011
2002  2057979  1963747

[63 rows x 2 columns]
#plotting the boy and girl time series plots
>>> birth_year.plot(); plt.show()
# when printing in white and black, using different lines to
# distinguish the two time series plots shown in Fig. 1.3
>>> boy,= plt.plot(birth_year['boys'],'b-', label='boys')
# 'b-' means blue and solid line
>>> girl,= plt.plot(birth_year['girls'],'r--',label='girls')
# 'r--' means red and dashed line
>>> plt.legend(handles=[boy, girl]); plt.show()
# plotting barcharts of the boy and girl series shown Fig.1.21
>>> fig = plt.figure()
>>> birth_year[:30].plot(kind='bar', color=['black','grey'],
    ax=fig.add_subplot(211))
>>> birth_year[30:].plot(kind='bar', color=['black','grey'],
    ax=fig.add_subplot(212))
>>> plt.show()
# plotting the boy and girl lag plots shown Fig. 1.22
>>> from pandas.plotting import lag_plot
# note the command "lag_plot" is not in the matplotlib
>>> fig = plt.figure()
>>> lag_plot(birth_year['boys'], ax=fig.add_subplot(211))
>>> plt.title('Lag plot of the boys')
>>> lag_plot(birth_year['girls'], ax=fig.add_subplot(212))
>>> plt.title('Lag plot of the girls')
>>> plt.show()
```

In this chapter we have illustrated the concept of time series by demonstrating the time series examples from various fields of theory and practice. We have introduced notions such as stationarity, autocorrelation, and so forth. We have also learned six visualization methods: time series plot, histogram, correlogram (ACF plot), boxplot, bar chart, and lag plot. They all are very helpful for us to understand and analyze a time series. There are still many other data visualization approaches. We will introduce them anytime they are needed.

## Problems

**1.1** Explain the concept of time series and give several everyday-life time series examples.

**1.2** Install Python and its extension packages discussed in this chapter on your computer, and then run all the Python code in this chapter again.

**1.3** The time series data "BitcoinPrice17-6-23-18-6-22" in the folder "Ptsadata" is the Bitcoin price from June 23, 2017, to June 22, 2018. Write the Python code to produce its histogram and lag plot and run it. Is the time series autocorrelated?

**1.4** The time series data "Yearly mean total sunspot number 1700–2017" in the folder "Ptsadata" is annual mean sunspot number from 1700 to 2017. With Python, produce its histogram, correlogram (ACF plot), and lag plot. Is the time series autocorrelated?

**1.5** Simulate a normal white noise sample of size 10 with the mean 2.3 and standard deviation 1.2, and then calculate the sample mean and sample standard deviation. Are you satisfied with your computed results? How about the sample size 10,000?

**1.6** Prove that if a time series $\{X_t\}$ follows the normal distribution family, then the weak stationarity is equivalent to the strick stationarity.

**1.7** The statement is that if the time series $\{X_t\}$ satisfies (1) for any time $t$, $\mathrm{E}(X_t^2) < \infty$; (2) $\mathrm{Cov}(X_t, X_{t+k}) = \gamma(k)$ for any time $t$ and $k$, then $\{X_t\}$ is stationary. Prove it if you think that it is true. Otherwise, construct a simple model as a counterexample.

**1.8** Suppose that $X_t = (-1)^t X$ where $X$ is a random variable with finite mean and variance. Find a necessary and sufficient condition on $X$ so that $\{X_t\}$ is (weakly) stationary.

**1.9** Let $\varepsilon_t \sim \mathrm{WN}(0, \sigma^2)$ be a white noise series. (1) Build a simplest model in terms of $\varepsilon_t$ so that it is mean stationary but not variance stationary. (2) Build a simplest model in terms of $\varepsilon_t$ so that it is variance stationary but not mean stationary.

# Chapter 2
# Exploratory Time Series Data Analysis

This chapter conducts exploratory time series data analysis with Python. In fact we have made some exploratory data analyses by means of time series plot, correlogram, boxplot, lag plot, and more in Chap. 1. In this chapter another correlation concept "partial autocorrelation function" is introduced which is helpful in modeling a time series. We consider how to statistically test whether a stationary time series is a white noise, which is indispensable in diagnosis of a resulting model. We also discuss effects of simple time series composition through simulation. Lastly we elaborate on methods and techniques for time series decomposition and smoothing.

## 2.1 Partial Autocorrelation Functions

In Sect. 1.3 we introduced the concept of autocorrelation function (ACF) and discussed its properties. Now we know how to use a time series sample to estimate its ACF and graph its ACF plot (correlogram). There is also another concept of correlation in time series analysis: partial autocorrelation function (PACF), which is as well helpful in modeling a time series. In this section, we explain and define PACF as well as discuss how to plot the PACF graph with Python.

### *2.1.1 Definition of PACF*

Let $\{X_t\}$ be a stationary time series with $\mathsf{E}(X_t) = 0$. Here the assumption $\mathsf{E}(X_t) = 0$ is for conciseness only. If $\mathsf{E}(X_t) = \mu \neq 0$, it is okay to replace $\{X_t\}$ by $\{X_t -$

 
C. Huang, A. Petukhina, *Applied Time Series Analysis and Forecasting with Python*, Statistics and Computing,
https://doi.org/10.1007/978-3-031-13584-2_2

$\mu\}$. Now consider the linear regression (prediction) of $X_t$ on $\{X_{t-k+1:t-1}\}$ for any integer $k \geq 2$. We use $\hat{X}_t$ to denote this regression (prediction):

$$\hat{X}_t = \alpha_1 X_{t-1} + \cdots + \alpha_{k-1} X_{t-k+1}$$

where $\{\alpha_1, \cdots, \alpha_{k-1}\}$ satisfy

$$\{\alpha_1, \cdots, \alpha_{k-1}\} = \arg\min_{\beta_1,\cdots,\beta_{k-1}} \mathsf{E}[X_t - (\beta_1 X_{t-1} + \cdots + \beta_{k-1} X_{t-k+1})]^2$$

That is, $\{\alpha_1, \cdots, \alpha_{k-1}\}$ are chosen by minimizing the mean squared error of prediction. Similarly, let $\hat{X}_{t-k}$ denote the regression (prediction) of $X_{t-k}$ on $\{X_{t-k+1:t-1}\}$:

$$\hat{X}_{t-k} = \eta_1 X_{t-1} + \cdots + \eta_{k-1} X_{t-k+1}$$

Note that if $\{X_t\}$ is stationary, then $\{\alpha_{1:k-1}\} = \{\eta_{1:k-1}\}$. Now let $\hat{Z}_{t-k} = X_{t-k} - \hat{X}_{t-k}$ and $\hat{Z}_t = X_t - \hat{X}_t$. Then $\hat{Z}_{t-k}$ is the residual of removing the effect of the intervening variables $\{X_{t-k+1:t-1}\}$ from $X_{t-k}$, and $\hat{Z}_t$ is the residual of removing the effect of $\{X_{t-k+1:t-1}\}$ from $X_t$.

**Definition 2.1** The partial autocorrelation function (PACF) at lag $k$ of a stationary time series $\{X_t\}$ with $\mathsf{E}(X_t) = 0$ is

$$\phi_{11} = \mathrm{Corr}(X_{t-1}, X_t) = \frac{\mathrm{Cov}(X_{t-1}, X_t)}{[\mathrm{Var}(X_{t-1})\mathrm{Var}(X_t)]^{1/2}} = \rho_1$$

and

$$\phi_{kk} = \mathrm{Corr}(\hat{Z}_{t-k}, \hat{Z}_t) = \frac{\mathrm{Cov}(\hat{Z}_{t-k}, \hat{Z}_t)}{[\mathrm{Var}(\hat{Z}_{t-k})\mathrm{Var}(\hat{Z}_t)]^{1/2}}, \ k \geq 2$$

According to the property of correlation coefficient (see, e.g., P172, Casella and Berger 2002), $|\phi_{kk}| \leq 1$. On the other hand, the following theorem paves the way to estimate the PACF of a stationary time series, and its proof can be seen in Fan and Yao (2003).

**Theorem 2.1** *Let $\{X_t\}$ be a stationary time series with $\mathsf{E}(X_t) = 0$, and $\{a_{1k}, \cdots, a_{kk}\}$ satisfy*

$$\{a_{1k}, \cdots, a_{kk}\} = \arg\min_{a_1,\cdots,a_k} \mathsf{E}(X_t - a_1 X_{t-1} - \cdots - a_k X_{t-k})^2$$

*Then $\phi_{kk} = a_{kk}$ for $k \geq 1$.*

*Example 2.1 (A Special Case: $\phi_{22}$)* Suppose that $\{X_t\}$ is a stationary time series with $\mathsf{E}(X_t) = 0$. Calculate $\phi_{22}$. At this point, we have

$$\hat{X}_1 = \rho_1 X_2, \ \hat{X}_3 = \rho_1 X_2$$

The proof is easy and left as an exercise. Thus

$$\begin{aligned} \text{Cov}(X_1 - \rho_1 X_2, X_3 - \rho_1 X_2) &= \gamma_2 - 2\rho_1\gamma_1 + \rho_1^2\gamma_0 \\ &= \gamma_0(\rho_2 - 2\rho_1^2 + \rho_1^2) \\ &= \gamma_0(\rho_2 - \rho_1^2) \end{aligned}$$

and

$$\begin{aligned} \text{Var}(X_1 - \rho_1 X_2) &= \text{Var}X_1 - 2\rho_1 \text{Cov}(X_1, X_2) + \rho_1^2 \text{Var}X_2 \\ &= \gamma_0(1 - 2\rho_1^2 + \rho_1^2) \\ &= \gamma_0(1 - \rho_1^2) \end{aligned}$$

Similarly, $\text{Var}(X_3 - \rho_1 X_2) = \gamma_0(1 - \rho_1^2)$. Therefore,

$$\begin{aligned} \phi_{22} &= \text{Corr}(X_1 - \rho_1 X_2, X_3 - \rho_1 X_2) \\ &= \frac{\text{Cov}(X_1 - \rho_1 X_2, X_3 - \rho_1 X_2)}{[\text{Var}(X_1 - \rho_1 X_2)\text{Var}(X_3 - \rho_1 X_2)]^{1/2}} \\ &= \frac{\rho_2 - \rho_1^2}{1 - \rho_1^2} \end{aligned}$$

The PACF clearly measures the correlation between $X_{t-k}$ and $X_t$ that is not explained by $\{X_{t-k+1:t-1}\}$. Moreover, it helps to identify an AR model. More discussion on the PACF can be found in Chaps. 3 and 4 of the book.

### 2.1.2 Sample PACF and PACF Plot

For a time series sample, there are a few ways to estimate the PACF. The details are discussed in Chaps. 3 and 4. Now we focus on how to use Python to implement PACF estimation and plot. Like the ACF plot, plotting $\phi_{kk}$ against lag $k$ forms the PACF plot. Before plotting PACF, there are two points to which we need to pay attention:

- For comparison, we additively define $\phi_{00} = 1$ and plot it too.
- Whether a time series is stationary or not, we still plot its PACF according to Definition 2.1 and then examine the PACF plot, just as we do for ACF.

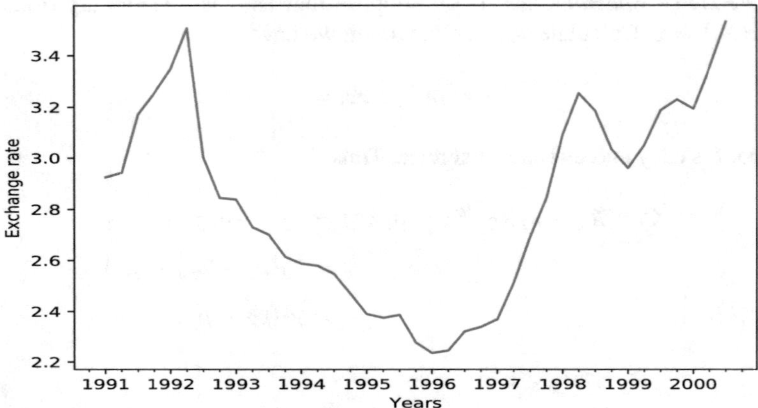


**Fig. 2.1** Time series plot of the quarterly exchange rates in Example 2.2

pyTSA_ExRate

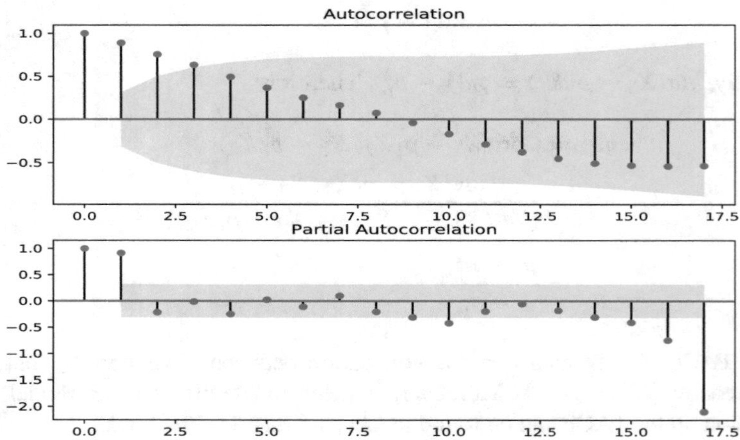


**Fig. 2.2** Statsmodels-plotted ACF and PACF of the quarterly exchange rates

pyTSA_ExRate

*Example 2.2 (Quarterly Exchange Rates of GBP to NZ Dollar)* The dataset "ExchRate NZ per UK" in the folder "Ptsadata" is from Cowpertwait and Metcalfe (2009). This time series is the quarterly exchange rates for British pounds sterling to New Zealand dollars from March 1991 to September 2000. Its time series plot is shown in Fig. 2.1. Its evolution pattern is relatively complex: roughly speaking, after an initial surge ending in 1992, a descent pattern leads to a minimum around 1996, which is followed by a reascent in the second half of the time series. Figure 2.2 is the ACF and PACF plots of the time series and plotted with the Python package `statsmodels` of version 0.12.1. In Fig. 2.2, we find two bugs: (1) $|\phi_{17,17}| > 1$. This is contradictory to the fact that $|\phi_{kk}| \leq 1$ for any $k \geq 0$; (2) The ticks on the

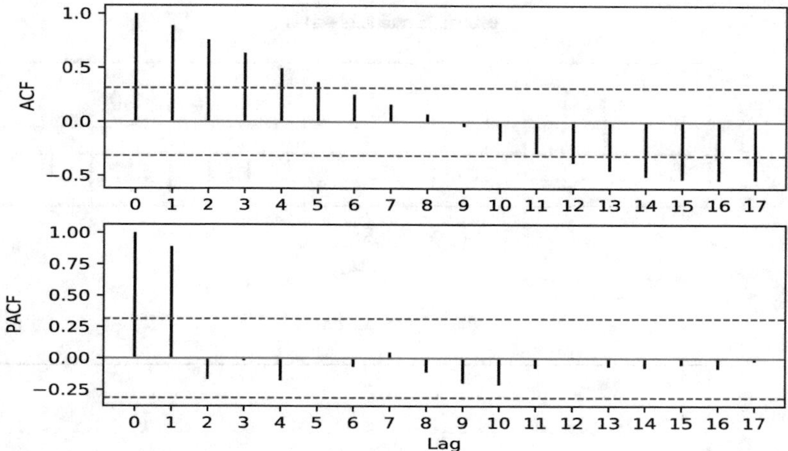


**Fig. 2.3** PythonTsa-plotted ACF and PACF of the quarterly exchange rates

pyTSA_ExRate

horizontal (time) axis become decimals such as 2.5, 7.5, and so on.[1] Now let us check Fig. 2.3 plotted with the Python package `PythonTsa`. It looks normal, and what is more, it is identical to Fig. 2.4 plotted with R.[2]

Below is the Python code for Example 2.2.

```
>>> import pandas as pd
>>> import numpy as np
>>> import matplotlib.pyplot as plt
>>> from PythonTsa.datadir import getdtapath
>>> dtapath=getdtapath()
>>> x=pd.read_csv(dtapath +'ExchRate NZ per UK.txt',header=0)
>>> dates = pd.date_range('1991', periods=len(x), freq='Q')
>>> x.index=dates; xts=pd.Series(x['xrate'])
>>> xts.plot(); plt.xlabel('Years')
>>> plt.ylabel('Exchange rate'); plt.show()
# The next is to plot ACF and PACF using statsmodels
>>> from statsmodels.graphics.tsaplots import plot_acf, plot_pacf
>>> fig = plt.figure()
>>> plot_acf(xts, lags=17, ax= fig.add_subplot(211))
>>> plot_pacf(xts, lags=17, ax= fig.add_subplot(212))
>>> plt.show()
# The next is to plot ACF and PACF using PythonTsa
```

[1] There is a warning in `statsmodels` of v. 0.13.1: FutureWarning: The default method "yw" can produce PACF values outside of the $[-1, 1]$ interval. After 0.13, the default will change to unadjusted Yule-Walker ("ywm"). You can use this method now by setting method ='ywm'. This approach gets right PACF values but the second bug still exists. We will report this to the developer.

[2] According to the R usage, the ticks on the horizontal (time) axis are in the number of periods, and $\phi_{00} = 1$ is not plotted out.

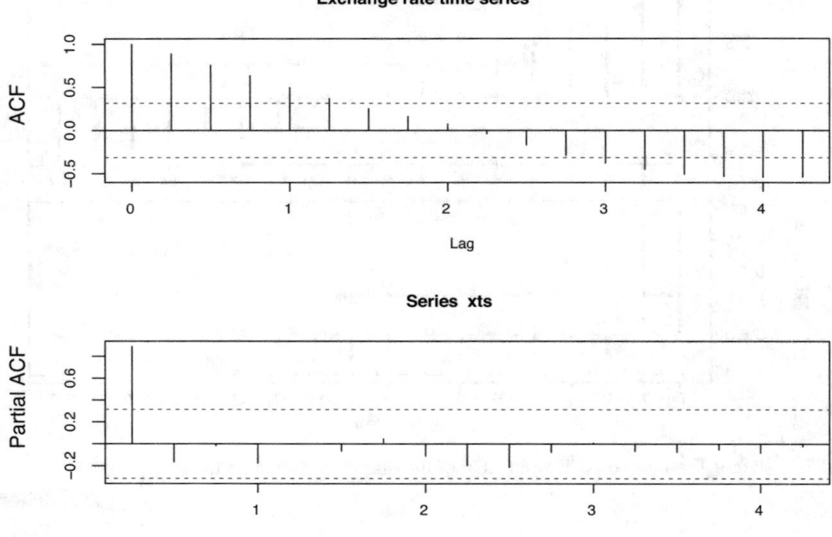


**Fig. 2.4** R-plotted ACF and PACF of the quarterly exchange rates

```
>>> from PythonTsa.plot_acf_pacf import acf_pacf_fig
>>> acf_pacf_fig(xts, both=True, lag=17)
>>> plt.show()
```

## 2.2 White Noise Test

We defined the white noise and discussed some properties of it in Sect. 1.3.4. An important issue is how to test whether a stationary time series is a white noise or not. In Sect. 1.3.4, we used the time series plot and the ACF plot of the stationary time series to decide if the time series is a white noise. In this section, we discuss how to statistically test whether a stationary time series is a white noise. For a stationary time series $\{X_t\}$, clearly, it is a white noise if and only if its autocorrelation function (ACF) $\rho_k = 0$ for any integer $k \neq 0$. There are several methods for testing whether a stationary time series is a white noise or not (see, e.g., Brockwell and Davis 1991, 2016 as well as Fan and Yao 2003). Now we focus on the portmanteau (Q) test, which has two types: Box-Pierce and Ljung-Box. We know that $r_k$ is the sample ACF of $\{X_t\}$ by Definition 1.7, and if $\{X_t\}$ is a white noise series, then $r_k$ is asymptotically normal with the mean 0 and the variance of $1/n$ where $n$ is the sample size. This naturally leads to the Box-Pierce test statistic (see Box and Pierce 1970)

$$Q_{BP}(m) = n \sum_{k=1}^{m} r_k^2$$

for the null hypothesis $H_0 : \rho_1 = \cdots = \rho_m = 0$ against the alternative hypothesis $H_1 : \rho_i \neq 0$ for some $i \in \{1 : m\}$ where $1 \leq m < T$ is any given integer and $T$ is the sample size. Under the null hypothesis that $\{X_t\}$ is a white noise ($H_0$ is true), $Q_{BP}(m)$ asymptotically follows a chi-squared distribution $\chi^2(m)$ with $m$ degrees of freedom. However, Ljung and Box (1978) argue that the chi-squared distribution does not provide a sufficiently accurate approximation to the distribution of the statistic $Q_{BP}(m)$ under the null hypothesis and propose a modified form of the statistic

$$Q_{LB}(m) = n(n+2) \sum_{k=1}^{m} \frac{r_k^2}{n-k} \tag{2.1}$$

which still asymptotically and better follows the chi-squared distribution $\chi^2(m)$ and is often called the Ljung-Box statistic and is preferred to the Box-Pierce statistic. Note that under the null hypothesis, for every integer $1 \leq m < T$, the $p$-value for $Q_{LB}(m)$ should be greater than 0.05 (the level of significance).

*Example 2.3 (Example 2.2 Continued)* The quarterly exchange rates of GBP to NZ dollar $\{X_t\}$ are a financial price time series. According to the way of thinking in Example 1.9, we take natural logarithms on $\{X_t\}$ and then difference $\{\log(X_t)\}$. So we obtain the return time series

$$Y_t = \log(X_t) - \log(X_{t-1})$$

Figure 2.5 is the time series plot of $\{Y_t\}$, and it seems to have neither trend nor seasonality. In addition, from Fig. 2.6, namely, the ACF plot of $\{Y_t\}$, we see that this ACF plot looks like the correlogram of a white noise except that the ACF value at lag 1 appears not to be zero. Then, is $\{Y_t\}$ a white noise or not after all? Now let us do the Ljung-Box test on $\{Y_t\}$. In the Python package `statsmodels`, there are a few functions for the Ljung-Box test, of which the function `acf` in the module `statsmodels.tsa.stattools` and the function `acorr_ljungbox` in the module `statsmodels.stats.diagnostic` are easy to use. For comparing their usage, we run the two functions at the same time. Below is the Python code, and from the results, it is clear that $\{Y_t\}$ cannot be a white noise due to the $p$-value for $Q_{LB}(1)$ is 0.0106 and the $p$-value for $Q_{LB}(2)$ is 0.0308. If we take difference on $\{Y_t\}$ or double difference on $\{\log(X_t)\}$, we obtain the new time series

$$Z_t = Y_t - Y_{t-1}$$

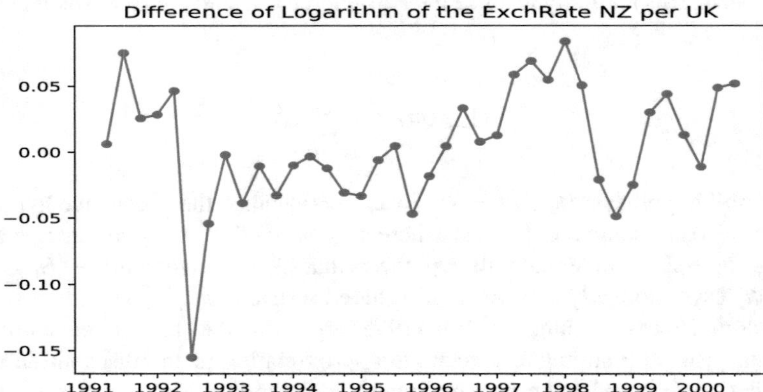


**Fig. 2.5** Time series plot of the differences of the log quarterly exchange rates

pyTSA_ExRateWN

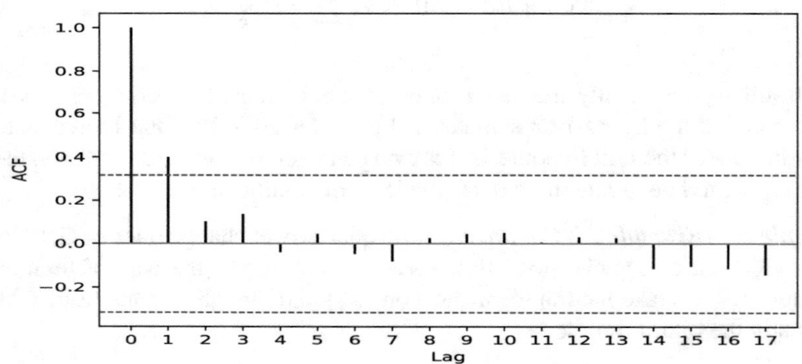


**Fig. 2.6** ACF plot of the differences of the log quarterly exchange rates

pyTSA_ExRateWN

and its correlogram shown in Fig. 2.7 is almost exactly the same as a white noise's one. Additionally, all the $p$-values for $Q_{LB}(m)$ are larger than 0.05. Hence the new time series can be seen as a white noise.

```
>>> logxts=np.log(xts)
>>> dlogxts=logxts.diff(1)
>>> dlogxts=dlogxts.dropna()  #delete "NaN"
>>> dlogxts.plot(marker='o', markersize=5)
>>> plt.title('Difference of Logarithm of the ExchRate NZ per UK')
>>> plt.show()
>>> acf_pacf_fig(dlogxts, both=False, lag=17)  #plot ACF only
>>> plt.show()
>>> from statsmodels.tsa.stattools import acf
>>> r,q,p=acf(dlogxts,nlags=35,qstat=True)
# r for ACF; q for Ljung-Box statistics; p for p-values
```

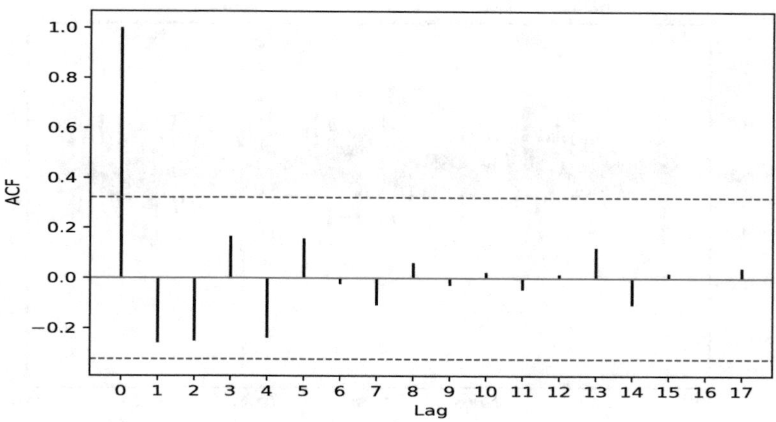


**Fig. 2.7** ACF plot of the double differences of the log quarterly exchange rates

pyTSA_ExRateWN

```
>>> p
array([0.01061724, 0.03083645, 0.05136147, 0.09819017, 0.16076553,
       ...])
>>> from statsmodels.stats.diagnostic import acorr_ljungbox
>>> q1,p1=acorr_ljungbox(dlogxts,lags=35,return_df=False,boxpierce=False)
>>> p1
array([0.01061724, 0.03083645, 0.05136147, 0.09819017, 0.16076553,
       ...])
>>> ddlogxts=dlogxts.diff(1) # difference again
>>> ddlogxts=ddlogxts.dropna()
>>> acf_pacf_fig(ddlogxts, both=False, lag=17)
>>> plt.show()
>>> r2,q2,p2=acf(ddlogxts,nlags=35,qstat=True)
>>> p2
array([0.10279916, 0.07226889, 0.09256769, 0.06353298, 0.0743523 ,
       0.12230988, 0.1563488 , 0.21311281, 0.28660097, 0.36751679,
       0.44441684, 0.52889322, 0.53656066, 0.55838455, 0.63132226,
       0.70003144, 0.75306299, 0.80635176, 0.79263419, 0.82537154,
       0.84889291, 0.74786537, 0.79541059, 0.72526094, 0.72807138,
       0.76359134, 0.76365521, 0.67031202, 0.69953091, 0.6323352 ,
       0.55188113, 0.58532784, 0.632728  , 0.61512833, 0.59652958])
```

*Example 2.4 (Monthly Returns of Procter and Gamble (PG) Stock from 1961 to 2016)* The time series data "monthly returns of Procter and Gamble stock from 1961 to 2016" ( denoted by the PG series ) is part of the dataset "monthly returns of Procter n Gamble stock n 3 market indexes 1961 to 2016" in the folder "Ptsadata." We first graph the time series plot of the PG series shown in Fig. 2.8 and use the function `describe( )` to compute descriptive statistics for the PG series such as $mean = 0.010342 > 0$ and $std = 0.055383$. So if you had purchased this stock, averagely you might made money. Besides, from Fig. 2.8 we find that both

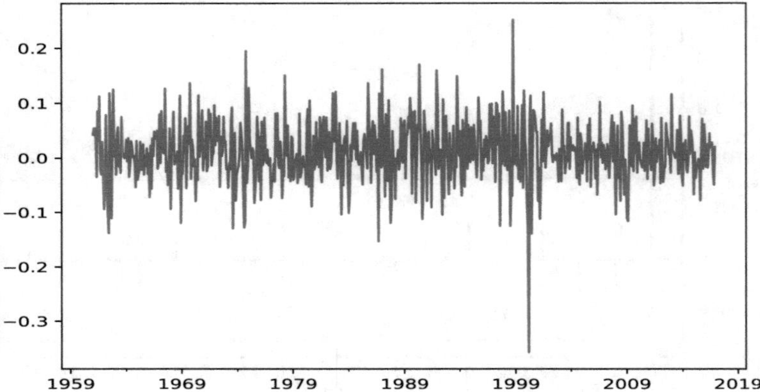


**Fig. 2.8** Time series plot of monthly returns of PG stock from 1961 to 2016

pyTSA_ReturnsPG

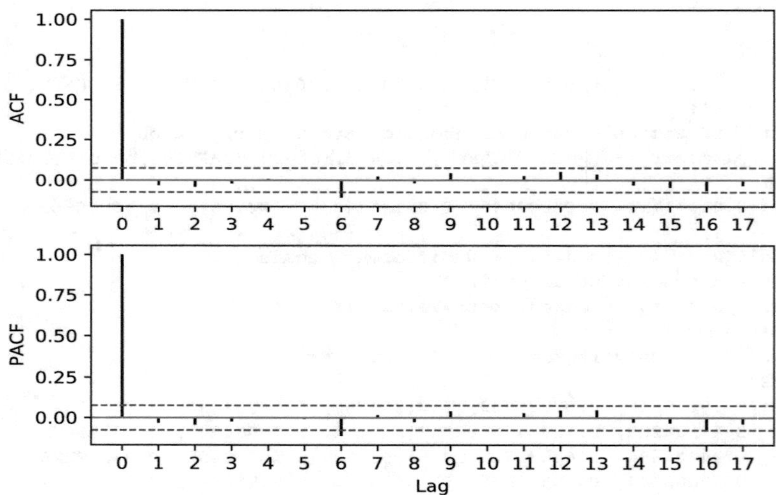


**Fig. 2.9** ACF and PACF plots of monthly returns of PG stock from 1961 to 2016

pyTSA_ReturnsPG

the maximum and the minimum in the PG series are around 1999. What is more, we are able to get their exact positions using the commands `idxmax` and `idxmin`. The positions of the maximum and minimum values are at October31 ,1998, and March 31, 2000, respectively. It was the turbulent period during which the Asian economic crisis occurred. Observing the Fig. 2.9 and the $p$-values for the Ljung-Box test, we conclude that the PG series is surely a white noise. The following is the Python code for this example.

```
>>> import pandas as pd
>>> import numpy as np
>>> from PythonTsa.datadir import getdtapath
>>> dtapath=getdtapath()
>>> x=pd.read_table(dtapath +'monthly returns of Procter
                n Gamble stock n 3 market indexes 1961
                to 2016.csv', sep=',', header=0)
>>> timeindex= pd.date_range('1961', periods=len(x), freq='M')
>>> x.index=timeindex
>>> yts=x['RET']
>>> import matplotlib.pyplot as plt
>>> yts.plot(); plt.show()
>>> yts.describe()
count    672.000000
mean       0.010342
std        0.055383
min       -0.357041
25%       -0.021283
50%        0.007384
75%        0.044261
max        0.250931
Name: RET, dtype: float64
>>> pd.Series.idxmax(yts) # get the position of the maximum
Timestamp('1998-10-31 00:00:00', freq='M')
>>> pd.Series.idxmin(yts) # get the position of the minimum
Timestamp('2000-03-31 00:00:00', freq='M')
>>> from PythonTsa.plot_acf_pacf import acf_pacf_fig
>>> acf_pacf_fig(yts, both=True, lag=17)
>>> plt.show()
>>> from statsmodels.tsa.stattools import acf
>>> r,q,p=acf(yts,qstat=True)
>>> p
array([0.41254176, 0.38900645, 0.54029397, 0.69709275, 0.81666593,
       0.11107219, 0.15000765, 0.20494966, 0.1976198 , 0.26606154,
       0.30642527, 0.25597948, 0.26708488, 0.29372717, 0.27787471,
       0.1902881 , 0.20381855, 0.24892409, 0.28338075, ...])
```

## 2.3 Simple Time Series Compositions

Before discussing how to decompose a time series in the next section, we first look at compositions of simple time series in this section. Given three very simple time series, we then compose them in different ways. Below is the three time series:

(1) $F_t = 0.2 + 0.1t$. It is obviously a linear increasing function. If a time series has one component such as $F_t$, then it possesses a *deterministic trend component*.
(2) $P_t = 2sin(2\pi t/50 + 0.3\pi)$. It is actually a periodic function with the period 50 or sine wave with the amplitude 2 and phase shift $0.3\pi$. If a time series has one component such as $P_t$, then it possesses *deterministic seasonality* or

*deterministic seasonal component*. But it is not necessary that the seasonal component is always a periodic function.

(3) $X_t$ = `np.random.normal(loc=0,scale=4.2,size=500)`. It is a random real number series sampled from the normal distribution with mean zero and standard deviation 4.2. What is more, it is a Gaussian white noise $X_t \sim \text{iidN}(0,4.2^2)$. Any time series except deterministic series has a component like $X_t$ or more complicated stochastic part, and this random component (part) is called *random (stochastic) variation component*.

These three time series plus addition (+) and multiplication (×) operations can be combined in 14 different ways, including the combinations of only two time series. All the compositions are as follows:

$$F_t + P_t,\ F_t \times P_t,\ F_t + X_t,\ F_t \times X_t,\ P_t + X_t,\ P_t \times X_t$$

and

$$F_t + P_t + X_t,\ (F_t + P_t) \times X_t,\ (F_t + X_t) \times P_t,\ (P_t + X_t) \times F_t$$

$$F_t \times P_t \times X_t,\ (F_t \times P_t) + X_t,\ (F_t \times X_t) + P_t,\ (P_t \times X_t) + F_t$$

where $F_t + P_t$ and $F_t \times P_t$ are deterministic time series and not considered. We graph the time series plots of the 12 compositions shown in Figs. 2.10, 2.11, and 2.12. Here we examine only 6 of the 12 compositions and the other is left as an exercise to the reader.

- $M_t = P_t + X_t$. The time series $M_t$ is a simple addition of the periodic function $P_t$ and the Gaussian white noise $X_t$. Its time series plot is shown in the third panel of Fig. 2.10. We observe that the time series is so contaminated by the white noise that we would hardly see the periodic (seasonal) component in it. Moreover, its mean is $2sin(2\pi t/50 + 0.3\pi)$ and not constant. But it is evidently variance stationary.
- $G_t = F_t + P_t + X_t$. The time series $G_t$ is formed by adding the linear increasing function $F_t$ to $M_t$. Its time series plot is shown in the top panel of Fig. 2.11. We expect that the time series has a linear increasing trend and it does. On the other hand, like $M_t$ we could not observe that there is a seasonal component in it. Besides, it is variance stationary although nonstationary.
- $S_t = F_t \times P_t \times X_t$. The time series $S_t$ is the result of interaction of $F_t$, $P_t$, and $X_t$. Its time series plot is shown in the top panel of Fig. 2.12. We cannot see that there exists any trend or seasonality in the time series plot. Both the trend $F_t$ and the seasonal component $P_t$ are masked. Interesting is that its mean is zero but variance is not constant. In other words, it is mean stationary although nonstationary.
- $V_t = F_t \times P_t + X_t$. The time series $V_t$ is shaped by multiplying the two deterministic components and then adding the random variation component to

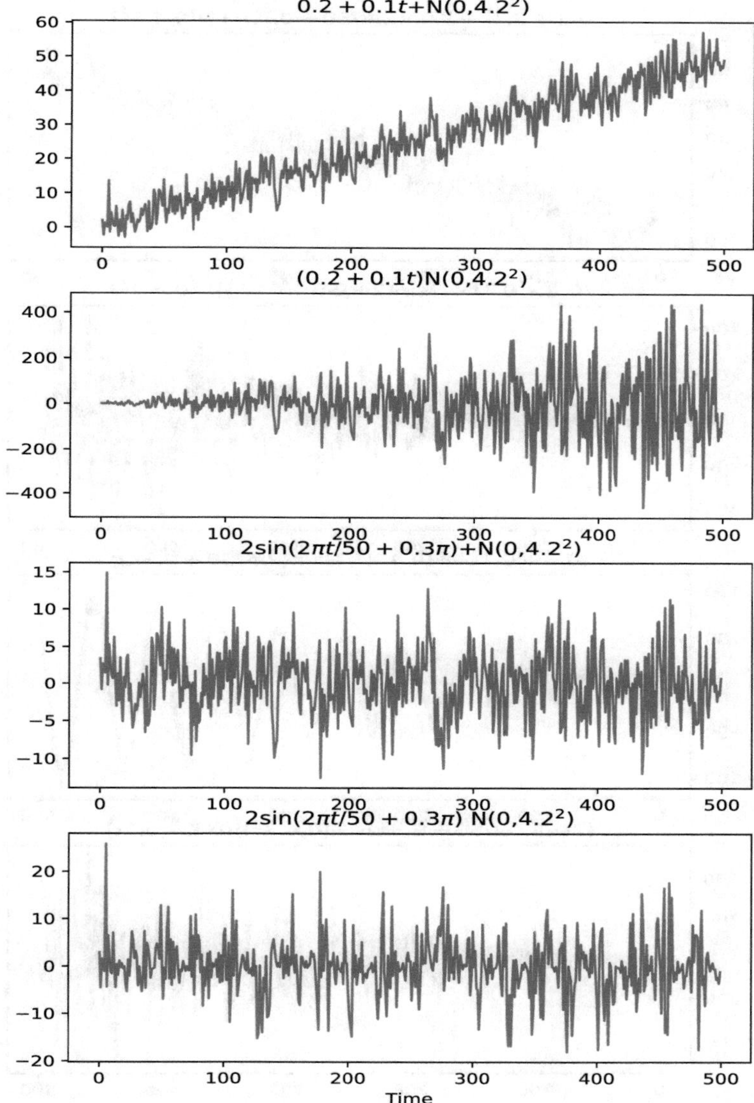


**Fig. 2.10** Time series plots of $F_t + X_t$, $F_t \times X_t$, $P_t + X_t$, and $P_t \times X_t$

pyTSA_TSComposition

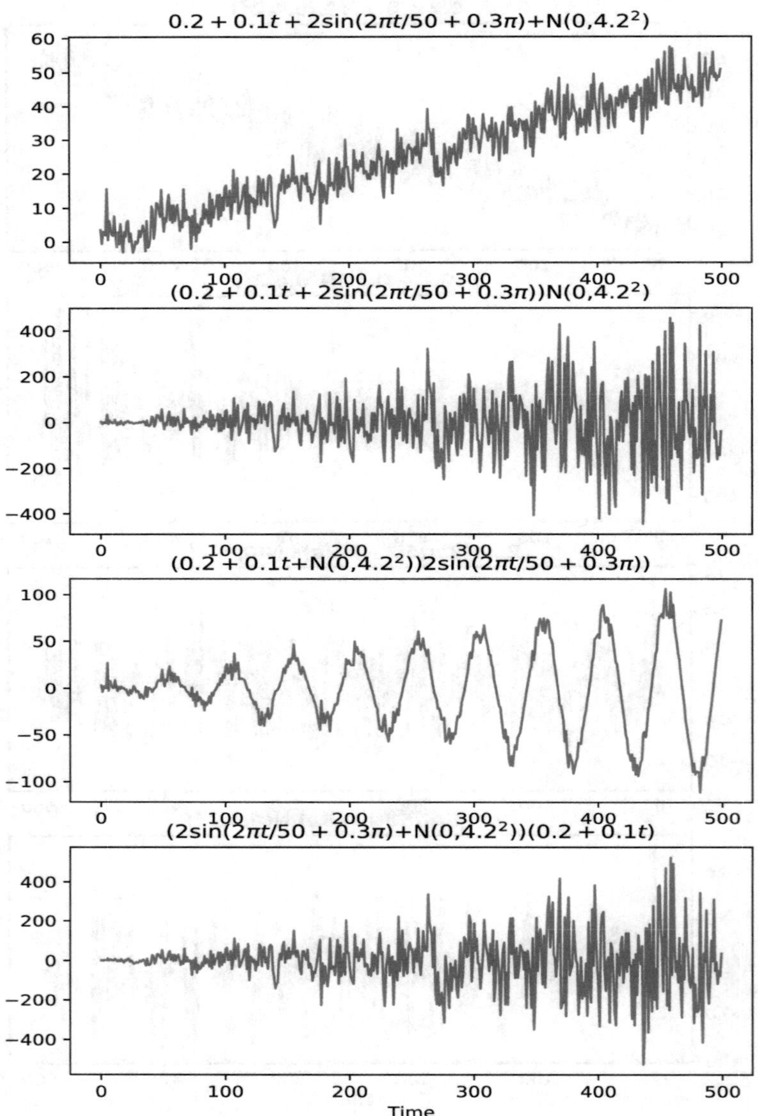


**Fig. 2.11** Time series plots of $F_t + P_t + X_t$, $(F_t + P_t)X_t$, $(F_t + X_t)P_t$, and $(P_t + X_t)F_t$ pyTSA_TSComposition

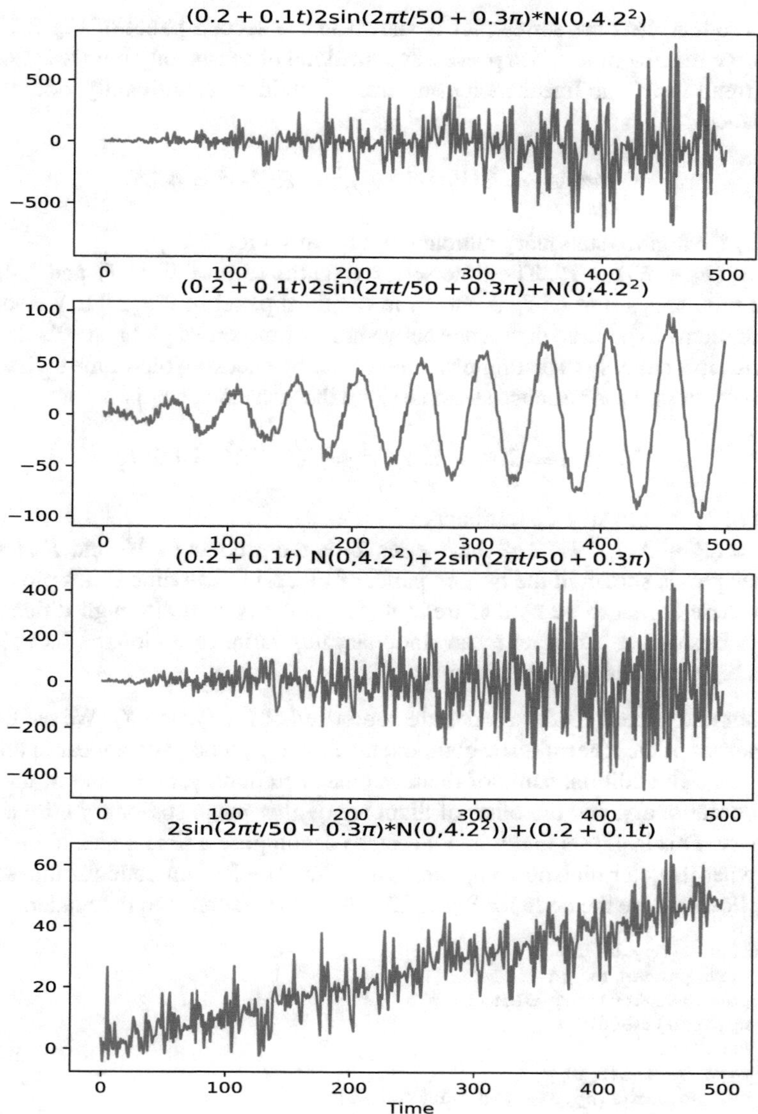


**Fig. 2.12** Time series plots of $F_t * P_t * X_t$, $F_t * P_t + X_t$, $F_t * X_t + P_t$, and $P_t * X_t + F_t$

pyTSA_TSComposition

the product. Its time series plot is shown in the second panel of Fig. 2.12. We observe that the time series possesses some kind of seasonality but we cannot see the trend in it. The linear increasing trend is hidden. Additionally, note that its variance

$$\mathrm{Var}(V_t) = \mathsf{E}(V_t - \mathsf{E}(V_t))^2 = \mathsf{E}(X_t)^2 = 4.2^2$$

So $V_t$ is variance stationary although not mean stationary.

- $Y_t = (F_t + X_t) \times P_t$. The time series $Y_t$ is the sum of $F_t \times P_t$ and $X_t \times P_t$. The time series plot of $Y_t$ is shown in the third panel of Fig. 2.11. We see that there seems to be little difference between $Y_t$'s time series plot and $V_t$'s. In other words, it is not easy to distinguish one another by checking their time series plots. Nevertheless, $V_t$ is variance stationary, but the variance of $Y_t$ is

$$\mathrm{Var}(Y_t) = \mathsf{E}(Y_t - \mathsf{E}(Y_t))^2 = \mathsf{E}(P_t X_t)^2 = 4.2^2 P_t^2$$

and so $Y_t$ is not variance stationary.

- $Z_t = (P_t + X_t) \times F_t$. The time series $Z_t$ is the product of $M_t$ and $F_t$. Its time series plot is shown in the bottom panel of Fig. 2.11. The time series plot shows that there seems to be neither trend nor seasonality in it although it does have both. Besides, it is neither mean stationary nor variance stationary. Its volatility gets bigger and bigger over time.

In summary, these six time series are comprised of $F_t$, $P_t$, and $X_t$. We see that the true features of the deterministic components $F_t$ and $P_t$ tend to disappear in the time series plots. In addition, some of them are mean stationary, and some of them are variance stationary, and the other of them are neither mean stationary nor variance stationary. This suggests that it is not easy to decompose a time series if we do not know what its deterministic components are like. The Python code for this section is as follows where the code for Fig. 2.12 is left as an exercise to the reader.

```
>>> import numpy as np
>>> import pandas as pd
>>> import matplotlib.pyplot as plt
>>> t=np.arange(500)
>>> f_t=0.2+0.1*t
>>> f_t=pd.Series(f_t)
>>> p_t=2*np.sin(2*np.pi*t/50 +0.3*np.pi)
>>> p_t=pd.Series(p_t)
>>> np.random.seed(1357)
>>> x_n=np.random.normal(loc=0, scale=4.2, size=500)
>>> x_n=pd.Series(x_n)
>>> fx_n=f_t+x_n; fx=f_t*x_n
>>> px_n=p_t+x_n; px=p_t*x_n
>>> fig = plt.figure()
>>> fx_n.plot(ax= fig.add_subplot(411))
>>> plt.title('$0.2+0.1t$+N(0,$4.2^2$)')
>>> fx.plot(ax= fig.add_subplot(412))
>>> plt.title('($0.2+0.1t$)N(0,$4.2^2$)')
>>> px_n.plot(ax= fig.add_subplot(413))
```

```
>>> plt.title('$ 2\sin(2\pi t/50+0.3\pi)$+N(0,$4.2^2$)')
>>> px.plot(ax= fig.add_subplot(414))
>>> plt.title('$ 2\sin(2\pi t/50 +0.3\pi)$ N(0,$4.2^2$)')
>>> plt.xlabel('Time'); plt.show()
>>> fApAx=f_t+p_t+x_n
>>> fAp_x=(f_t+p_t)*x_n
>>> fAx_p=(f_t+x_n)*p_t
>>> pAx_f=(p_t+x_n)*f_t
>>> fig = plt.figure()
>>> fApAx.plot(ax= fig.add_subplot(411))
>>> plt.title('$0.2+0.1t+2\sin(2\pi t/50 +0.3\pi)$+N(0,$4.2^2$)')
>>> fAp_x.plot(ax= fig.add_subplot(412))
>>> plt.title('($0.2+0.1t+2\sin(2\pi t/50+0.3\pi)$)N(0,$4.2^2$)')
>>> fAx_p.plot(ax= fig.add_subplot(413))
>>> plt.title('($0.2+0.1t$+N(0,$4.2^2$))$2\sin(2\pi t/50+0.3\pi)$)')
>>> pAx_f.plot(ax= fig.add_subplot(414))
>>> plt.title('($2\sin(2\pi t/50+0.3\pi)$+N(0,$4.2^2$))($0.2+0.1t$)')
>>> plt.xlabel('Time'); plt.show()
```

## 2.4 Time Series Decomposition and Smoothing

In practice, many of realistic time series possess either deterministic seasonality (component) or a deterministic trend (component). Some of them may have both. We have observed several time series examples with a trend or/and seasonality in Chap. 1. After extracting the trend and seasonal components from a time series, the remainder is its random (variation) component. Decomposing a time series is helpful to better understand it and improve forecast accuracy. In this section, we further discuss how to split a time series into their components. Knowing the background of a time series is helpful to find its components. For example, in most cases, sales data possess deterministic seasonality, and macroeconomic time series have a deterministic trend.

### *2.4.1 Deterministic Components and Decomposition Models*

The deterministic trend (component) of a time series $\{X_t\}$ should possess the following features:

- It reflects systematic change in the time series and such change is relatively slow.
- It can be expressed by an aperiodic function of time $T(t) = T_t$.

The deterministic seasonality (component) of a time series $\{X_t\}$ should have the following features:

- It reflects systematic change in the time series.
- The seasonality is of a fixed frequency. The fixed frequency is usually called the (minimum) seasonal period.

- The movement pattern in one period is similar to the movement pattern in the next period. In other words, the seasonal effects are reasonably stable with respect to timing, direction, and magnitude.

The seasonality is also known as the seasonal variation; seasonal effect; seasonal component; and so forth. It tends to be caused by the changes of the calendar, the weather, the timing of decisions, and so on. Besides, in some literature, a component called the cycle is decomposed from a time series. In this book we combine the cycle with the trend (sometimes called trend-cycle component), and hence, there are at most three decomposable components in a time series.

There are two decomposition models used often in practice:

(1) the additive model $X_t = T_t + S_t + R_t$; (2) the multiplicative model $X_t = T_t S_t R_t$ where $T_t$, $S_t$, and $R_t$, respectively, denote the trend, seasonality, and random component. As long as the movements in the time series are almost repetitive from one period to another or the seasonal component $S_t$ looks like a mathematical periodic function of time $t$, the two decomposition models fit well. What is more, generally speaking, the additive decomposition model is more appropriate if the magnitude of the seasonal variation does not vary with the level of the time series. When the seasonal variation appears to be proportional to the level of the time series, then the multiplicative decomposition model is more appropriate. Note that the values in the time series data are required to be nonzero for the multiplicative decomposition model.

*Example 2.5 (Australian Employed Total Persons)* It does matter to have a job! The data set "AustraliaEmployedTotalPersons" in the folder Ptsadata is from the Australian Bureau of Statistics (http://www.abs.gov.au/). It is the time series of Australian monthly employed total persons from February 1978 to November 2018 (in thousands). Its time series plot is shown in Fig. 2.13. We see that there exists a linearly increasing trend in it, but it seems to have no seasonality. In order to more carefully check the time series, we graph a new time series plot shown in Fig. 2.14 which time window is from January 2013 to January 2017. At this point we clearly observe that there is seasonality in it and the period is 12 months or 1 year. Moreover, we are able to graph seasonal plots (viz., seasonal subseries plots) which can clearly exhibit every seasonal pattern and the changes in seasonality for a time series. The seasonal plots of this example are shown in Fig. 2.15 where the horizontal red lines indicate the means for each monthly season. It is easy to see that all the seasonal patterns are almost the same and there are hardly the changes in seasonality.

Now we choose the additive model to decompose this time series, and the resulting plot is shown in Fig. 2.16 (the multiplicative decomposition is left as an exercise to the reader). From the residual time series plot in the bottom panel of Fig. 2.16, the decomposition is not bad. Let us have a look at the ACF plot of the decomposition residual series shown in Fig. 2.17. It looks like an ACF plot of a stationary time series except the ACF values at lag 12 and lag 24. These two ACF values suggest that the time series should have seasonal correlation which will be discussed in detail in Chap. 5. Besides, we take the time window as 36 months

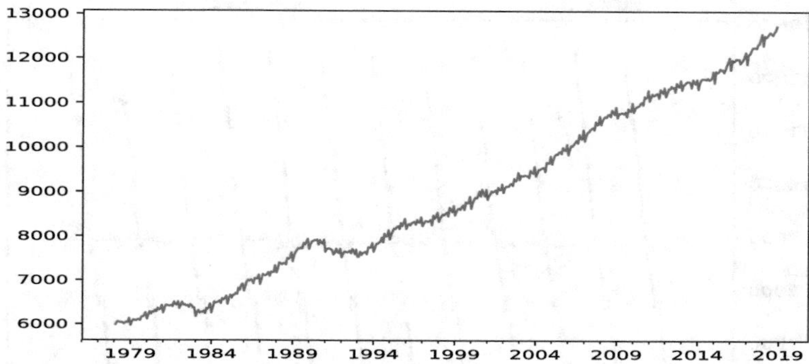


**Fig. 2.13** Time series plot of Australian employed persons from 1978.2 to 2018.11

pyTSA_AustralianLabour

**Fig. 2.14** Time series plot of Australian employed persons from 2013.1 to 2017.1

pyTSA_AustralianLabour

and plot the rolling means and rolling standard deviations shown in Fig. 2.18. We observe that the rolling means and rolling standard deviations are almost constant which further verifies that the residual series should be stationary. Below is the Python code for this example.

```
>>> import pandas as pd
>>> import matplotlib.pyplot as plt
>>> from PythonTsa.datadir import getdtapath
>>> dtapath=getdtapath()
>>> aul=pd.read_excel(dtapath +
        'AustraliaEmployedTotalPersons.xlsx', header=0)
>>> timeindex= pd.date_range('1978-02',periods=len(aul),freq='M')
>>> aul.index=timeindex
>>> aults=aul['EmployedP']
>>> aults.plot(); plt.show()
# Graph time series plot from 2013.1 to 2017.1
>>> aults['2013-01':'2017-01'].plot(); plt.show()
```

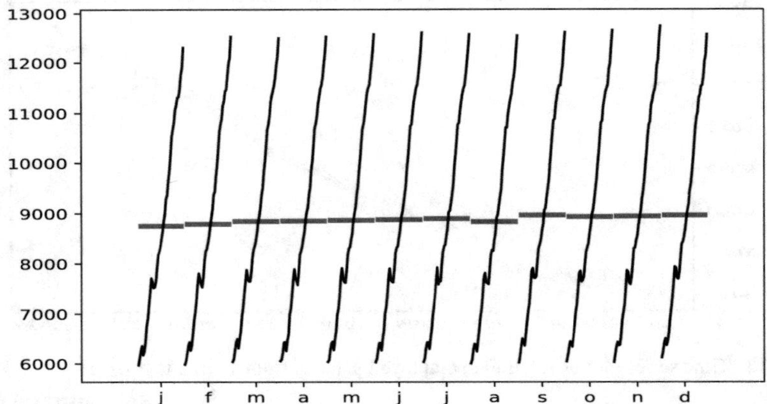


**Fig. 2.15** Seasonal plots of Australian employed persons from 1978.2 to 2018.11

pyTSA_AustralianLabour

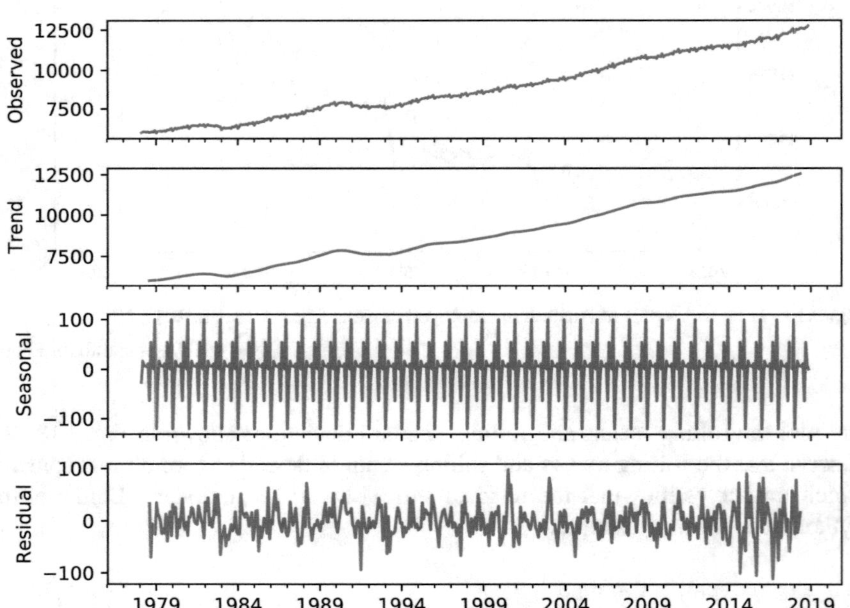


**Fig. 2.16** Additive decomposition of the time series in Example 2.5

pyTSA_AustralianLabour

```
>>> from statsmodels.graphics.tsaplots import month_plot
>>> month_plot(aults); plt.show() #Plot seasonal plots
>>> from statsmodels.tsa.seasonal import seasonal_decompose
>>> aultsdeca=seasonal_decompose(aults, model='additive')
>>> aultsdeca.plot(); plt.show()
```

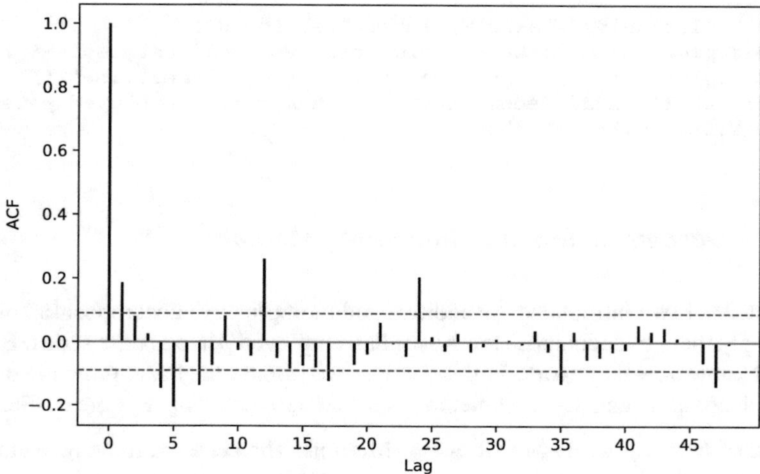


**Fig. 2.17** ACF plot of add. decom. resid. of the time series in Example 2.5

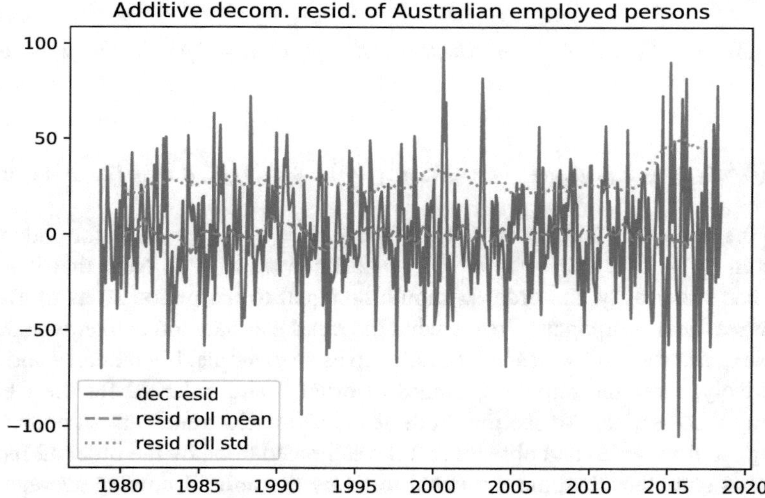


**Fig. 2.18** Rolling means and stds of add. decom. resid. of the series in Example 2.5

pyTSA_AustralianLabour

```
>>> aultsdecaResid=aultsdeca.resid.dropna()
>>> from PythonTsa.plot_acf_pacf import acf_pacf_fig
>>> acf_pacf_fig(aultsdecaResid, both=False, lag=48)
>>> plt.show()
>>> ar=aultsdecaResid
>>> rolm=pd.Series.rolling(ar, window=36, center=True).mean()
>>> rolstd=pd.Series.rolling(ar, window=36, center=True).std()
```

```
>>> plt.plot(aultsdecaResid, label='dec resid')
>>> plt.plot( rolm, label='resid roll mean', linestyle='--' )
>>> plt.plot(rolstd, label='resid roll std', linestyle=':')
>>> plt.title('Add. decom. resid. of Australian employed persons')
>>> plt.legend(); plt.show()
```

### *2.4.2 Decomposition and Smoothing Methods*

In Sect. 2.4.1, we introduced the additive and multiplicative decomposition models and apply them to decompose the Australian employed persons time series. But we do not know how they work. In this section, we discuss the basic principle of time series decomposition. The first method is the centered moving average or filtering.

**Centered Moving Average (Filtering) Method** The centered moving average or filtering approach is usually used to smooth or filter a time series $\{X_{1:n}\}$ so as to estimate its deterministic trend. The centered moving average of order $p$ is defined as

$$T_t = (X_{t-d} + X_{t-d+1} + \cdots + X_{t+d-1} + X_{t+d})/p, \; d = (p-1)/2, \; p \text{ is odd.}$$

and

$$T_t = (0.5X_{t-d} + X_{t-d+1} + \cdots + X_{t+d-1} + 0.5X_{t+d})/p, \; d = p/2, \; p \text{ is even.}$$

where the range of $t$ is from $(p+1)/2$ to $n-d = n-(p-1)/2$ for odd $p \geq 3$ and from $(p+2)/2$ to $n-d = n-p/2$ for even $p \geq 2$. Note that if a time series has seasonality, the order $p$ should be equal to the period so as to average out the seasonal component. To see what the trend estimate looks like, we take the Chinese quarterly GDP series in Example 1.6 as an example. Figures 2.19 and 2.20, respectively, show the moving averages of order 3 and order 12 for the Chinese quarterly GDP series. We see that both of the estimated trends are smoother than the original time series and observe that the estimated trend by the order 12 moving average is smoother than the estimated trend by the order 3 moving average. The pity is that we have to lose $2d = p-1$ data for odd $p$ or $2d = p$ data for even $p$ when doing the centered moving average of order $p$. The Python code for Figs. 2.19 and 2.20 is left as an exercise to the reader.

**Holt-Winters Smoothing** The Holt-Winters smoothing is proposed by Holt (1957) and Winters (1960) and uses exponentially weighted moving averages to update estimates of three components of a time series with a trend and seasonality. These three components are the (seasonally adjusted) level, slope, and seasonal component. Note that in some literature, the slope is called "trend," but it is obviously different from the concept of trend in Sect. 2.4.1. It is easy to understand that the seasonal pattern in one period can be different from the seasonal pattern

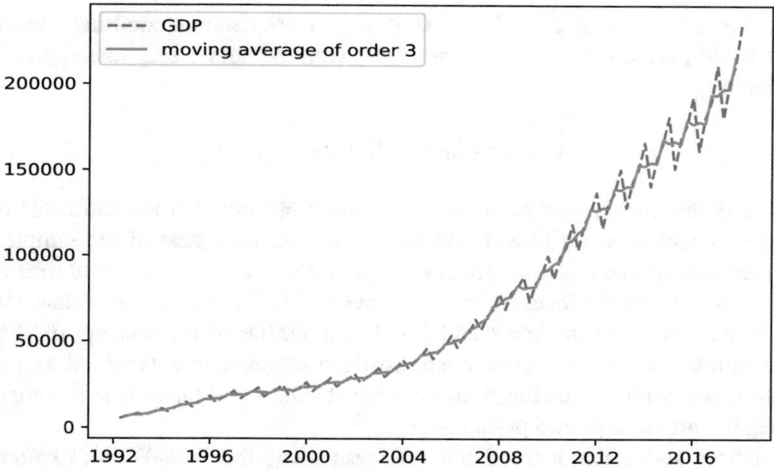


**Fig. 2.19** Moving average of order 3 for Chinese quarterly GDP from 1992 to 2017

**Fig. 2.20** Moving average of order 12 for Chinese quarterly GDP (1992–2017)

in another period due to vagaries of fashion, variation in climate, and so forth. The advantage of the Holt-Winters smoothing is that it allows for such change of seasonal patterns. Now we focus on two often used Holt-Winters smoothing approaches. The first is the additive Holt-Winters smoothing and its smoothing equations are

$$
\begin{aligned}
L_t &= \alpha(X_t - S_{t-p}) + (1-\alpha)(L_{t-1} + B_{t-1}) \\
B_t &= \beta(L_t - L_{t-1}) + (1-\beta)B_{t-1} \\
S_t &= \gamma(X_t - L_t) + (1-\gamma)S_{t-p}
\end{aligned}
$$

where $0 \leq \alpha \leq 1$, $0 \leq \beta \leq 1$, and $0 \leq \gamma \leq 1$ are called smoothing parameters and $p$ is the period length. At this point the Holt-Winters $h$ step ahead prediction function is

$$\hat{X}_{t+h|t} = L_t + hB_t + S_{t+h-p(k+1)}$$

where $k$ is the integer part of $(h-1)/p$, which ensures that the estimates of the seasonal indices used for forecasting come from the final year of the sample. The level estimate at time $t$ is a weighted average of the seasonally adjusted time series value at time $t$ and the level estimate at time $t-1$. The slope is a certain kind of level increment between time $t$ and $t-1$ and similar to the concept of slope in mathematics. The seasonal component estimate at time $t$ is a weighted average of the raw observation minus the level estimate at time $t$ and the seasonal component value at the previous period point $t-p$.

If a time series has a trend but no seasonality, the Holt-Winters smoothing becomes the Holt trend smoothing or double (parameter) exponential smoothing:

$$\begin{aligned} L_t &= \alpha X_t + (1-\alpha)(L_{t-1} + B_{t-1}) \\ B_t &= \beta(L_t - L_{t-1}) + (1-\beta)B_{t-1} \end{aligned}$$

If a time series has neither a trend nor seasonality, the Holt-Winters smoothing turns to the simple exponential smoothing or single (parameter) exponential smoothing:

$$L_t = \alpha X_t + (1-\alpha)L_{t-1} \tag{2.2}$$

It is easy to prove that Eq. (2.2) is equivalent to

$$L_t = \alpha X_t + \alpha(1-\alpha)X_{t-1} + \cdots + \alpha(1-\alpha)^{t-2}X_2 + \alpha(1-\alpha)^{t-1}X_1$$

and also equivalent to

$$L_t = \sum_{i=0}^{\infty} \alpha(1-\alpha)^i X_{t-i} = \alpha X_t + \alpha(1-\alpha)X_{t-1} + \alpha(1-\alpha)^2 X_{t-2} + \cdots \tag{2.3}$$

Since the weights in Eq. (2.3) decrease exponentially fast, this type of smoothing is called *exponential smoothing* or *exponentially weighted moving average*.

The second is the multiplicative Holt-Winters smoothing and its smoothing equations are

$$\begin{aligned} L_t &= \alpha(X_t/S_{t-p}) + (1-\alpha)(L_{t-1} + B_{t-1}) \\ B_t &= \beta(L_t - L_{t-1}) + (1-\beta)B_{t-1} \\ S_t &= \gamma(X_t/L_t) + (1-\gamma)S_{t-p} \end{aligned}$$

and now the Holt-Winters $h$ step ahead prediction function is

$$\hat{X}_{t+h|t} = (L_t + hB_t)S_{t+h-p(k+1)}$$

Special cases of the multiplicative Holt-Winters smoothing can be discussed too. For example, if a time series has seasonality but no trend, then the multiplicative Holt-Winters smoothing is reduced to the following:

$$L_t = \alpha(X_t/S_{t-p}) + (1-\alpha)L_{t-1}$$
$$S_t = \gamma(X_t/L_t) + (1-\gamma)S_{t-p}$$

For more about the Holt-Winters smoothing method, it can be found in Hyndman and Athanasopoulos (2018) and Ord et al. (2017). In the next subsection, as a case study, we will consider decomposing a real-life time series.

### 2.4.3 *Example*

*Example 2.6 (Example 1.6 Continued)* We clearly see that the Chinese quarterly GDP from 1992 to 2017 has both an increasing trend and seasonality from Example 1.6. Now we decompose it with the Python function `seasonal_decompose`. The additive decomposition result plots, and multiplicative decomposition result plots are, respectively, shown in Figs. 2.21 and 2.22. By comparing the additive decomposition residual time plot shown in the bottom panel of Fig. 2.21 with the multiplicative decomposition residual time plot shown in the bottom panel of Fig. 2.22 as well as the two residual series lag plots shown in Fig. 2.23, it turns out that the multiplicative decomposition is a little better than the additive decomposition. However, from Fig. 2.24, we observe that the correlogram of the multiplicative decomposition residual series is not like one of a stationary time series. In other words, there should be a certain structure in the Chinese quarterly GDP series which has not been revealed. Therefore, unlike Example 2.5, here both of the decomposition results are not satisfactory.

Now we turn to use the additive Holt-Winters smoothing approach to smooth and predict the Chinese quarterly GDP series. The smoothing result plots is shown in Fig. 2.25. We see that the slope plot shown in the third panel of Fig. 2.25 is clearly different from the trend plots shown in the second panels of Figs. 2.21 and 2.22. Besides, the seasonal component time series plot shown in the bottom panel of Fig. 2.25 exhibits the periodic changes. By observing the time series plot shown in Fig. 2.26 of the residuals of the Chinese quarterly GDP Holt-Winters smoothing, the lag plot of the residuals shown in Fig. 2.27, and the ACF plot of the residuals shown in Fig. 2.28, we can say that the trend and seasonal components in the Chinese quarterly GDP series have been completely removed, and the residual series looks like a stationary time series. Additionally, in the ACF plot, we see that the ACF value

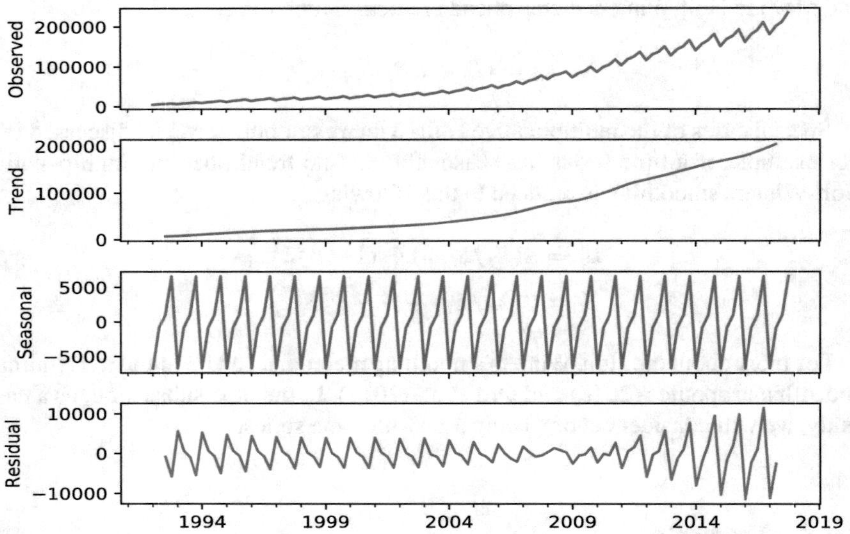


**Fig. 2.21** Additive decomposition of Chinese quarterly GDP from 1992 to 2017

pyTSA_GDPChinaDecomposition

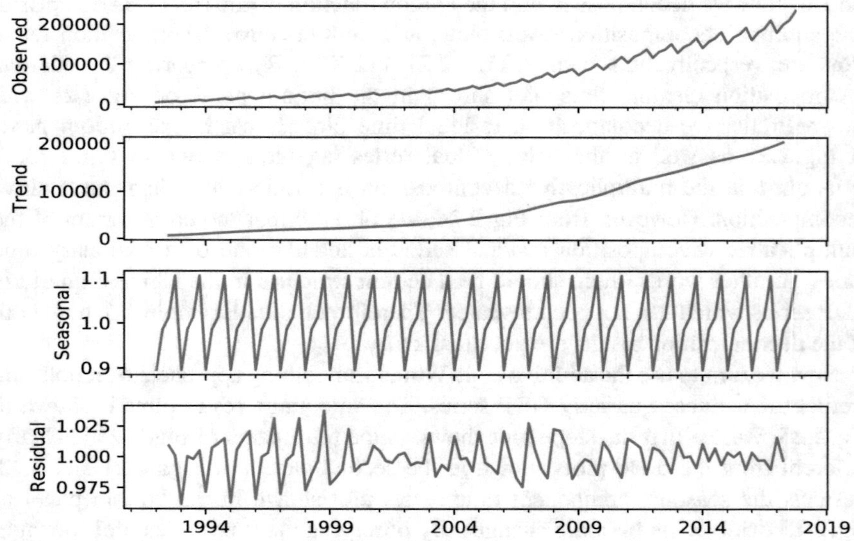


**Fig. 2.22** Multiplicative decom. of Chinese quarterly GDP from 1992 to 2017

pyTSA_GDPChinaDecomposition

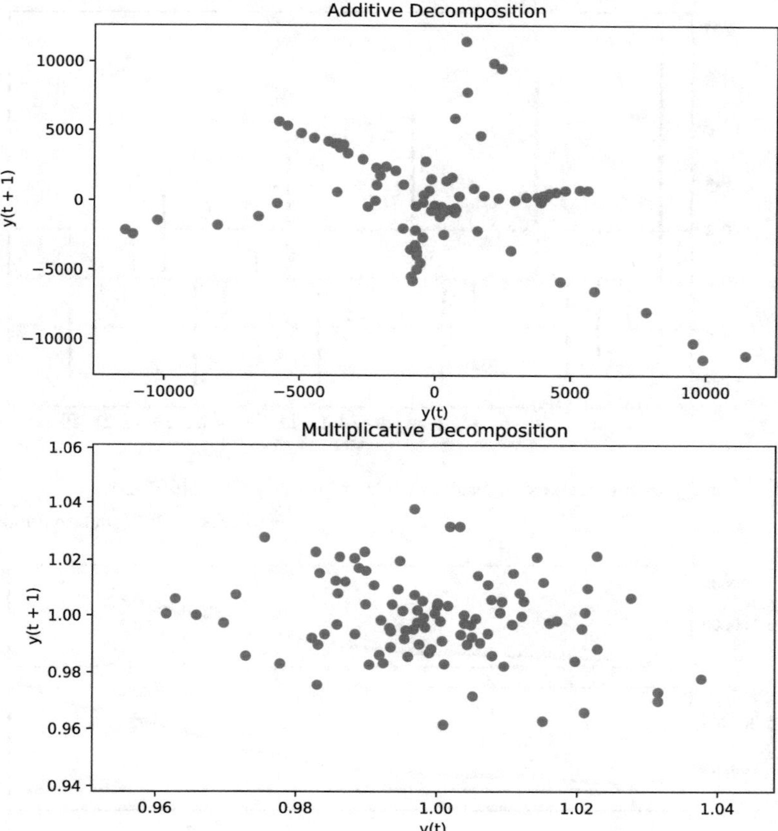


**Fig. 2.23** Lag plots of decom. residuals of Chinese quarterly GDP (1992–2017)

pyTSA_GDPChinaDecomposition

at lag 12 is significantly nonzero, which suggests that the residual series should have seasonal correlation (this term is traditional and should perhaps be called periodic correlation). Now we are able to predict the future values of the Chinese quarterly GDP and obtain the out-of-sample forecast values for the year 2018:

$$202178.3, \ 222846.7, \ 234333.5, \ 254601.4$$

which are, respectively, for the first, second, third, and fourth quarter of 2018. On the other hand, the actual Chinese GDP of the 4 quarters of 2018 are, respectively,

$$198783.1, \ 220178.0, \ 231937.7, \ 253598.6$$

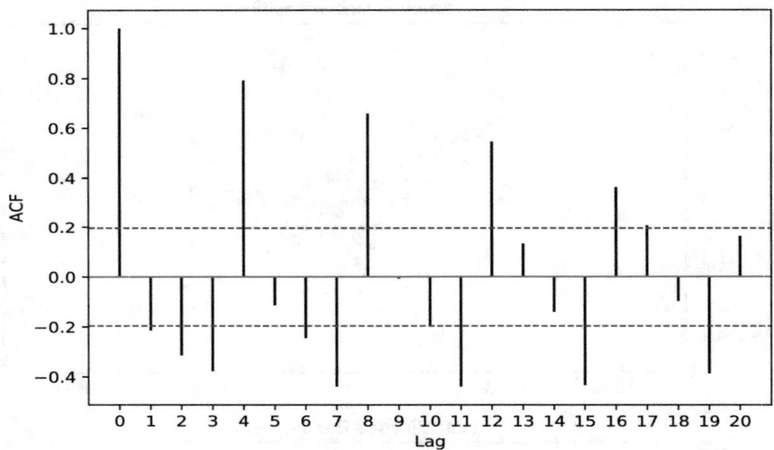


**Fig. 2.24** ACF plot of mult. decom. resid. of Chinese quarterly GDP (1992–2017)

pyTSA_GDPChinaDecomposition

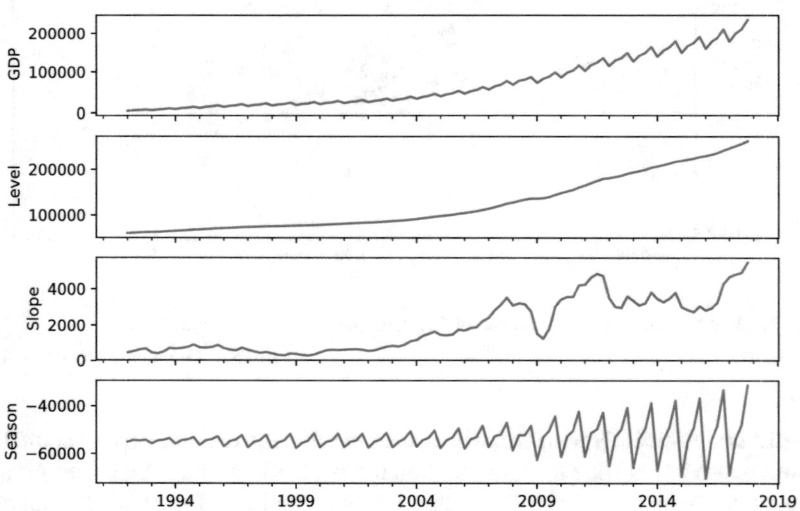


**Fig. 2.25** Holt-Winters smoothing result plots of Chinese quarterly GDP

pyTSA_GDPChinaDecomposition

which are available from The Chinese National Bureau of Statistics website: http://www.stats.gov.cn/. Comparing the forecast values with the actual values knows that the out-of-sample forecasting result by the additive Holt-Winters smoothing is satisfactory in terms of this example. Generally speaking, the Holt-Winters smoothing method is quite suitable for macroeconomic data. In addition, Fig. 2.29 displays the comparison of Chinese quarterly GDP series with its Holt-Winters out-

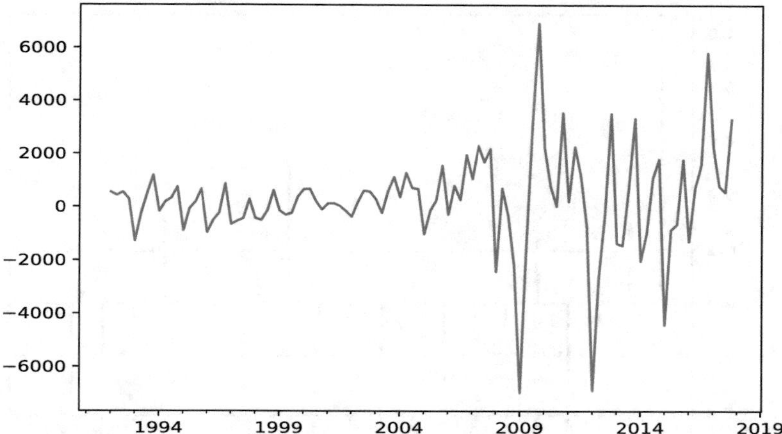


**Fig. 2.26** Time plot of resid. of Chinese quarterly GDP Holt-Winters smoothing
pyTSA_GDPChinaDecomposition

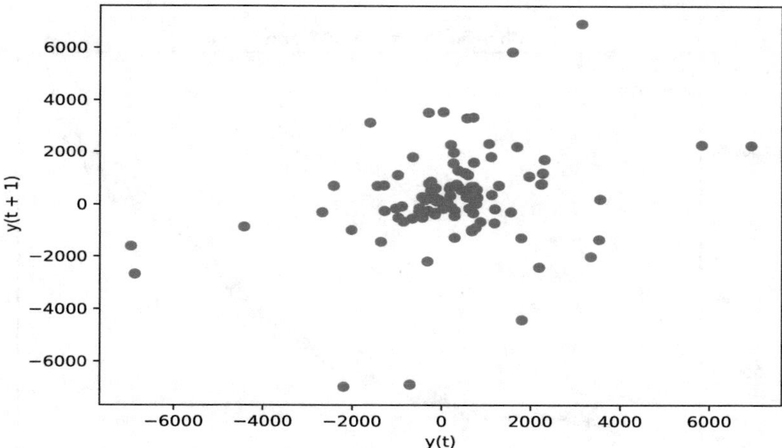


**Fig. 2.27** Lag plot of residuals of Chinese quarterly GDP Holt-Winters smoothing
pyTSA_GDPCinaSmoothing

of-sample and in-sample predict. Lastly, we do a Ljung-Box test on the residual series of the Chinese quarterly GDP Holt-Winters smoothing. We find that all the $p$-values are tiny for lag $m \leq 20$. This suggests that the residual series possesses seasonal and/or autocorrelation, which is the topic that we will discuss in Chaps. 3 and 4.

Below is the Python code for Example 2.6:

```
>>> import pandas as pd
```

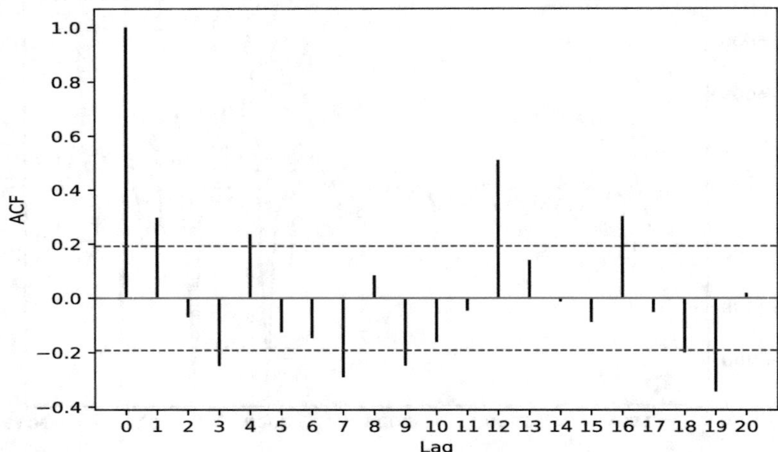


**Fig. 2.28** ACF plot of residuals of Chinese quarterly GDP Holt-Winters smoothing

pyTSA_GDPChinaDecomposition

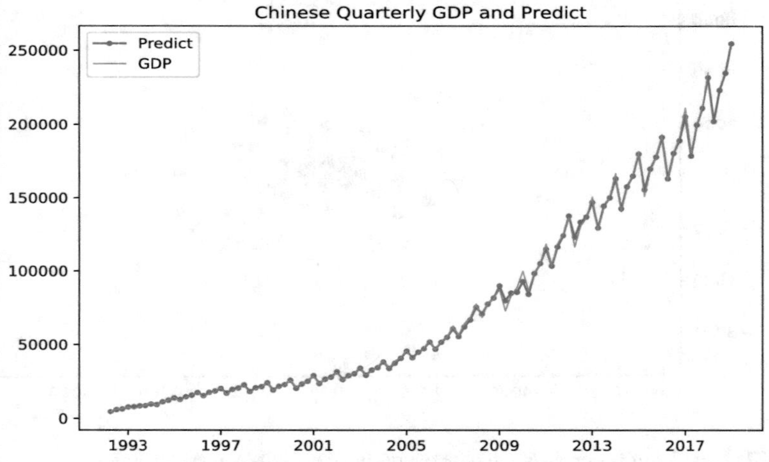


**Fig. 2.29** Comparison of Chinese quarterly GDP with its Holt-Winters predict

pyTSA_GDPChinaDecomposition

```
>>> import numpy as np
>>> from PythonTsa.datadir import getdtapath
>>> dtapath=getdtapath()
>>> x=pd.read_csv(dtapath +
                'gdpquarterlychina1992.1-2017.4.csv',header=0)
>>> dates = pd.date_range(start='1992',periods=len(x),freq='Q')
>>> x.index=dates
>>> x=pd.Series(x['GDP'])
>>> from statsmodels.tsa.seasonal import seasonal_decompose
```

```
>>> xdeca=seasonal_decompose(x, model='additive')
>>> xdecm=seasonal_decompose(x, model='multiplicative')
>>> import matplotlib.pyplot as plt
>>> xdeca.plot(); plt.show()
>>> xdecm.plot(); plt.show()
>>> from pandas.plotting import lag_plot
>>> fig = plt.figure()
>>> lag_plot(xdeca.resid, ax=fig.add_subplot(211))
>>> plt.title('Additive Decomposition')
>>> lag_plot(xdecm.resid, ax=fig.add_subplot(212))
>>> plt.title('Multiplicative Decomposition')
>>> plt.show()
>>> xdecmResid=xdecm.resid
>>> xdecmResid=xdecmResid.dropna()
>>> from PythonTsa.plot_acf_pacf import acf_pacf_fig
>>> acf_pacf_fig(xdecmResid, both=False, lag=20)
>>> plt.show()
>>> from statsmodels.tsa.holtwinters import ExponentialSmoothing
>>> xhwfit=ExponentialSmoothing(x,trend='add',seasonal='add',
    seasonal_periods=4).fit(method='L-BFGS-B')
>>> ax1=plt.subplot(411);x.plot();plt.setp(ax1.get_xticklabels(),
        visible=False); plt.ylabel('GDP')
>>> ax2=plt.subplot(412,sharex=ax1);xhwfit.level.plot();plt.setp
      (ax2.get_xticklabels(),visible=False); plt.ylabel('Level')
>>> ax3=plt.subplot(413,sharex=ax1);xhwfit.slope.plot();plt.setp
      (ax3.get_xticklabels(),visible=False); plt.ylabel('Slope')
# now "slope" is removed, use "trend" instead.
>>> ax4=plt.subplot(414, sharex=ax1); xhwfit.season.plot();
        plt.ylabel('Season')
>>> plt.show()
>>> xhwfit.resid.plot(); plt.show()
>>> lag_plot( xhwfit.resid); plt.show()
>>> acf_pacf_fig(xhwfit.resid, both=False, lag=20)
>>> plt.show()
>>> y=xhwfit.forecast(4)  #out-of-sample forecasting
>>> y
2018-03-31    202178.288835
2018-06-30    222846.650597
2018-09-30    234333.458081
2018-12-31    254601.425695
Freq: Q-DEC, dtype: float64
>>> z=xhwfit.predict(start='1992-03-31',end='2018-12-31')
#In-sample prediction and out-of-sample forecasting
>>> z=pd.DataFrame(z,columns={'Predict'})
>>> zx=z.join(x)
>>> Predict,=plt.plot(zx['Predict'],marker='.',label='Predict')
>>> GDP,=plt.plot(zx['GDP'], linewidth=1.0, label='GDP')
>>> plt.title('Chinese Quarterly GDP and Predict')
>>> plt.legend(handles=[Predict, GDP]); plt.show()
>>> from statsmodels.tsa.stattools import acf
>>> r,q,p=acf(xhwfit.resid,qstat=True, nlags=20)
>>> p
array([2.09574355e-03, 6.72877624e-03, 7.87531414e-04, 1.30115533e-04,
       1.58990371e-04, 1.35621751e-04, 5.24844206e-06, 9.11896120e-06,
       1.05247148e-06, 6.89591301e-07, 1.42558449e-06, 5.86893273e-12,
```

```
5.61549021e-12, 1.44169367e-11, 2.39456667e-11, 4.62614486e-13,
9.99850424e-13, 2.82861168e-13, 1.07451681e-15, 2.64968588e-15])
```

There are other decomposition approaches such as SEATS, STL, X11, and so on. The reader can find them in Dagum and Bianconcini (2016) as well as Hyndman and Athanasopoulos (2018). There are also other smoothing methods such as kernel smoothing, local polynomial fitting, smoothing spline method, and so forth. The reader can refer to Ghosh (2018), Fan and Yao (2003), as well as Chacón and Duong (2018).

## Problems

**2.1** Examine the composition time series that are not examined in Sect. 2.3.

**2.2** Write and then run the Python code for Fig. 2.12.

**2.3** Make a multiplicative decomposition of the Australian employed total persons in Example 2.5, and then explain properties of the decomposition residuals.

**2.4** Write and then run the Python code for Figs. 2.19 and 2.20 (Hint: use the function `rolling().mean()` ).

**2.5** Find the command for graphing quarterly seasonal plots in the Python package `statsmodels` , and then graph the seasonal plots for the dataset in Example 2.6 as well as give your comment on the plot.

**2.6** Use the multiplicative Holt-Winters smoothing method to analyze the dataset "gdpquarterlychina1992.1-2017.4.csv," and compare your results with the results in Example 2.6.

**2.7** The time series data "AustraliaUnemployedTotalPersons" in the folder "Ptsa-data" is from the Australian Bureau of Statistics. Make an exploratory data analysis of it.

**2.8** For the interpretation of economic statistics such as unemployment data or GDP data, it is important to identify the presence of seasonal components and to remove them so as to eliminate the effect of seasonal influence on the raw time series and reveal the true trend. This process is called the *seasonal adjustment*, and if the seasonal components are removed from the original time series data, the resulting values are known as the *seasonally adjusted time series*. Using the `seasonal_decompose` function, make a seasonal adjustment of the time series data "gdpquarterlychina1992.1-2017.4.csv," and describe what is the difference between the seasonally adjusted series and the trend component.

**2.9** By the Internet, download the Bitcoin price at the end of everyday from July 1, 2018, to the present. Then make an exploratory time series data analysis of the downloaded dataset using the approaches introduced in Chaps. 1 and 2.

# Chapter 3
# Stationary Time Series Models

This chapter first introduces the backshift operator, which is widely used for model simplicity and differencing, which is one way to make a nonstationary time series stationary. Then we present a statistical test on stationarity—the KPSS stationarity test. Third, we define MA, AR, and ARMA models and discuss their properties, including invertibility, causality, and more. We also distinguish the ARMA model from the ARMA process.

## 3.1 Backshift Operator, Differencing, and Stationarity Test

### *3.1.1 Backshift Operator*

The backshift (or lag) operator is like a matrix in mathematics but operates on a time series$\{X_t\}$. Using $B$ for backshift, we define the backshift operator by

$$BX_t = X_{t-1}$$

That is, $B$ has the effect of shifting the data back one step when operating on $\{X_t\}$. Generally, we define

$$B^n X_t = B^{n-1}(BX_t) = X_{t-n}$$

for any integer $n \geq 1$ and $B^0 X_t = X_t$. The operating properties of $B$ itself are similar to the ones of the matrix. For example, we have that for any real numbers $\alpha$ and $\beta$, and any time series $\{X_t\}$ and $\{Y_t\}$,

$$B(\alpha X_t \pm \beta Y_t) = B(\alpha X_t) \pm B(\beta Y_t) = \alpha X_{t-1} \pm \beta Y_{t-1}$$

 
C. Huang, A. Petukhina, *Applied Time Series Analysis and Forecasting with Python*, Statistics and Computing,
https://doi.org/10.1007/978-3-031-13584-2_3

and

$$(1-B)^n X_t = \sum_{i=0}^{n}(-1)^i \frac{n!}{i!(n-i)!} B^i X_t = \sum_{i=0}^{n}(-1)^i \frac{n!}{i!(n-i)!} X_{t-i}$$

Using the backshift operator $B$, the random walk can be written as $(1-B)X_t = W_t$ where $W_t \sim \text{WN}(0, \sigma^2)$. In the following chapters, we will see that the backshift operator is widely used for model expression simplicity.

### 3.1.2 *Differencing and Stationarity*

In Example 1.9, we have seen that for a nonstationary time series, by computing the differences between consecutive time series values, we can make it stationary. This method of making a nonstationary time series stationary is called *differencing*. It can help stabilize the mean and variance of a time series, therefore removing (or reducing) trend and seasonality. Formally we give the following definition.

**Definition 3.1** Using the backshift operator $B$, (1) the differencing of order $d$ (or $d$th order differencing) is defined as

$$\nabla^1 X_t = \nabla X_t = (1-B)X_t = X_t - X_{t-1}, \quad \nabla^d X_t = (1-B)^d X_t$$

It is also called the difference operator of order $d$. (2) The differencing of lag $k$ is defined as

$$\nabla_1 X_t = \nabla X_t = (1-B)X_t = X_t - X_{t-1}, \quad \nabla_k X_t = (1-B^k)X_t = X_t - X_{t-k}$$

(3) If the lag $k$ in (2) is the number of seasons (or the period of a time series), the lag $k$ differencing is known as the seasonal differencing.

Pay attention to the following remarks about this definition:

- We do differencing in order to make a nonstationary time series stationary.
- If a time series possesses trend but no seasonality, in order to make it stationary, just use the $d$th order differencing.
- If a time series has seasonality, we should firstly do the seasonal differencing and then consider whether the $d$th order differencing is needed. It is because the seasonally differenced series will sometimes be stationary.
- $\nabla^d \neq \nabla_d$ except for $d = 1$.

Note that stationarizing a time series through differencing is an important part of the process of fitting an ARIMA model, as discussed in the next chapters. Examples of making a nonstationary time series stationary by differencing are given in the next subsection.

### 3.1.3 KPSS Stationarity Test

The stationarity test has been an important issue in time series analysis. In practice, only by using a time series dataset, can we test stationarity of the time series. In the previous two chapters, we decide whether a time series is stationary or not by examining its time series plot and correlogram. If a time series has a clear trend or seasonality, it is obviously nonstationary. For a nonstationary time series, we do everything possible to eliminate its trend and/or seasonality so that we can further analyze it. However, the difficulty is how to test whether a time series that has neither obvious trend nor clear seasonality is stationary. Kwiatkowski, Phillips, Schmidt, and Shin (KPSS, 1992) proposed a test of the null hypothesis that an observable time series is (level) stationary or trend stationary (stationary around a deterministic trend). The time series can first be expressed as the sum of deterministic trend, random walk, and stationary error, and the test is the Lagrange multiplier test of the hypothesis that the random walk has zero variance. More details about the KPSS stationarity test will discussed in Chap. 9. For application of the KPSS test, in the module `statsmodels.tsa.stattools`, there is a test function `kpss` that is the KPSS stationarity test where the argument `regression='c'` means that the function tests the stationarity of a time series without clear trend and obvious seasonality.

*Example 3.1 (Southern Hemisphere Temperature Volatility Data Series)* The dataset "Southtemperature" in the folder Ptsadata is the monthly temperature volatility series (Jan. 1850–Dec. 2007) for the southern hemisphere, which were extracted from the database maintained by the University of East Anglia Climatic Research Unit. This dataset is read from left to right. In order to make it belong to the `Series` class in `pandas`, we use the function `pd.concat` to connect every row in it. As it has 158 rows, we use the `for` loop to complete this connection. The time series plot of the temperature data is shown in Fig. 3.1, and we see that it has trend but no seasonality. Hence we take a first difference on the series and then graph the time series plot of the differenced temperature series shown in Fig. 3.2, which exhibits no trend (the trend is eliminated). We continue to plot the ACF of the first differenced temperature data, which is shown in Fig. 3.3, and we observe that for any $k \geq 2$, the ACF value at lag $k$ is almost zero. Both the time series plot and ACF plot show that the differenced temperature series should be stationary. Now we further use the KPSS stationarity test to test the stationarity of the differenced temperature series. The testing result shows that the $p$-value of the test statistic is greater than 0.1, which gives a powerful evidence to support that the differenced temperature series is stationary. In summary, the first differenced temperature series is stationary although the original temperature series is nonstationary. The Python code for this example is as follows.

```
>>> import numpy as np
>>> import pandas as pd
>>> from PythonTsa.datadir import getdtapath
>>> dtapath=getdtapath()
```

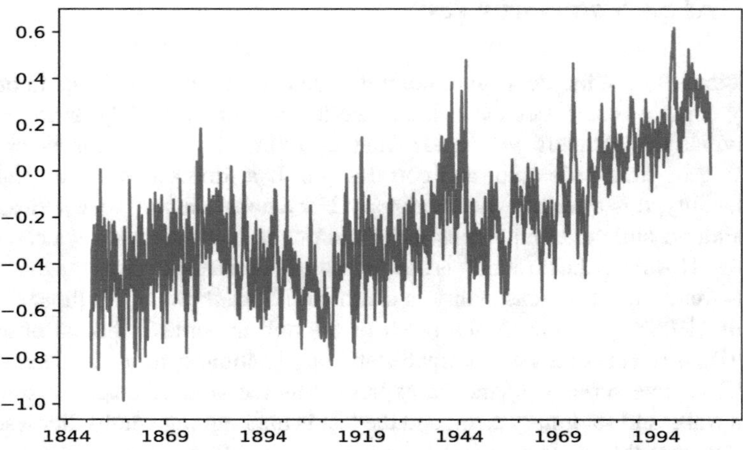


**Fig. 3.1** Time series plot of the temperature data for the southern hemisphere

pyTSA_TemperatureSH

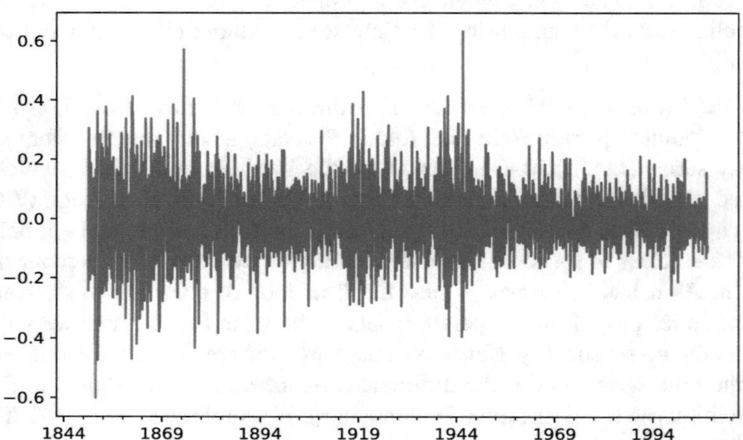


**Fig. 3.2** Time plot of the first differenced temp. data for the southern hemisphere

pyTSA_TemperatureSH

```
>>> tem=pd.read_csv(dtapath +'Southtemperature.txt',
                    header=None,sep='\s+')
# now read_table is deprecated, use read_csv instead.
>>> temts=pd.concat([tem.loc[0],tem.loc[1]],ignore_index='true')
>>> for i in range(2,158):
        temts=pd.concat([temts,tem.loc[i]],ignore_index='true')
>>> type(temts)
<class 'pandas.core.series.Series'>
>>> dates = pd.date_range('1850', periods= len(temts), freq='M')
>>> temts.index=dates
```

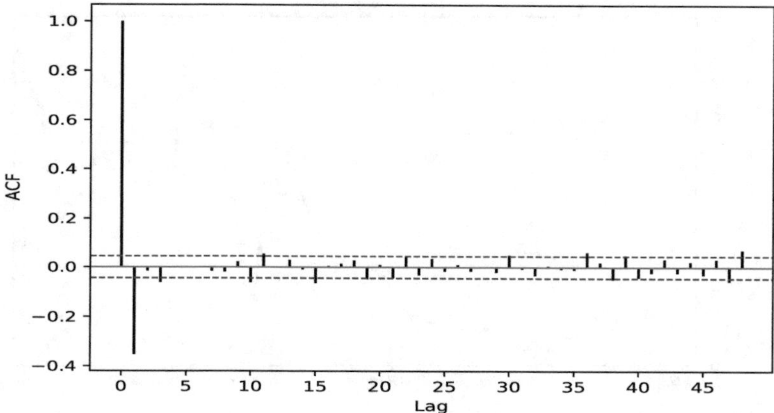


**Fig. 3.3** ACF plot of the first differenced temp. data for the southern hemisphere

pyTSA_TemperatureSH

```
>>> import matplotlib.pyplot as plt
>>> temts.plot(); plt.show()
>>> dt=temts.diff(1)  # the first differencing
>>> dt=dt.dropna()
>>> dt.plot(); plt.show()
>>> from PythonTsa.plot_acf_pacf import acf_pacf_fig
>>> acf_pacf_fig(dt, both=False, lag=48)
>>> plt.show()
>>> from statsmodels.tsa.stattools import kpss
>>> kpss(dt,regression='c', nlags='auto')
InterpolationWarning: p-value is greater than the indicated p-value
(0.026191175282707216, 0.1, 56,
{'10%': 0.347, '5%': 0.463, '2.5%': 0.574, '1%': 0.739})
```

*Example 3.2 (The Chinese Quarterly GDP)* We have discussed the Chinese quarterly GDP series from 1992 to 2017 in the previous chapters. From Fig. 1.7, we see that the time series has both trend and seasonality. Since it is the quarterly data, the number of seasons is 4 naturally. Now we seasonally difference it with the lag 4 and graph the time series plot of the seasonally differenced series shown in Fig. 3.4. From Fig. 3.4, we observe that the seasonality in the quarterly GDP series has been removed. Additionally, the seasonally differenced series unexpectedly reveals the truth that the GDP values around 2009 sharply decrease. This fact cannot be seen from the original time series plot shown in Fig. 1.7 due to the impact of seasonality. It is well known that the 2008 subprime mortgage crisis that happened in the United States triggered the world economic crisis. Like other economies, China's economy was also affected by the global economic crisis around 2009. The time series plot of the seasonally differenced series demonstrates this big event well. The pity is that the seasonally differenced series is still nonstationary. We continue to take a first

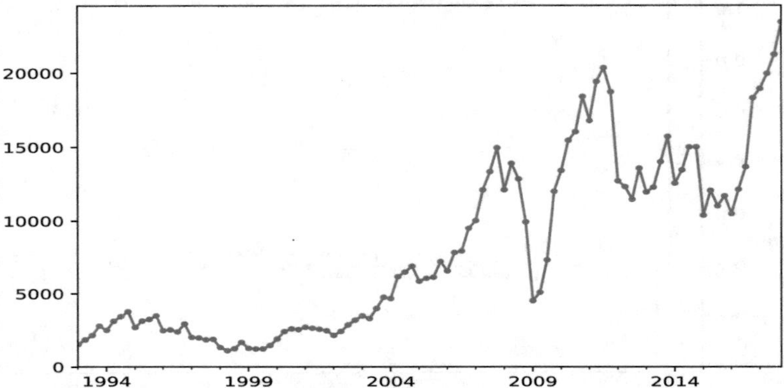


**Fig. 3.4** Time series plot of the seasonally differenced Chinese quarterly GDP

pyTSA_GDPChinaDiff

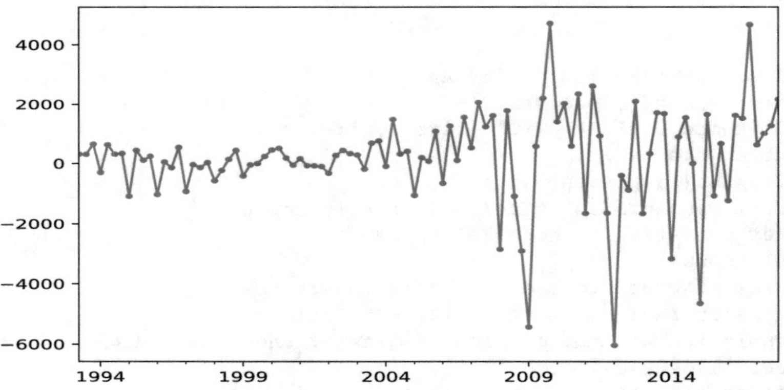


**Fig. 3.5** Time plot of the seasonal and first differences of Chinese quarterly GDP

pyTSA_GDPChinaDiff

difference on the seasonally differenced series and obtain the seasonally and firstly differenced series. From its time series plot shown in Fig. 3.5, we see that the trend in the original series or seasonally differenced series has been eliminated, and the resulting series looks stationary. From its ACF plot shown in Fig. 3.6, we observe that the ACF values at lags 4, 12, and 16 as well as 5, 7, and 19 are significantly nonzero. This indicates that there are both autocorrelation and seasonal correlation in the resulting series. In other words, the correlations of a time series cannot be differenced off. Lastly, the KPSS test result verifies that the seasonally and firstly differenced series is stationary. The following is the Python code for this example.

```
>>> import pandas as pd
>>> import numpy as np
>>> from PythonTsa.datadir import getdtapath
```

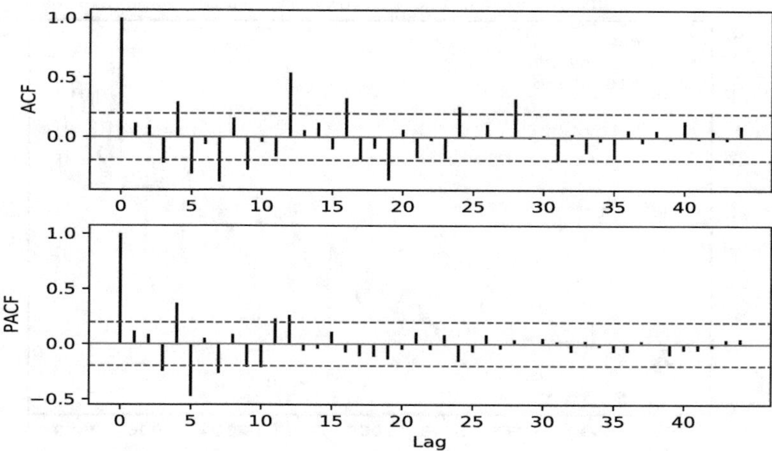


**Fig. 3.6** ACF plot of the seasonally and firstly differenced Chinese quarterly GDP
pyTSA_GDPChinaDiff

```
>>> dtapath=getdtapath()
>>> x=pd.read_csv(dtapath +
                'gdpquarterlychina1992.1-2017.4.csv',header=0)
>>> dates = pd.date_range(start='1992',periods=len(x),freq='Q')
>>> x.index=dates; x=pd.Series(x['GDP'])
>>> dx=x.diff(4)  # seasonal differencing
>>> dx=dx.dropna()
>>> import matplotlib.pyplot as plt
>>> dx.plot(marker='o',ms=3) # ms means marker size
>>> plt.show()
>>> d1dx=dx.diff(1)
>>> d1dx=d1dx.dropna()
>>> d1dx.plot(marker='o',ms=3); plt.show()
>>> from PythonTsa.plot_acf_pacf import acf_pacf_fig
>>> acf_pacf_fig(d1dx, both=True, lag=44)
>>> plt.show()
>>> from statsmodels.tsa.stattools import kpss
>>> kpss(d1dx, regression='c', nlags='auto')
InterpolationWarning: p-value is greater than the indicated p-value
(0.10664288385692093, 0.1, 3,
{'10%': 0.347, '5%': 0.463, '2.5%': 0.574, '1%': 0.739})
```

*Example 3.3 (Monthly Anti-Diabetic Drug Sales in Australia)* The dataset "AntidiabeticDrugSales" in the folder Ptsadata is the monthly anti-diabetic drug sales in Australia from July 1991 to June 2008. It is from the Medicare Australia. Here we use the notation $a10$ to denote it. In order to stabilize the variance function of the series $a10$, we take a logarithm on $a10$ and obtain the log series of $a10$ ($loga10$). Comparing Fig. 3.8 with Fig. 3.7, we can find that the rolling standard deviation line of $loga10$ is almost horizontal, while the rolling standard deviation line of $a10$ is

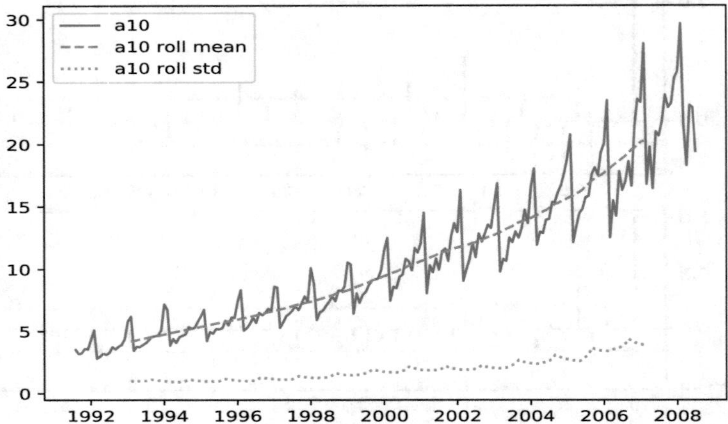


**Fig. 3.7** Time series plot and rolling std of the anti-diabetic drug sales in Australia
pyTSA_AntiDiabetSales fv

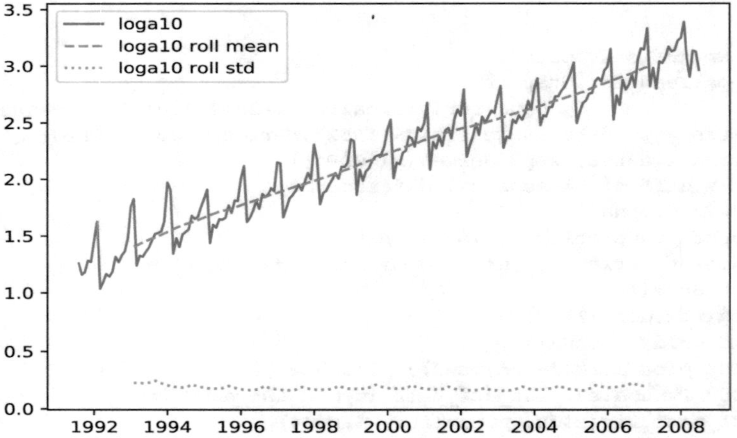


**Fig. 3.8** Time plot and rolling std of the log anti-diabetic drug sales in Australia
pyTSA_AntiDiabetSales

ascending. This indicates that the log transform has really stabilized the variance of the series $a10$ although $loga10$ still has both trend and seasonality. Now we seasonally difference $loga10$ with lag 12 and get the resulting series $dloga10$. Its time series plot is shown in Fig. 3.9 and looks well stationary. The KPSS test further affirms its stationarity. Note that for the series $loga10$ with trend and seasonality, the lag 12 seasonal differencing removes both seasonality and trend in it. The first differencing is no more needed. The following is the code for this example.

```
>>> import pandas as pd
>>> import numpy as np
```

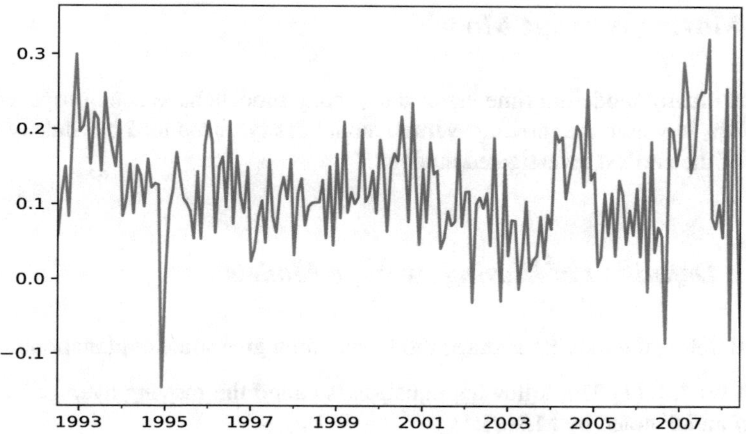


**Fig. 3.9** Seasonal differences of the log anti-diabetic drug sales in Australia

pyTSA_AntiDiabetSales

```
>>> from PythonTsa.datadir import getdtapath
>>> dtapath=getdtapath()
>>> a10=pd.read_csv(dtapath +
                    'AntidiabeticDrugSales.csv',header=None)
>>> dates=pd.date_range(start='1991-07',periods=len(a10),freq='M')
>>> a10.index=dates; a10=pd.Series(a10[0])
>>> import matplotlib.pyplot as plt
>>> rolma10=pd.Series.rolling(a10, window=36, center=True).mean()
>>> rolstda10=pd.Series.rolling(a10, window=36, center=True).std()
>>> from pandas.plotting import register_matplotlib_converters
>>> register_matplotlib_converters()
# explicitly register matplotlib converters.
>>> import matplotlib.pyplot as plt
>>> plt.plot(a10, label='a10')
>>> plt.plot(rolma10, label='a10 roll mean', linestyle='--')
>>> plt.plot(rolstda10, label='a10 roll std', linestyle=':')
>>> plt.legend(); plt.show()
>>> loga10=np.log(a10)
>>> rolmloga10=pd.Series.rolling(loga10,window=36,center=True).mean()
>>> rolstdloga10=pd.Series.rolling(loga10,window=36,center=True).std()
>>> plt.plot(loga10, label='loga10')
>>> plt.plot(rolmloga10, label='loga10 roll mean', linestyle='--')
>>> plt.plot(rolstdloga10, label='loga10 roll std', linestyle=':')
>>> plt.legend(); plt.show()
>>> dloga10=loga10.diff(12)
>>> dloga10=dloga10.dropna()
>>> dloga10.plot(); plt.show()
>>> from statsmodels.tsa.stattools import kpss
>>> kpss(dloga10, regression='c', nlags='auto')
InterpolationWarning: p-value is greater than the indicated p-value
(0.1634206184632204, 0.1, 7,
{'10%': 0.347, '5%': 0.463, '2.5%': 0.574, '1%': 0.739})
```

## 3.2 Moving Average Models

For statistically modeling time series data, many models have been proposed since the 1920s, in which the moving average model firstly introduced by Slutsky (1927) is one of the earliest investigated models.

### *3.2.1 Definition of Moving Average Models*

We first define the moving average model and then give some explanations.

**Definition 3.2** (1) The following equation is called the moving average model of order $q$ and denoted by MA($q$):

$$X_t = \mu + \varepsilon_t + \theta_1\varepsilon_{t-1} + \theta_2\varepsilon_{t-2} + \cdots + \theta_q\varepsilon_{t-q} \tag{3.1}$$

where $\{\varepsilon_t\} \sim \text{WN}(0, \sigma_\epsilon^2)$, that is, $\{\varepsilon_t\}$ is a white noise series and $\mu, \theta_1, \cdots, \theta_q$ are real-valued parameters (coefficients) with $\theta_q \neq 0$. (2) If a time series $\{X_t\}$ is stationary and satisfies such an equation as (3.1), then we call it an MA($q$) process.

Note the following remarks about this definition:

- For simplicity, we often assume that the intercept (const term) $\mu = 0$; otherwise, we can consider $\{X_t - \mu\}$.
- We distinguish the concept of MA models from the concept of MA processes.
- Sometimes $\varepsilon_t$ in Eq. (3.1) is known as the *innovation* term or *shock* term.
- The series $\{X_t\}$ generated by Eq. (3.1) or the MA($q$) model is obviously always stationary.

Additionally, using the backshift operator $B$, the MA($q$) model can be rewritten as

$$X_t = \theta(B)\varepsilon_t$$

where $\theta(z) = 1 + \theta_1 z + \cdots + \theta_q z^q$ is the MA polynomial.

**Definition 3.3** If the time series $\{X_t\}$ generated by a model is stationary, then we say that the model is stationary.

Note that this definition is not just for the MA model but for any model that can generate a time series. According to it, the MA($q$) model is always stationary. This is an important property of the MA($q$) model.

*Example 3.4 (An MA(2) Model)* Given an MA(2) model as follows

$$X_t = \varepsilon_t + 0.6\varepsilon_{t-1} - 0.3\varepsilon_{t-2}$$

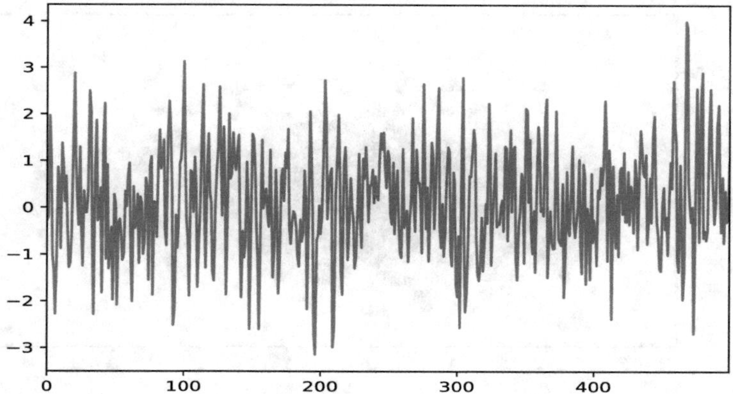


**Fig. 3.10** Time series plot of the simulated MA(2) sample in Example 3.4

pyTSA_SimMA2

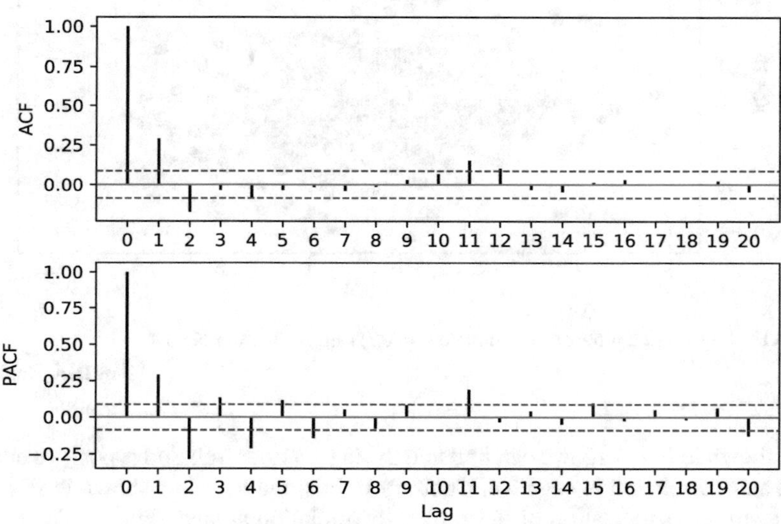


**Fig. 3.11** ACF and PACF plots of the simulated MA(2) sample in Example 3.4

pyTSA_SimMA2

where $\varepsilon_t \sim$ iidN(0,1), simulate a sample of size (length) 500 from the MA(2) model, and then graph the time series, ACF and PACF plots of it. The Python function `arma_generate_sample` can generate this time series sample. Figure 3.10 is the time series plot, which has neither trend nor seasonality and looks very stationary. Figure 3.11 is the ACF and PACF plots of the simulated sample. From it, we observe that after lag 2 almost the ACF values are zero except at lag 11 while the PACF values at 3, 4, 6, and 12 seem to be all nonzero. Furthermore, from the lag

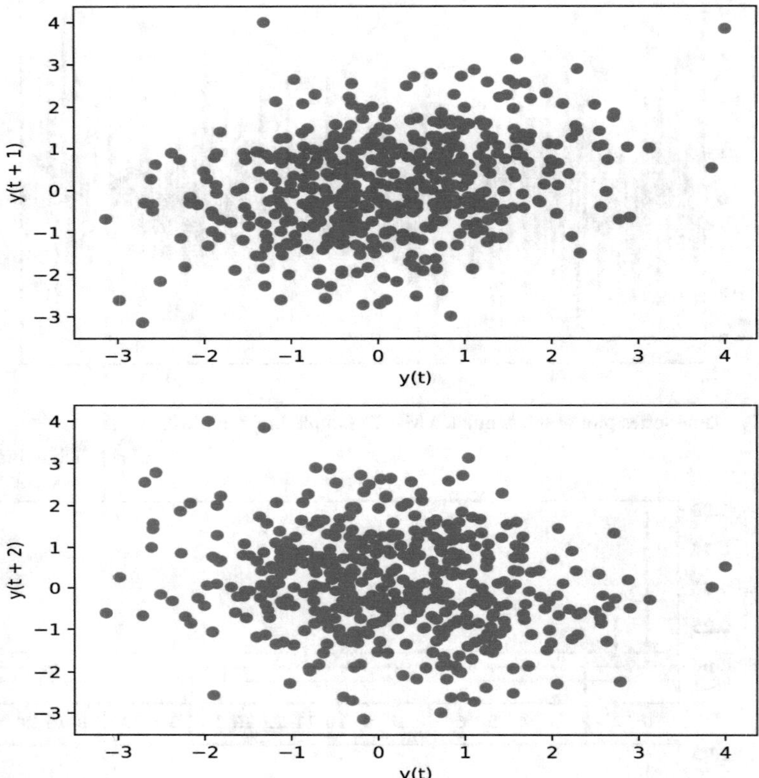


**Fig. 3.12** Lag 1 and 2 plots of the simulated MA(2) sample in Example 3.4

pyTSA_SimMA2

plots shown in Fig. 3.12, we can find that the lag 1 autocorrelation is positive and the lag 2 autocorrelation is negative. Contrary to this, the lag plots shown in Fig. 3.13 give a strong visualization of the zero autocorrelation at lags 3 and 4. The Python code for this example is as follows.

```
>>> import numpy as np
>>> import pandas as pd
>>> from statsmodels.tsa.arima_process import arma_generate_sample
>>> ma=np.array([1, 0.6,-0.3])
>>> np.random.seed(123457)
>>> x= arma_generate_sample(ar=[1], ma=ma, nsample=500)
# ar=[1] means no ar part in the model
>>> type(x)
<class 'numpy.ndarray'>  # x is not a series
>>> x=pd.Series(x)
>>> type(x)
<class 'pandas.core.series.Series'>  # now x is a series
```

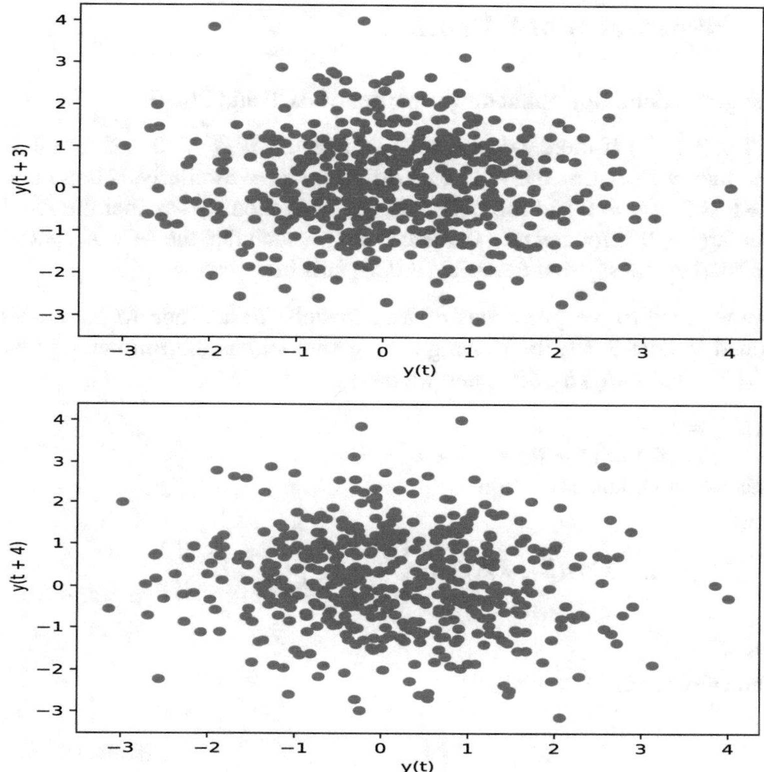


**Fig. 3.13** Lag 3 and 4 plots of the simulated MA(2) sample in Example 3.4

pyTSA_SimMA2

```
>>> import matplotlib.pyplot as plt
>>> x.plot(); plt.show()
>>> from PythonTsa.plot_acf_pacf import acf_pacf_fig
>>> acf_pacf_fig(x, both=True, lag=20)
>>> plt.show()
>>> from pandas.plotting import lag_plot
>>> fig = plt.figure()
>>> lag_plot(x, lag=1, ax=fig.add_subplot(211))
>>> lag_plot(x, lag=2, ax=fig.add_subplot(212))
>>> plt.show()
>>> fig = plt.figure()
>>> lag_plot(x, lag=3, ax=fig.add_subplot(211))
>>> lag_plot(x, lag=4, ax=fig.add_subplot(212))
>>> plt.show()
```

### 3.2.2 Properties of MA Models

We first give a definition about the properties of ACF and PACF:

**Definition 3.4** (1) If there exists $h$ such that the ACF $\rho_h \neq 0$ and for all $k > h$, $\rho_k = 0$, then we say that the ACF cuts off after lag $h$. Similarly, if there is $h$ such that the PACF $\phi_{hh} \neq 0$ and for all $k > h$, $\phi_{kk} = 0$, then we say that the PACF cuts off after lag $h$. (2) If for any $n > 0$, there is $m \geq n$ such that the ACF $\rho_m \neq 0$ (PACF $\phi_{mm} \neq 0$), then we say that the ACF (PACF) tails off.

Now we turn to the properties of MA models. In addition to the stationarity mentioned in Sect. 3.2.1, the moving average model has the following properties. Let $\{X_t\}$ be the MA($q$) model. Then we have

(1) $\mathsf{E}(X_t) = \mu$.
(2) $\gamma_0 = \text{Var}(X_t) = (1 + \theta_1^2 + \cdots + \theta_q^2)\sigma_\epsilon^2$.
(3) The autocovariance function

$$\gamma_k = \text{Cov}(X_t, X_{t+k}) = \begin{cases} 0, & \text{if } k > q, \\ \sigma_\epsilon^2 \sum_{i=0}^{q-k} \theta_i\theta_{i+k}, & \text{if } 0 \leq k \leq q \end{cases}$$

where $\theta_0 = 1$.
(4) The autocorrelation function

$$\rho_k = \text{Corr}(X_t, X_{t+k}) = \begin{cases} 1, & \text{if } k = 0, \\ 0, & \text{if } k > q, \\ \sum_{i=0}^{q-k} \theta_i\theta_{i+k} / \sum_{i=0}^{q} \theta_i^2, & \text{if } 0 < k \leq q. \end{cases}$$

The properties above confirm that all the moving average models are stationary. Note that for all $k > q$, $\rho_k = 0$, that is, the ACF of MA($q$) models cuts off after lag $q$, which is very helpful in building a MA model.

*Example 3.5 (Properties of the MA(1) Model)* Let $X_t = \varepsilon_t + \theta_1\varepsilon_{t-1}$. Then we have the autocorrelation function (ACF)

$$\rho_k = \text{Corr}(X_t, X_{t+k}) = \begin{cases} 1, & \text{if } k = 0, \\ 0, & \text{if } k > 1, \\ \theta_1/(1 + \theta_1^2), & \text{if } k = 1. \end{cases}$$

The partial autocorrelation(PACF) at lag 2 for the MA(1) mode] is

$$\phi_{22} = \frac{-\theta_1^2}{1 + \theta_1^2 + \theta_1^4}.$$

In fact, now $\rho_1 = \theta_1/(1+\theta_1^2)$, $\rho_2 = 0$, and according to Example 2.1, we have

$$\phi_{22} = \frac{\rho_2 - \rho_1^2}{1-\rho_1^2} = \frac{-\rho_1^2}{1-\rho_1^2} = \frac{-\theta_1^2}{1+\theta_1^2+\theta_1^4}.$$

In general, we have

$$\phi_{kk} = \frac{(-1)^{k+1}\theta_1^k(1-\theta_1^2)}{1-\theta_1^{2(k+1)}} = \frac{(-1)^{k+1}\theta_1^k}{1+\theta_1^2+\theta_1^4+\cdots+\theta_1^{2k}}. \tag{3.2}$$

The proof is left to the reader as an exercise.

### 3.2.3 *Invertibility*

Consider two MA(1) models: (1) $X_t = \varepsilon_t + \theta\varepsilon_{t-1}$ and (2) $Y_t = \varepsilon_t + \theta^{-1}\varepsilon_{t-1}$. According to Example 3.5, it is easily seen that both of them have the same autocorrelation function. Thus there exists a question: if using the sample ACF to estimate the coefficients or choose the two models, which one is preferable? Besides, we know that the innovation series $\{\varepsilon_t\}$ is unobservable. Hence, we want to use the data $\{X_t\}$ or $\{Y_t\}$ to represent the innovation series so as to estimate them.

We rewrite model (1) as $\varepsilon_t = X_t - \theta\varepsilon_{t-1}$ for all $t$. Then we have

$$\varepsilon_t = X_t - \theta\varepsilon_{t-1} = X_t - \theta(X_{t-1} - \theta\varepsilon_{t-2}) = X_t - \theta X_{t-1} + \theta^2\varepsilon_{t-2}$$

We may continue this substitution and if and only if $|\theta| < 1$, we obtain

$$\varepsilon_t = X_t + (-\theta)X_{t-1} + (-\theta)^2X_{t-2} + (-\theta)^3X_{t-3} + \cdots$$

Similarly model (2) can be formally rewritten as

$$\varepsilon_t = Y_t + (-\frac{1}{\theta})Y_{t-1} + (-\frac{1}{\theta})^2Y_{t-2} + (-\frac{1}{\theta})^3Y_{t-3} + \cdots$$

But this series diverges if $|\theta| < 1$. Therefore if $|\theta| < 1$, we naturally choose model (1) and abandon model (2). At this point, model (1) is said to be invertible because the innovation term $\varepsilon_t$ can be inversely expressed by the data $\{X_t\}$. Note that the MA polynomial for model (1) is $1 + \theta z$ and its root is $z = -1/\theta$. So the absolute value of the root $|z| > 1$ if and only if the one of the coefficient $|\theta| < 1$.

In general, we define the invertibility for a time series as follows.

**Definition 3.5** (1) A time series $\{X_t\}$ is said to be invertible if there exist coefficients $\pi_j$ such that

$$\varepsilon_t = X_t + \pi_1 X_{t-1} + \pi_2 X_{t-2} + \pi_3 X_{t-3} + \cdots = \sum_{j=0}^{\infty} \pi_j X_{t-j}, \sum_{j=0}^{\infty} |\pi_j| < \infty$$

where $\pi_0 = 1$. At this point, we say that the time series $\{X_t\}$ has an AR($\infty$) representation. (2) We say that a model is invertible if the time series generated by it is invertible.

From now on, any resulting MA model is required to be invertible. In order to decide whether an MA model is invertible or not, the following theorem can be used.

**Theorem 3.1 (Invertibility Theorem)** *An MA($q$) model defined by Eq. (3.1) is invertible if and only if the roots of its MA polynomial $\theta(z) = 1 + \theta_1 z + \cdots + \theta_q z^q$ exceed 1 in modulus or lie outside the unit circle on the complex plane.*

Pay attention to the following remarks:

- In the light of the fundamental theorem of algebra (see, e.g., Waerden 2003), a polynomial of degree $q$ has $q$ roots in the complex field, and if it has complex roots, they must be conjugate complex numbers.
- Using the relationship between the roots and the coefficients of the degree 2 polynomial $\theta(z) = 1 + \theta_1 z + \theta_2 z^2$, it may be proved that both of the roots of the polynomial exceed 1 in modulus if and only if

$$\begin{cases} |\theta_2| < 1, \\ \theta_2 + \theta_1 > -1, \\ \theta_2 - \theta_1 > -1. \end{cases}$$

  Thus, we can conveniently use the three inequations of the coefficients to decide whether a MA(2) model is invertible or not.
- It may also be proved that given an autocorrelation function, there exists only one invertible MA process whose autocorrelation function is the given one.
- It may be shown that for the MA($q$) model, the coefficients $\{\pi_j\}$ in Definition 3.5 satisfy $\pi_0 = 1$ and

$$\pi_j = -\sum_{k=1}^{j} \theta'_k \pi_{j-k},\ j \geq 1 \text{ where } \theta'_k = \theta_k \text{ if } k \leq q \text{ and } \theta'_k = 0 \text{ if } k > q.$$

- Equation (3.2) suggests that the PACF of the invertible MA(1) model decays at exponential rate but tails off (viz., always nonzero). So does the PACF of the general invertible MA($q$) model.

*Example 3.6 (Invertibility)*

(1) Given two MA models

$$X_t = \varepsilon_t - 2\varepsilon_{t-1} + 0.6\varepsilon_{t-2} - \varepsilon_{t-3} + \varepsilon_{t-4} + 2\varepsilon_{t-5} \qquad (a)$$

$$X_t = \varepsilon_t - 0.2\varepsilon_{t-1} + 0.6\varepsilon_{t-2} - 0.1\varepsilon_{t-3} + 0.1\varepsilon_{t-4} + 0.2\varepsilon_{t-5} \qquad (b)$$

which of them is invertible? Both MA models are of order 5 and so their corresponding MA polynomials are of degree 5. It is well known that there are no closed-form root expressions for general polynomials of degree $q \geq 5$. We have to find their numerical solutions. Fortunately, Python gives the commands of solving a polynomial equation. With Python we get a root of model (a) corresponding MA polynomial: `0.6293124+0.11862764j` and its absolute value (modulus) is 0.64039567 $<$ 1. Therefore model (a) is not invertible. Note that the imaginary unit is denoted by $j$ not by $i$ in Python. For model (b), all the root absolute values of its corresponding MA polynomial are greater than 1. So model (b) is invertible.

(2) Consider the MA(2) model:

$$X_t = \varepsilon_t - \frac{2}{3}\varepsilon_{t-1} + \frac{4}{9}\varepsilon_{t-2} \qquad (c)$$

where $\theta_1 = -2/3, \theta_2 = 4/9$. We have

$$\begin{cases} |\theta_2| = \frac{4}{9} < 1, \\ \theta_2 + \theta_1 = \frac{4}{9} - \frac{2}{3} = -\frac{2}{9} > -1, \\ \theta_2 - \theta_1 = \frac{4}{9} + \frac{2}{3} = \frac{10}{9} > -1. \end{cases}$$

Thus, model (c) is invertible. Furthermore, we can use the Python function `arma2ar` to estimate the coefficients $\pi_j$ in Definition 3.5. We obtain

$$\varepsilon_t = X_t + 0.6667X_{t-1} + 0.0000X_{t-2} - 0.2963X_{t-3} + \cdots$$

The code for this example is as follows.

```
>>> import numpy as np
>>> p1=[2, 1, -1, 0.6, -2, 1]     # MA model (a)
>>> r1=np.roots(p1)
>>> r1
array([-1.44125189+0.j          , -0.15868646+0.9059506j ,
       -0.15868646-0.9059506j , 0.6293124 +0.11862764j,
        0.6293124 -0.11862764j])
>>> abs(r1)
array([1.44125189, 0.91974337, 0.91974337, 0.64039567, 0.64039567])
>>> p2=[0.2, 0.1, -0.1, 0.6, -0.2, 1]    # MA model (b)
```

```
>>> r2=np.roots(p2)
>>> r2
array([-1.97706234+0.j          ,  0.9495042 +0.98246025j,
        0.9495042 -0.98246025j, -0.21097303+1.14465167j,
       -0.21097303-1.14465167j])
>>> abs(r2)
array([1.97706234, 1.36630391, 1.36630391, 1.16393173, 1.16393173])
>>> from statsmodels.tsa.arima_process import arma2ar
>>> arma2ar(ar=[1], ma=[1, -2/3, 4/9], lags=16)
array([ 1.        ,  0.66666667,  0.        , -0.2962963 ,
       -0.19753086,  0.        ,  0.0877915 ,  0.05852766,
        0.        , -0.02601229, -0.01734153,  0.        ,
        0.00770735,  0.00513823,  0.        , -0.00228366])
# ar=[1] means no ar part in the model
```

## 3.3 Autoregressive Models

The autoregressive (AR) model was firstly presented by Yule who used the AR model to fit the sunspot number series in Yule (1927). It is one of the fundamental models in time series analysis.

### *3.3.1 Definition of Autoregressive Models*

We first give the definition of AR models and then make some explanations.

**Definition 3.6** (1) The following equation is called the autoregressive model of order $p$ and denoted by AR($p$):

$$X_t = \varphi_0 + \varphi_1 X_{t-1} + \varphi_2 X_{t-2} + \cdots + \varphi_p X_{t-p} + \varepsilon_t \tag{3.3}$$

where $\{\varepsilon_t\} \sim \text{WN}(0, \sigma_\epsilon^2)$, $\text{E}(X_s\varepsilon_t) = 0$ if $s < t$ and $\varphi_0, \varphi_1, \cdots, \varphi_p$ are real-valued parameters (coefficients) with $\varphi_p \neq 0$. (2) If a time series $\{X_t\}$ is stationary and satisfies such an equation as (3.3), then we call it an AR($p$) process.

Note the following remarks about this definition:

- For simplicity, we often assume that the intercept (const term) $\varphi_0 = 0$; otherwise, we can consider $\{X_t - \mu\}$ where $\mu = \varphi_0/(1 - \varphi_1 - \cdots - \varphi_p)$.
- We distinguish the concept of AR models from the concept of AR processes. AR models may or may not be stationary and AR processes must be stationary.
- $\text{E}(X_s\varepsilon_t) = 0 (s < t)$ means that $X_s$ in the past has nothing to do with $\varepsilon_t$ at the current time $t$.

- Like the definition of MA models, sometimes $\varepsilon_t$ in Eq. (3.3) is called the innovation or shock term.

In addition, using the backshift operator $B$, the AR($p$) model can be rewritten as

$$\varphi(B)X_t = \varepsilon_t$$

where $\varphi(z) = 1 - \varphi_1 z - \cdots - \varphi_p z^p$ is called the *(corresponding) AR polynomial.* Besides, in the Python package |`StatsModels`|, $\varphi(B)$ is called the AR lag polynomial.

*Example 3.7 (An AR(2) Model)* Given an AR(2) model as follows

$$X_t = 0.8X_{t-1} - 0.3X_{t-2} + \varepsilon_t$$

where $\varepsilon_t \sim$ iidN(0,1), simulate a sample of size (length) 500 from the AR(2) model, and then do some analysis of the simulated time series data. With the Python function `arma_generate_sample`, we can get this time series sample. Its time series plot is shown in Fig. 3.14, which exhibits neither trend nor seasonality and looks stationary. Its ACF and PACF plots are shown in Fig. 3.15. From it, we observe that the PACF values are almost zero after lag 3 but the ACF value at lag 11 still appears nonzero. Furthermore, from the lag plot shown in Fig. 3.16, it seems to been seen that there exists positive lag 11 autocorrelation in the sample series.

If an AR(2) model is given as follows

$$X_t = 0.8X_{t-1} - 1.3X_{t-2} + \varepsilon_t$$

then it is nonstationary. This test is left to the reader as an exercise. The following is the Python code for this example.

```
>>> import numpy as np
>>> import pandas as pd
>>> from statsmodels.tsa.arima_process import arma_generate_sample
>>> ar=np.array([1, -0.8, 0.3])
>>> np.random.seed(123457)
>>> x= arma_generate_sample(ar=ar, ma=[1], nsample=500)
# ma=[1] means no ma part in the model
>>> x=pd.Series(x)
>>> import matplotlib.pyplot as plt
>>> x.plot(); plt.show()
>>> from PythonTsa.plot_acf_pacf import acf_pacf_fig
>>> acf_pacf_fig(x, both=True, lag=20)
>>> plt.show()
>>> from pandas.plotting import lag_plot
>>> lag_plot(x, lag=11);plt.show()
```

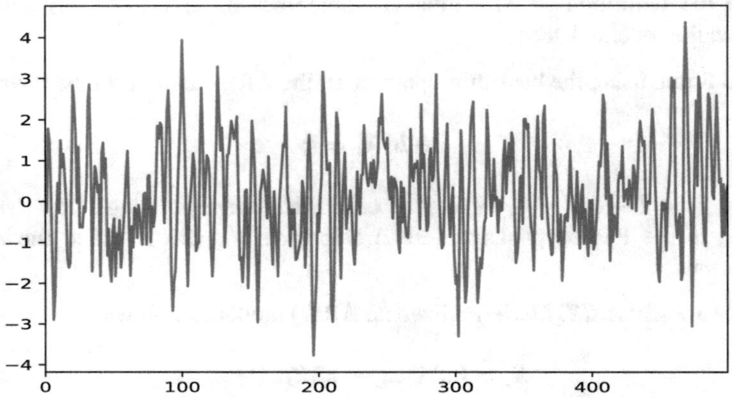


**Fig. 3.14** Time series plot of the simulated AR(2) sample in Example 3.7

pyTSA_SimAR2

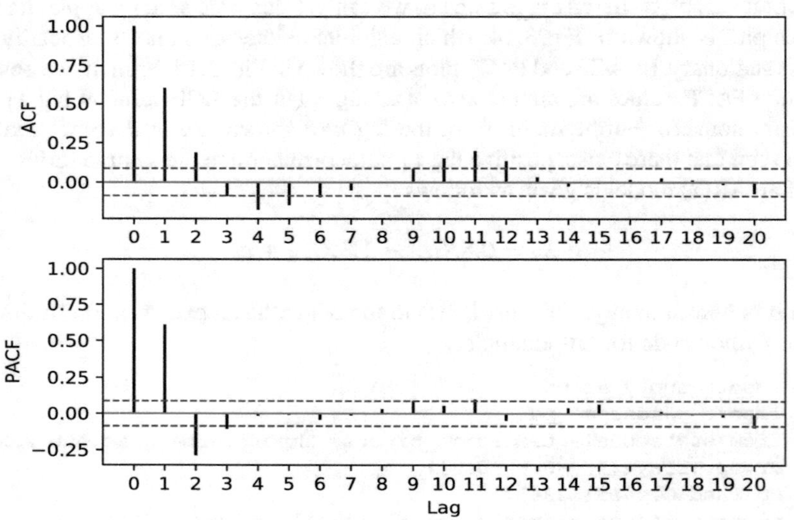


**Fig. 3.15** ACF and PACF plots of the simulated AR(2) sample in Example 3.7

pyTSA_SimAR2

### 3.3.2 *Durbin-Levinson Recursion Algorithm*

In Sect. 2.1, we introduce the concept of partial autocorrelations (PACF) and draw the PACF plot with Python. Actually, there are other ways to define this concept. Suppose that $\{X_t\}$ is any stationary time series with the mean zero. It may be shown that for a given lag $k \geq 1$, the partial autocorrelation $\phi_{kk}$ is just the last coefficient

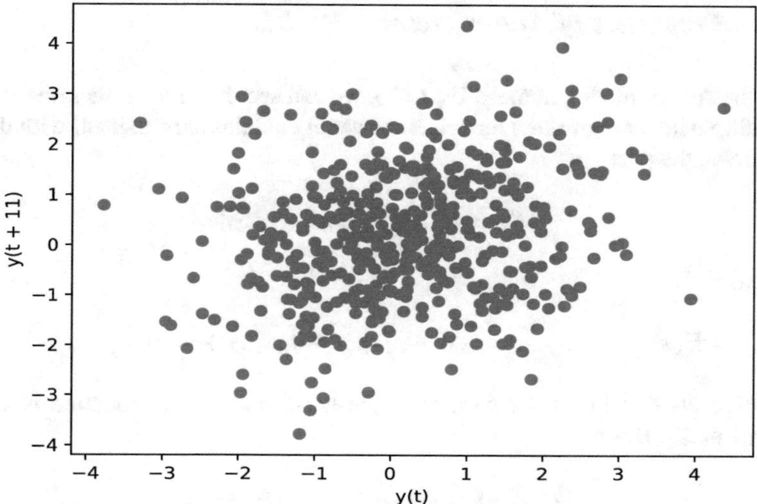


**Fig. 3.16** Lag 11 plot of the simulated AR(2) sample in Example 3.7

pyTSA_SimAR2

in the AR($k$) model (viz., the regression of $X_t$ on $\{X_{t-k:t-1}\}$):

$$X_t = \phi_{k1}X_{t-1} + \phi_{k2}X_{t-2} + \cdots + \phi_{kk}X_{t-k} + \varepsilon_{kt}$$

For $1 \leq j \leq k$, multiplying this equation by $X_{t-j}$, taking expectations, and dividing by $\gamma_0$, we obtain (noting $\rho_n = \rho_{-n}$)

$$\rho_j = \phi_{k1}\rho_{j-1} + \phi_{k2}\rho_{j-2} + \cdots + \phi_{kk}\rho_{j-k} \tag{3.4}$$

If $\{\rho_{1:k}\}$ are given, then we can solve these equations for $\phi_{kk}$.

Additionally, Durbin (1960) and Levinson (1947) give a method for solving Eqs. (3.4) (now called Durbin-Levinson recursion algorithm). They show that Eqs. (3.4) can recursively be solved as follows:

$$\phi_{kk} = \frac{\rho_k - \sum_{j=1}^{k-1}\phi_{k-1,j}\rho_{k-j}}{1 - \sum_{j=1}^{k-1}\phi_{k-1,j}\rho_j}, \quad \phi_{11} = \rho_1$$

where $k > 1$ and for $1 \leq j \leq k-1$, $\phi_{kj} = \phi_{k-1,j} - \phi_{kk}\phi_{k-1,k-j}$. For example, $\phi_{22} = (\rho_2 - \rho_1^2)/(1 - \rho_1^2)$, $\phi_{21} = \phi_{11}(1 - \phi_{22})$ and

$$\phi_{33} = \frac{\rho_3 - \phi_{21}\rho_2 - \phi_{22}\rho_1}{1 - \phi_{21}\rho_1 - \phi_{22}\rho_2}.$$

Thus, given sample ACF, namely, the estimates for $\rho_k$, we can obtain the estimates $\hat{\phi}_{kk}$ of $\phi_{kk}$.

### 3.3.3 Properties of Autoregressive Models

From the AR($p$) model, namely, Eq. (3.3), we can see that it is in the same form as the multiple linear regression model. However, it explains current itself with its own past. Given the past

$$\{X_{(t-p):(t-1)}\} = \{x_{(t-p):(t-1)}\}$$

we have

$$\mathsf{E}(X_t|X_{(t-p):(t-1)}) = \varphi_0 + \varphi_1 x_{t-1} + \varphi_2 x_{t-2} + \cdots + \varphi_p x_{t-p}$$

This suggests that given the past, the right-hand side of this equation is a good estimate of $X_t$. Besides

$$\text{Var}(X_t|X_{(t-p):(t-1)}) = \text{Var}(\varepsilon_t) = \sigma_\epsilon^2$$

Now we suppose that the AR($p$) model, namely, Eq. (3.3), is stationary; then we have

(1) The model mean $\mathsf{E}(X_t) = \mu = \varphi_0/(1 - \varphi_1 - \cdots - \varphi_p)$. Thus, the model mean $\mu = 0$ if and only if $\varphi_0 = 0$.
(2) If the mean is zero or $\varphi_0 = 0$ ((3) and (4) below have the same assumption), noting that $\mathsf{E}(X_t\varepsilon_t) = \sigma_\epsilon^2$, we multiply Eq. (3.3) by $X_t$, take expectations, and then get

$$\text{Var}(X_t) = \gamma_0 = \varphi_1\gamma_1 + \varphi_2\gamma_2 + \cdots + \varphi_p\gamma_p + \sigma_\epsilon^2$$

Furthermore

$$\gamma_0 = \sigma_\epsilon^2/(1 - \varphi_1\rho_1 - \varphi_2\rho_2 - \cdots - \varphi_p\rho_p).$$

(3) For all $k > p$, the partial autocorrelation $\phi_{kk} = 0$, that is, the PACF of AR($p$) models cuts off after lag $p$, which is very helpful in identifying an AR model. In fact, at this point, the predictor or regression of $X_t$ on $\{X_{t-k+1:t-1}\}$ is

$$\hat{X}_t = \varphi_1 X_{t-1} + \cdots + \varphi_{k-1} X_{t-k+1}$$

Thus, $X_t - \hat{X}_t = \varepsilon_t$. Moreover, $X_{t-k} - \hat{X}_{t-k}$ is a function of $\{X_{t-k:t-1}\}$, and $\varepsilon_t$ is uncorrelated to everyone in $\{X_{t-k:t-1}\}$. Therefore

$$\text{Cov}(X_{t-k} - \hat{X}_{t-k}, X_t - \hat{X}_t) = \text{Cov}(X_{t-k} - \hat{X}_{t-k}, \varepsilon_t) = 0.$$

By Definition 2.1, $\phi_{kk} = 0$.

(4) We multiply Eq. (3.3) by $X_{t-k}$, take expectations, divide by $\gamma_0$, and then obtain the recursive relationship between the autocorrelations:

$$\text{for } k \geq 1, \ \rho_k = \varphi_1\rho_{k-1} + \varphi_2\rho_{k-2} + \cdots + \varphi_p\rho_{k-p} \tag{3.5}$$

For Eq. (3.5), let $k = 1, 2, \cdots, p$. Then we arrive at a set of difference equations, which is known as the *Yule-Walker equations*. If the ACF $\{\rho_{1:p}\}$ are given, then we can solve the Yule-Walker equations to obtain the estimates for $\{\varphi_{1:p}\}$, and the solutions are called the *Yule-Walker estimates* .

(5) Since the model is a stationary AR($p$) now, naturally it satisfies $X_t = \varphi_1 X_{t-1} + \varphi_2 X_{t-2} + \cdots + \varphi_p X_{t-p} + \varepsilon_t$. Hence $\phi_{pp} = \varphi_p$. If the AR($p$) model is further Gaussian and a sample of size $T$ is given, then (a) $\hat{\phi}_{pp} \to \varphi_p$ as $T \to \infty$; (b) according to Quenouille (1949), for $k > p$, $\sqrt{T}\hat{\phi}_{kk}$ asymptotically follows the standard normal(Gaussian) distribution N(0, 1), or $\hat{\phi}_{kk}$ is asymptotically distributed as N(0, $1/T$).

*Example 3.8 (ACF and PACF of Stationary AR(2) Models)* Putting $p = 2$ in Eq. (3.5), we have that $\rho_k = \varphi_1\rho_{k-1} + \varphi_2\rho_{k-2}$. (1) Letting $k = 1$ gets $\rho_1 = \varphi_1\rho_0 + \varphi_2\rho_{-1}$. Knowing $\rho_{-1} = \rho_1$ and $\rho_0 = 1$ makes $\rho_1 = \varphi_1 + \varphi_2\rho_1$. Thus $\rho_1 = \varphi_1/(1 - \varphi_2)$. (2) Letting $k = 2$ gets

$$\rho_2 = \varphi_1\rho_1 + \varphi_2\rho_0 = \frac{\varphi_1^2 + \varphi_2 - \varphi_2^2}{1 - \varphi_2}.$$

Putting $k = 1$ in Eq. (3.4), we have $\rho_1 = \phi_{11}\rho_0$. So $\phi_{11} = \rho_1 = \varphi_1/(1 - \varphi_2)$. Putting $k = 2$ in Eq. (3.4), we have

$$\begin{cases} \rho_1 = \phi_{21}\rho_0 + \phi_{22}\rho_1 \\ \rho_2 = \phi_{21}\rho_1 + \phi_{22}\rho_0 \end{cases}$$

By Gramer's rule in linear algebra, we get

$$\phi_{22} = \frac{\begin{vmatrix} \rho_0 & \rho_1 \\ \rho_1 & \rho_2 \end{vmatrix}}{\begin{vmatrix} \rho_0 & \rho_1 \\ \rho_1 & \rho_0 \end{vmatrix}} = \frac{\rho_2 - \rho_1^2}{1 - \rho_1^2}$$

Therefore

$$\phi_{22} = \frac{\frac{\varphi_1^2 + \varphi_2 - \varphi_2^2}{1 - \varphi_2} - (\frac{\varphi_1}{1 - \varphi_2})^2}{1 - (\frac{\varphi_1}{1 - \varphi_2})^2} = \frac{(1 - \varphi_2)(\varphi_1^2 + \varphi_2 - \varphi_2^2) - \varphi_1^2}{(1 - \varphi_2)^2 - \varphi_1^2} = \varphi_2$$

### 3.3.4 Stationarity and Causality of AR Models

Consider the AR(1) model:

$$X_t = \varphi X_{t-1} + \varepsilon_t,\ \varepsilon_t \sim \mathrm{WN}(0, \sigma_\epsilon^2) \tag{3.6}$$

For $|\varphi| < 1$, let $X_{1t} = \sum_{j=0}^{\infty} \varphi^j \varepsilon_{t-j}$ and for $|\varphi| > 1$, let $X_{2t} = -\sum_{j=1}^{\infty} \varphi^{-j} \varepsilon_{t+j}$. It is easy to show that both $\{X_{1t}\}$ and $\{X_{2t}\}$ are stationary and satisfy Eq. (3.6). That is, both are the stationary solution of Eq. (3.6). This gives rise to a question: which one of both is preferable? Obviously, $\{X_{2t}\}$ depends on future values of unobservable $\{\varepsilon_t\}$, and so it is unnatural. Hence we take $\{X_{1t}\}$ and abandon $\{X_{2t}\}$. In other words, we require that the coefficient $\varphi$ in Eq. (3.6) is less 1 in absolute value. At this point, the AR(1) model is said to be causal and its causal expression is $X_t = \sum_{j=0}^{\infty} \varphi^j \varepsilon_{t-j}$. In general, the definition of causality is given below.

**Definition 3.7** (1) A time series $\{X_t\}$ is causal if there exist coefficients $\psi_j$ such that

$$X_t = \sum_{j=0}^{\infty} \psi_j \varepsilon_{t-j},\ \sum_{j=0}^{\infty} |\psi_j| < \infty$$

where $\psi_0 = 1,\ \{\varepsilon_t\} \sim \mathrm{WN}(0, \sigma_\epsilon^2)$. At this point, we say that the time series $\{X_t\}$ has an MA($\infty$) representation. (2) We say that a model is causal if the time series generated by it is causal.

Causality suggests that the time series $\{X_t\}$ is caused by the white noise (or innovations) from the past up to time $t$. Besides, the time series $\{X_{2t}\}$ is an example that is stationary but not causal. In order to determine whether an AR model is causal, similar to the invertibility for the MA model, we have the following theorem.

**Theorem 3.2 (Causality Theorem)** *An AR model defined by Eq. (3.3) is causal if and only if the roots of its AR polynomial $\varphi(z) = 1 - \varphi_1 z - \cdots - \varphi_p z^p$ exceed 1 in modulus or lie outside the unit circle on the complex plane.*

Note the following remarks:

- In the light of the existence and uniqueness on page 75 of Brockwell and Davis (2016), an AR model defined by Eq. (3.3) is stationary if and only if its AR polynomial $\varphi(z) = 1 - \varphi_1 z - \cdots - \varphi_p z^p \neq 0$ for all $|z| = 1$ or all the roots of the AR polynomial do not lie on the unit circle. Hence for the AR model defined by Eq. (3.3), its stationarity condition is weaker than its causality condition.
- A causal time series is surely a stationary one. So an AR model that satisfies the causal condition is naturally stationary. But a stationary AR model is not necessarily causal.

- If the time series $\{X_t\}$ generated by Eq. (3.3) is not from the remote past, namely,

$$t \in T = \{\cdots, -n, \cdots, -1, 0, 1, \cdots, n, \cdots\}$$

but starts from an initial value $X_0$, then it may be nonstationary, not to mention causality. For example, $X_t = 0.2X_{t-1} + \varepsilon_t$ with the initial value $X_0 \equiv 0$ is nonstationary. The proof is left to the reader as an exercise.
- According to the relationship between the roots and the coefficients of the degree 2 polynomial $\varphi(z) = 1 - \varphi_1 z - \varphi_2 z^2$, it may be proved that both of the roots of the polynomial exceed 1 in modulus if and only if

$$\begin{cases} |\varphi_2| < 1, \\ \varphi_2 + \varphi_1 < 1, \\ \varphi_2 - \varphi_1 < 1. \end{cases}$$

Thus, we can conveniently use the three inequations to decide whether a AR(2) model is causal or not.
- It may be shown that for an AR($p$) model defined by Eq. (3.3), the coefficients $\{\psi_j\}$ in Definition 3.7 satisfy $\psi_0$=1 and

$$\psi_j = \sum_{k=1}^{j} \varphi'_k \psi_{j-k},\ j \geq 1 \text{ where } \varphi'_k = \varphi_k \text{ if } k \leq p \ \text{ and } \varphi'_k = 0 \ \text{ if } k > p.$$

Pertaining to the causality theorem and the remarks above, we can refer to, for example, pages 74–75 of Brockwell and Davis (2016), pages 31–32 of Fan and Yao (2003), and pages 87 and 495 of Shumway and Stoffer (2017).

*Example 3.9 (Causality)* We have two AR models:

$$X_t = 0.2X_{t-1} + 0.6X_{t-2} - 0.1X_{t-3} + 0.1X_{t-4} + 0.2X_{t-5} + \varepsilon_t \qquad (i)$$

$$X_t = 0.2X_{t-1} - 0.6X_{t-2} - 0.1X_{t-3} + 0.1X_{t-4} + 0.2X_{t-5} + \varepsilon_t \qquad (ii)$$

The model (i) corresponding AR polynomial is

$$\varphi(z) = 1 - 0.2z - 0.6z^2 + 0.1z^3 - 0.1z^4 - 0.2z^5$$

which has a root $z = 1$ (called unit root). Hence model (i) is nonstationary, not to mention causality. The time series plot of model (i) is shown in Fig. 3.17 and exhibits clear nonstationarity. The model (ii) corresponding AR polynomial is

$$\varphi(z) = 1 - 0.2z + 0.6z^2 + 0.1z^3 - 0.1z^4 - 0.2z^5$$

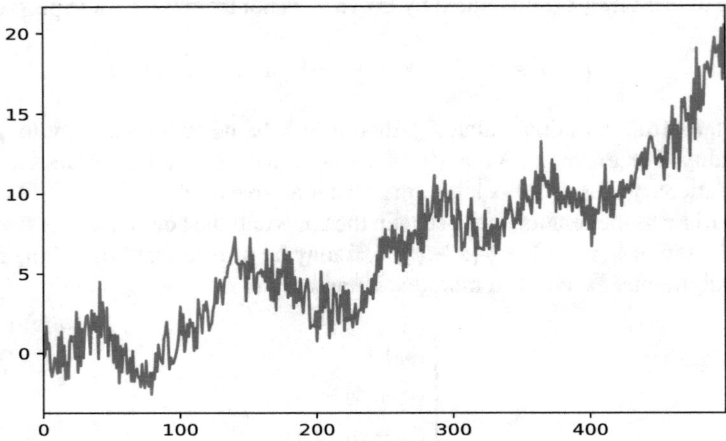


**Fig. 3.17** Time series plot of the noncausal AR(5) model in Example 3.9

pyTSA_Causality

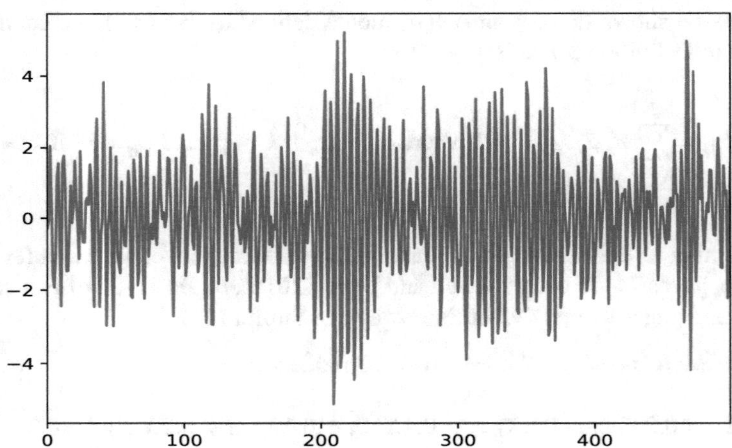


**Fig. 3.18** Time series plot of the causal AR(5) model in Example 3.9

pyTSA_Causality

Its roots are all greater than 1 in modulus. Thus model (ii) is causal and its time series plot shown in Fig. 3.18 looks very stationary. The Python code for this example is as follows.

```
>>> import numpy as np
>>> import pandas as pd
>>> import matplotlib.pyplot as plt
>>> p1=[-0.2, -0.1, 0.1, -0.6, -0.2, 1]    # AR model (i)
>>> r1=np.roots(p1)
>>> r1
```

```
array([-1.38071518+0.42051849j, -1.38071518-0.42051849j,
        0.63071518+1.41504072j,  0.63071518-1.41504072j,
        1.        +0.j        ])
>>> abs(r1)
array([1.44333302, 1.44333302, 1.54923912, 1.54923912, 1.        ])
>>> p2=[-0.2, -0.1, 0.1, 0.6, -0.2, 1]    # AR model (ii)
>>> r2=np.roots(p2)
>>> r2
array([-1.27880602+1.13255338j, -1.27880602-1.13255338j,
        1.5769574 +0.j        ,  0.24032732+1.01430874j,
        0.24032732-1.01430874j])
>>> abs(r2)
array([1.70822188, 1.70822188, 1.5769574 , 1.04239121, 1.04239121])
>>> from statsmodels.tsa.arima_process import arma_generate_sample
>>> ar1=np.array([1, -0.2, -0.6, 0.1, -0.1, -0.2])
>>> np.random.seed(123457)
>>> x1= arma_generate_sample(ar=ar1, ma=[1], nsample=500)
>>> x1=pd.Series(x1)
>>> x1.plot(); plt.show()
>>> ar2=np.array([1, -0.2, 0.6, 0.1, -0.1, -0.2])
>>> np.random.seed(123457)
>>> x2= arma_generate_sample(ar=ar2, ma=[1], nsample=500)
>>> x2=pd.Series(x2)
>>> x2.plot(); plt.show()
```

*Example 3.10 (MA($\infty$) Representation)* Given the AR(2) model

$$X_t = 0.8X_{t-1} - 0.3X_{t-2} + \varepsilon_t$$

where $\varphi_1 = 0.8$, $\varphi_2 = -0.3$. Thus

$$\begin{cases} |\varphi_2| = 0.3 < 1, \\ \varphi_2 + \varphi_1 = 0.5 < 1, \\ \varphi_2 - \varphi_1 = -1.1 < 1. \end{cases}$$

The AR(2) model is causal. Its MA($\infty$) representation can be obtained with the Python function arma2ma :

$$X_t = \varepsilon_1 + 0.800\varepsilon_2 + 0.340\varepsilon_3 + 0.032\varepsilon_4 - 0.076\varepsilon_5 + \cdots$$

```
>>> from statsmodels.tsa.arima_process import arma2ma
>>> arma2ma(ar=[1, -0.8, 0.3], ma=[1], lags=15)
array([ 1.00000000e+00,  8.00000000e-01,  3.40000000e-01,
        3.20000000e-02, -7.64000000e-02, -7.07200000e-02,
       -3.36560000e-02, -5.70880000e-03,  5.52976000e-03,
        6.13644800e-03,  3.25023040e-03,  7.59249920e-04,
       -3.67669184e-04, -5.21910323e-04, -3.07227503e-04])
```

## 3.4 Autoregressive Moving Average Models

In this section we introduce an important class of time series models known as autoregressive moving average (ARMA) models. From their name, we can imagine that ARMA models are formed by combining the AR and MA models. If we use only an AR model or MA model to fit a time series, we sometimes have to face a long AR or MA model in order to capture the complex structure of this time series. On the other hand, we want a parsimonious model to fit the data. That is why we combine the AR and MA models to form a new and parsimonious model.

### 3.4.1 Definitions

Now we give the definition of ARMA models as follows.

**Definition 3.8** (1) The following equation is called the autoregressive moving average model of order $(p, q)$ and denoted by ARMA$(p, q)$:

$$X_t = \varphi_0 + \varphi_1 X_{t-1} + \cdots + \varphi_p X_{t-p} + \varepsilon_t + \theta_1 \varepsilon_{t-1} + \cdots + \theta_q \varepsilon_{t-q} \tag{3.7}$$

where $\{\varepsilon_t\} \sim \mathrm{WN}(0, \sigma_\epsilon^2)$, $\mathrm{E}(X_s \varepsilon_t) = 0$ if $s < t$, and $\{\varphi_k\}$ and $\{\theta_k\}$ are real-valued parameters (coefficients) with $\varphi_p \neq 0$ and $\theta_q \neq 0$. (2) If a time series $\{X_t\}$ is stationary and satisfies such an equation as (3.7), then we call it an ARMA$(p, q)$ process.

We often assume the intercept (const term) $\varphi_0 = 0$. Using the backshift operator $B$, the ARMA$(p, q)$ model can be rewritten as

$$\varphi(B) X_t = \theta(B) \varepsilon_t$$

where $\varphi(z) = 1 - \varphi_1 z - \cdots - \varphi_p z^p$ is the AR polynomial and $\theta(z) = 1 + \theta_1 z + \cdots + \theta_q z^q$ is the MA polynomial. We always assume that $\varphi(z)$ and $\theta(z)$ have no common factors. Besides, $\varphi(B) X_t = \varepsilon_t$ and $X_t = \theta(B) \varepsilon_t$ are, respectively, called the *AR part* and *MA part* of the ARMA$(p, q)$ model. Of course, both the AR model and MA model are two special cases of the ARMA model: AR$(p)$=ARMA$(p, 0)$ and MA$(q)$= ARMA$(0, q)$.

*Example 3.11 (An ARMA(2,2) Model)* Consider the following ARMA(2,2) model:

$$X_t = 0.8 X_{t-1} - 0.6 X_{t-2} + \varepsilon_t + 0.7 \varepsilon_{t-1} + 0.4 \varepsilon_{t-2}$$

We can simulate it with Python. Its time series plot is shown in Fig. 3.19 and looks stationary. In addition, its ACF and PACF plots are shown in Fig. 3.20. Both ACF and PACF seem to tail off, namely, for lag $k \geq 3$, and many ACF and PACF values are still nonzero. The Python code for this example is as follows.

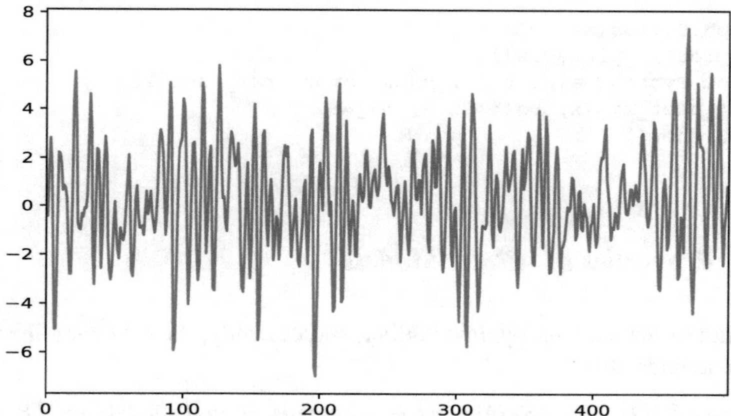


**Fig. 3.19** Time series plot of the ARMA(2,2) model in Example 3.11

pyTSA_ARMA

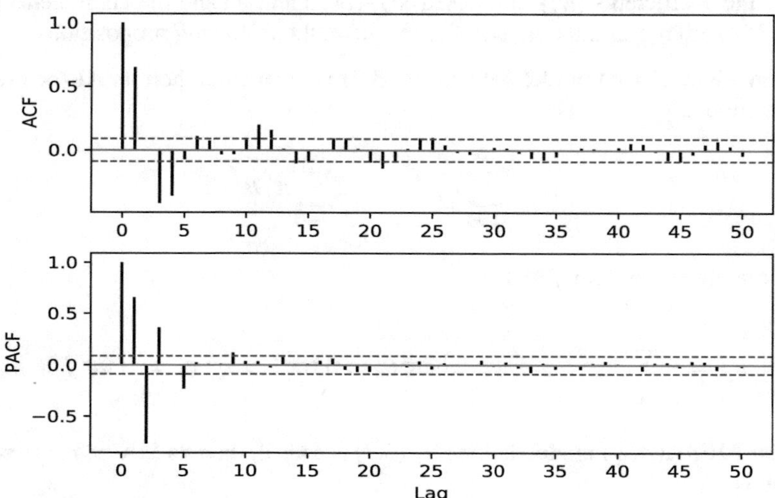


**Fig. 3.20** ACF and PACF plots of the ARMA(2,2) model in Example 3.11

pyTSA_ARMA

```
>>> import numpy as np
>>> import pandas as pd
>>> import matplotlib.pyplot as plt
>>> from statsmodels.tsa.arima_process import arma_generate_sample
>>> ar=np.array([1, -0.8, 0.6])
>>> ma=np.array([1, 0.7, 0.4])
>>> np.random.seed(123457)
>>> x= arma_generate_sample(ar=ar, ma=ma, nsample=500)
```

```
>>> x=pd.Series(x)
>>> x.plot(); plt.show()
>>> from PythonTsa.plot_acf_pacf import acf_pacf_fig
>>> acf_pacf_fig(x, both=True, lag=50)
>>> plt.show()
```

### 3.4.2 Properties of ARMA Models

Pertaining to the stationarity, invertibility, and causality, the following theorem is not hard to understand.

**Theorem 3.3** *(1) The ARMA$(p, q)$ is stationary if and only if its AR part is stationary. (2) The ARMA$(p, q)$ is invertible if and only if its MA part is invertible. (3) The ARMA$(p, q)$ is causal if and only if its AR part is causal.*

For the coefficients $\{\pi_j\}$ in the AR($\infty$) representation and the coefficients $\{\psi_j\}$ in the MA($\infty$) representation, it is easy to prove the following propositions.

- If an ARMA$(p, q)$ model defined by (3.7) is invertible, then its AR($\infty$) representation is

$$\varepsilon_t = \sum_{j=0}^{\infty} \pi_j X_{t-j} = (\sum_{j=0}^{\infty} \pi_j B^j) X_t$$

where $\pi_0 = 1$ and for $j \geq 1$

$$\pi_j = -\sum_{k=1}^{j} \theta_k \pi_{j-k} - \varphi_j \text{ where } \theta_k = 0 \text{ for } k > q,\ \varphi_j = 0 \text{ for } j > p.$$

- If an ARMA$(p, q)$ model defined by (3.7) is causal, then its MA($\infty$) representation is

$$X_t = \sum_{j=0}^{\infty} \psi_j \varepsilon_{t-j} = (\sum_{j=0}^{\infty} \psi_j B^j) \varepsilon_t$$

where $\psi_0 = 1$ and for $j \geq 1$

$$\psi_j = \sum_{k=1}^{j} \varphi_k \psi_{j-k} + \theta_j \text{ where } \varphi_k = 0 \text{ for } k > p,\ \theta_j = 0 \text{ for } j > q.$$

If an ARMA$(p, q)$ model (3.7) is causal and $\{\varepsilon_t\} \sim \text{iid}(0, \sigma_\epsilon^2)$, for each natural integer $h$, write $\boldsymbol{r}(h) = (r_1, \cdots, r_h)$ and $\boldsymbol{\rho}(h) = (\rho_1, \cdots, \rho_h)$, and then $\boldsymbol{r}(h)$ asymptotically follows the normal distribution $\text{N}(\boldsymbol{\rho}(h), n^{-1}W)$ where $n$ is the sample size and $W$ is the covariance matrix. This is an important large-sample property of the sample ACF for ARMA models. For details, you can find in Anderson and Walker (1964) as well as Brockwell and Davis (1991).

*Example 3.12 (Causality and Invertibility of ARMA Models)*

(1) Given an ARMA(2, 2) model:

$$X_t = 0.8X_{t-1} - 0.6X_{t-2} + \varepsilon_t + 0.7\varepsilon_{t-1} + 0.4\varepsilon_{t-2}$$

where $\varphi_1 = 0.8, \varphi_2 = -0.6$ and $\theta_1 = 0.7, \theta_2 = 0.4$; it is causal because

$$|\varphi_2| = 0.6 < 1, \varphi_2 + \varphi_1 = -0.6 + 0.8 = 0.2 < 1, \varphi_2 - \varphi_1 = -0.6 - 0.8 = -1.4 < 1$$

It is also invertible due to

$$|\theta_2| = 0.4 < 1, \theta_2 + \theta_1 = 0.4 + 0.7 = 1.1 > -1, \theta_2 - \theta_1 = 0.4 - 0.7 = -0.3 > -1$$

Hence it has both MA($\infty$) representation and AR($\infty$) representation. Furthermore, its MA($\infty$) representation is

$$X_t = \varepsilon_t + 1.5000\varepsilon_{t-1} + 1.0000\varepsilon_{t-2} - 0.1000\varepsilon_{t-3} - 0.6800\varepsilon_{t-4} + \cdots$$

and its AR($\infty$) representation is

$$\varepsilon_t = X_t - 1.5000X_{t-1} + 1.2500X_{t-2} - 0.2750X_{t-3} - 0.3075X_{t-4} + \cdots$$

The corresponding Python code is below.

```
>>> from statsmodels.tsa.arima_process import arma2ma
>>> arma2ma(ar=[1, -0.8, 0.6], ma=[1, 0.7, 0.4], lags=9)
array([ 1.        ,  1.5       ,  1.        , -0.1       , -0.68 ,
       -0.484    ,  0.0208    ,  0.30704 ,  0.233152])
>>> from statsmodels.tsa.arima_process import arma2ar
>>> arma2ar(ar=[1, -0.8, 0.6], ma=[1, 0.7, 0.4], lags=9)
array([ 1.        , -1.5        ,  1.25        , -0.275     , -0.3075 ,
        0.32525 , -0.104675 , -0.0568275 ,  0.08164925])
```

(2) Consider the following ARMA model:

$$X_t = 0.2X_{t-1} + 0.6X_{t-2} - 0.1X_{t-3} + 0.3X_{t-4} + \varepsilon_t + 0.4\varepsilon_{t-1} - 0.8\varepsilon_{t-2} + 0.9\varepsilon_{t-3} \quad (a)$$

Model (a) is an ARMA(4,3) model. Its AR polynomial is

$$\varphi(z) = 1 - 0.2z - 0.6z^2 + 0.1z^3 - 0.3z^4$$

and has a root $z = 1$. Thus Model (a) is neither causal nor stationary. On the other hand, its MA polynomial is

$$\theta(z) = 1 + 0.4z - 0.8z^2 + 0.9z^3$$

and has a root $z = -0.70674545$ inside the unit circle. So Model (a) is not invertible. Besides, we can use Python functions to directly check whether an ARMA model is stationary and/or invertible. Below is the corresponding Python code.

```
>>> import numpy as np
>>> import statsmodels.api as sm
>>> map=[0.9, -0.8, 0.4,  1]  # MA polynomial for model (a)
>>> maroot=np.roots(map)
>>> maroot
array([ 0.79781717+0.96728461j,  0.79781717-0.96728461j,
       -0.70674545+0.j        ])
>>> ar=[1, -0.2, -0.6, 0.1, -0.3] # coefficients of AR polynomial
>>> ma=[1, 0.4, -0.8, 0.9]
>>> arma_process = sm.tsa.ArmaProcess(ar, ma)
>>> arma_process.isstationary     # check stationarity
True
# wrong! We will report this to the developer.
>>> arma_process.isinvertible     # check invertibility
False
>>> from PythonTsa.CheckStationarynInvertible
    import isstationary, isinvertible
>>> isstationary(arma_process)
False
>>> isinvertible(arma_process)
False
```

Now we discuss the behavior of the true (theoretical) ACF and PACF of ARMA models. Let us start with an example.

*Example 3.13 (ACF of ARMA(1,1) Model)* Consider the following causal ARMA(1,1) model:

$$X_t = \varphi X_{t-1} + \varepsilon_t + \theta\varepsilon_{t-1} \tag{3.8}$$

Multiplying it by $X_{t-k}$ and then taking expectations, we obtain

$$\gamma_k = \varphi\gamma_{k-1} + \mathsf{E}(X_{t-k}\varepsilon_t) + \theta\mathsf{E}(X_{t-k}\varepsilon_{t-1}) \tag{3.9}$$

For $k \geq 2$, Eq. (3.9) becomes $\gamma_k = \varphi\gamma_{k-1}$. Due to $\mathsf{E}(X_t\varepsilon_{t-1}) = (\varphi + \theta)\sigma_\epsilon^2$ and $\gamma_{-k} = \gamma_k$, putting $k = 0$ in Eq. (3.9) gets

$$\gamma_0 = \varphi\gamma_1 + \sigma_\epsilon^2 + \theta(\varphi + \theta)\sigma_\epsilon^2$$

and putting $k = 1$ in Eq. (3.9) gets

$$\gamma_1 = \varphi\gamma_0 + \theta\sigma_\epsilon^2$$

Solving the two equations above, we have

$$\gamma_0 = \frac{1+\theta^2+2\varphi\theta}{1-\varphi^2}\sigma_\epsilon^2$$

and

$$\gamma_1 = \frac{(\varphi+\theta)(1+\varphi\theta)}{1-\varphi^2}\sigma_\epsilon^2.$$

Furthermore, we obtain

$$\gamma_k = \varphi\gamma_{k-1} = \cdots = \varphi^{k-1}\gamma_1 = \frac{(\varphi+\theta)(1+\varphi\theta)}{1-\varphi^2}\sigma_\epsilon^2\varphi^{k-1}, \quad k \geq 2$$

and the ACF

$$\rho_k = \frac{\gamma_k}{\gamma_0} = \frac{(\varphi+\theta)(1+\varphi\theta)}{1+\theta^2+2\varphi\theta}\varphi^{k-1}, \quad k \geq 1.$$

In general, for a causal ARMA$(p, q)$ model $\varphi(B)X_t = \theta(B)\varepsilon_t$, it is easy to obtain the Yule-Walker equations:

$$\rho_k - \varphi_1\rho_{k-1} - \varphi_2\rho_{k-2} - \cdots - \varphi_p\rho_{k-p} = 0, \quad k > q$$

It may be shown that the general solution of this equation is

$$\rho_k = c_1 z_1^{-k} + \cdots + c_p z_p^{-k}$$

where $\{c_{1:p}\}$ are constants(not all zero) and $\{z_{1:p}\}$ are the roots of the AR polynomial $\varphi(z)$. Noting $|z_i| > 1$ due to the causality of the ARMA$(p, q)$ model, it follows that $\rho_k$ decays at an exponential rate but tails off.

For the PACF $\phi_{kk}$ of ARMA$(p, q)$ models, according to the Durbin-Levinson recursion algorithm, using the solved $\rho_k$ above, we can thus calculate as many values for $\phi_{kk}$ as we want. Moreover, because an invertible ARMA model has an AR$(\infty)$ representation, the PACF $\phi_{kk}$ will tail off. In summary, we have Table 3.1 about the behavior of the ACF and PACF of a causal and invertible ARMA model, which are useful in identifying models.

**Table 3.1** Behavior of the ACF and PACF of ARMA models

| | MA($q$) | AR($p$) | ARMA($p, q$) ($p > 0, q > 0$) |
|---|---|---|---|
| ACF | Cuts off after lag $q$ | Tails off | Tails off |
| PACF | Tails off | Cuts off after lag $p$ | Tails off |

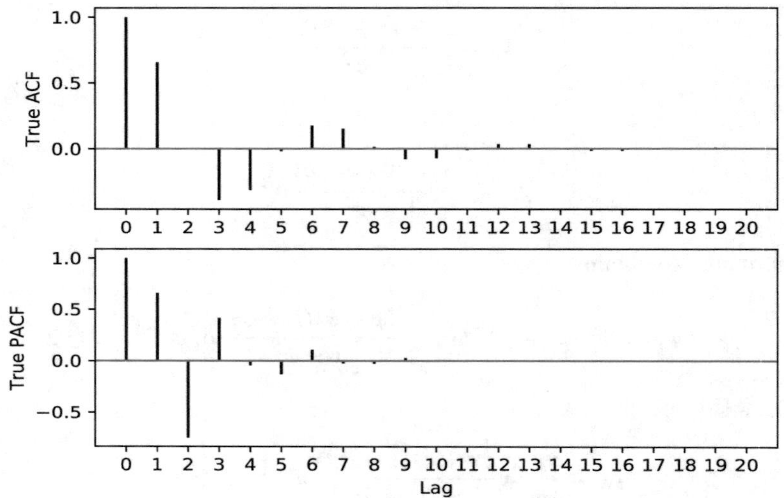


**Fig. 3.21** True ACF and PACF plots of the ARMA(2,2) model in Example 3.14

pyTSA_ARMA2

*Example 3.14 (ARMA Model True ACF and PACF Tailing Off)* Given an ARMA(2, 2) model:

$$X_t = 0.8X_{t-1} - 0.6X_{t-2} + \varepsilon_t + 0.7\varepsilon_{t-1} + 0.4\varepsilon_{t-2}$$

Using the function |`Tacf_pacf_fig`| in the Python package |`PythonTsa`|, we can draw the true (theoretical or model) ACF and PACF plots of the ARMA(2, 2) model shown in Fig. 3.21. We see that both true ACF and PACF decay to zero fast. And at the same time, many ACF and PACF values are still nonzero for lag $k \geq 3$, that is, both true ACF and PACF tail off.

```
>>> import matplotlib.pyplot as plt
>>> ar=[1, -0.8, 0.6]
>>> ma=[1, 0.7, 0.4]
>>> from PythonTsa.True_acf import Tacf_pacf_fig
>>> Tacf_pacf_fig(ar,ma, both=True, lag=20)
>>> plt.show()
```

This chapter has defined MA, AR, and ARMA models as well as elaborated on their properties. Now we are ready for fitting an ARMA model to a time series sample that comes from a stationary process. The next chapter will discuss how to build an ARMA model for the stationary time series.

## Problems

**3.1** Choose a country in which you are interested and analyze its quarterly GDP series using the methods introduced in this chapter.

**3.2** Given an AR(2) model as follows

$$X_t = 0.8X_{t-1} - 1.3X_{t-2} + \varepsilon_t$$

where $\varepsilon_t \sim$ iidN(0,1), first, simulate a sample of length 100 from the AR(2) model, and then graph its time series plot. Third, find the roots of its AR polynomial and determine whether it is stationary.

**3.3** Show that for AR models, if the mean is zero, then $\mathsf{E}(X_t\varepsilon_t) = \sigma_\epsilon^2$.

**3.4** Suppose that $X_t = \varphi_1 X_{t-1} + \varepsilon_t$ is stationary. Show that (1) $\gamma_0 = \text{Var}(X_t) = \sigma_\epsilon^2/(1-\varphi_1^2)$; (2) $\gamma_k = \varphi_1^k\sigma_\epsilon^2/(1-\varphi_1^2)$; (3) $\phi_{11} = \varphi_1$.

**3.5** Suppose that $X_t = \varphi_1 X_{t-1} + \varphi_2 X_{t-2} + \varepsilon_t$ is stationary. Prove that

$$\gamma_0 = \text{Var}(X_t) = \frac{(1-\varphi_2)\sigma_\epsilon^2}{(1+\varphi_2)[(1-\varphi_2)^2 - \varphi_1^2]}.$$

**3.6** Let $X_t = \varepsilon_t + \theta_1\varepsilon_{t-1}$. Show that for $k \geq 1$ we have

$$\phi_{kk} = \frac{(-1)^{k+1}\theta_1^k(1-\theta_1^2)}{1-\theta_1^{2(k+1)}} = \frac{(-1)^{k+1}\theta_1^k}{1+\theta_1^2+\theta_1^4+\cdots+\theta_1^{2k}}.$$

**3.7** For $|\varphi| < 1$, let $X_t = \sum_{j=0}^{\infty}\varphi^j\varepsilon_{t-j}$, $\varepsilon_t \sim \text{WN}(0,\sigma_\epsilon^2)$. Prove that$\{X_t\}$ is stationary.

**3.8** Prove that $X_t = 0.2X_{t-1} + \varepsilon_t$ with the initial value $X_0 \equiv 0$ is nonstationary.

**3.9** For a causal ARMA$(p,q)$ model $\varphi(B)X_t = \theta(B)\varepsilon_t$, show that

$$\gamma(k) = \text{Cov}(X_{t+k}, X_k) = \sigma_\epsilon^2\sum_{j=0}^{\infty}\psi_j\psi_{j+k}$$

where $\{\psi_j\}$ are the coefficients in the MA$(\infty)$ representation.

**3.10** Determine which of the following models are causal and/or invertible:

(1) $X_t = 1.8X_{t-1} + 0.8X_{t-2} + \varepsilon_t + 0.2\varepsilon_{t-1} - 0.6\varepsilon_{t-2}$.
(2) $X_t = 0.8X_{t-1} - 0.8X_{t-2} + 0.4X_{t-3} + \varepsilon_t + 0.2\varepsilon_{t-1} - 0.6\varepsilon_{t-2} + 0.3\varepsilon_{t-3}$.
(3) $X_t = 0.7X_{t-1} - 0.8X_{t-2} + 0.6X_{t-3} + \varepsilon_t - 0.2\varepsilon_{t-1} - 0.6\varepsilon_{t-2} + 0.3\varepsilon_{t-3}$.

**3.11** Given three ARMA models

(1) $X_t = \varepsilon_t + 0.2\varepsilon_{t-1} - 0.6\varepsilon_{t-2} - 0.3\varepsilon_{t-3}$.
(2) $X_t = 0.7X_{t-1} - 0.8X_{t-2} + 0.6X_{t-3} + \varepsilon_t$.
(3) $X_t = 0.7X_{t-1} - 0.8X_{t-2} + 0.6X_{t-3} + \varepsilon_t - 0.2\varepsilon_{t-1} - 0.6\varepsilon_{t-2} + 0.3\varepsilon_{t-3}$.

graph their true ACF and PACF plots, respectively, and then observe the behaviors of the ACF and PACF.

# Chapter 4
# ARMA and ARIMA Modeling and Forecasting

In this chapter, given a time series sample (data) from an ARMA process, we consider how to build an ARMA model that sufficiently characterizes the process. We also consider how to fit an ARIMA model to a time series that has trend but no seasonality. We introduce the methods of moments, (conditional) least squares, as well as maximum likelihood; elaborate on a few order determination approaches and how to use Python and its extension packages to select model order; and estimate the model. Then we present residual analysis and diagnosis of models. At last, using the resulting model forecasts the future values of the time series.

## 4.1 Model Building Problems

For building an ARMA model, a time series dataset is required to be stationary. Thus before estimating the ARMA model, we should check if the time series data is stationary using those procedures in Sect. 3.1.

In Chap. 3, we have given the definition of the ARMA model and elaborated on its properties. Now we know that Eq. (3.7) is the ARMA$(p, q)$ model:

$$X_t = \varphi_0 + \varphi_1 X_{t-1} + \cdots + \varphi_p X_{t-p} + \varepsilon_t + \theta_1 \varepsilon_{t-1} + \cdots + \theta_q \varepsilon_{t-q} \tag{4.1}$$

where $\{\varepsilon_t\} \sim \text{WN}(0, \sigma_\epsilon^2)$, $\varphi_p \neq 0$, $\theta_q \neq 0$ and $\varphi_0$ is the intercept (const). Another expression of the ARMA$(p, q)$ model is

$$\varphi(B)X_t = \varphi_0 + \theta(B)\varepsilon_t$$

where $\varphi(z) = 1 - \varphi_1 z - \cdots - \varphi_p z^p$, $\theta(z) = 1 + \theta_1 z + \cdots + \theta_q z^q$. Hence the number of parameters to be estimated is $p + q + 2$. Additionally, in practice, order$(p, q)$ is also unknown and needs to be determined.

 
C. Huang, A. Petukhina, *Applied Time Series Analysis and Forecasting with Python*, Statistics and Computing,
https://doi.org/10.1007/978-3-031-13584-2_4

There are two important procedures for selecting ARMA model order $(p, q)$. One is to firstly compute ACF as well as PACF and then by Table 3.1 to choose order$(p, q)$. Another is a few information criteria such as AIC, BIC, HQIC, and so on. We will present these methods in detail in Sect. 4.3.

If $X_t$ has no seasonality and by differencing, we have that $Y_t = \nabla^d X_t = (1 - B)^d X_t$ is stationary (sometimes we need firstly transform the original series), then we can build an ARMA$(p, q)$ model for $Y_t$ as follows:

$$\varphi(B)Y_t = \varphi_0 + \theta(B)\varepsilon_t.$$

Expressing this model in terms of the original time series $X_t$, we have

$$\varphi(B)(1 - B)^d X_t = \varphi_0 + \theta(B)\varepsilon_t. \tag{4.2}$$

**Definition 4.1** (1) Equation (4.2) where $Y_t = (1 - B)^d X_t$ is stationary is called an ARIMA$(p, d, q)$ model. (2) A time series $X_t$ that satisfies Eq. (4.2) is said to be an ARIMA$(p, d, q)$ process or follows an ARIMA$(p, d, q)$ model.

Evidently, as long as $d \geq 1$, the ARIMA$(p, d, q)$ process is nonstationary, and when $d = 0$, the ARIMA$(p, d, q)$ process becomes an ARMA$(p, q)$ process. Besides, note that ARIMA means "AutoRegressive Integrated Moving Average," and we build an ARIMA$(p, d, q)$ model by building an ARMA$(p, q)$ model.

In summary, there are a few basic steps to fitting an ARMA model to a given time series dataset (sample):

- Plotting the time series data (time plot, ACF, and PACF) and performing a stationarity test in order to check if it is stationary. For nonstationary data, transforming and\or differencing the data makes them stationary. A generic data transformation is the *Box-Cox transformation*: if $\lambda \neq 0$, $Y_t = \lambda^{-1}(X_t^{\lambda} - 1)$; otherwise $Y_t = \log(X_t)$. And the latter is most used.
- Selecting order $(p, q)$ of the model (viz., identifying the model).
- Estimating parameters.
- Diagnosing and then accepting the model or considering a new model.

## 4.2 Estimation Methods

This section considers a few common approaches to estimate parameters in building an ARMA model given a stationary time series sample $\{X_{1:n}\}$. They all have a long history but are still vibrant. For simplicity, we assume that the const term $\varphi_0 = 0$ in this section. Otherwise, let $Z_t = X_t - \mu$ where $\mu = \mathrm{E}(X_t)$. Then Equation(4.1) turns into

$$Z_t = \varphi_1 Z_{t-1} + \cdots + \varphi_p Z_{t-p} + \varepsilon_t + \theta_1 \varepsilon_{t-1} + \cdots + \theta_q \varepsilon_{t-q}$$

where the const term is zero or no const term exists in it. In addition, $\mu$ can be estimated with sample mean $\sum_{t=1}^{n} X_t/n$ and $\varphi_0 = \mu(1 - \varphi_1 - \cdots - \varphi_p)$. Hence, as long as we can obtain the estimates of $\varphi_1, \cdots, \varphi_p$, then the estimate of $\varphi_0$ is also arrived at.

### 4.2.1 The Innovations Algorithm

The innovations algorithm is a recursive one and widely applied in time series modeling and forecasting. Note that the algorithm to be discussed in this subsection is applicable to all time series with finite second moments, regardless of whether they are stationary or not.

Assume that $\{X_t\}$ is a time series with mean zero and $\mathsf{E}(X_t^2) < \infty, t \geq 1$. Denote the autocovariance function by $\kappa(i, j) = \mathsf{E}(X_i X_j)$. Let $\hat{X}_1 = 0$ and $\hat{X}_j = \mathsf{E}(X_j|X_1, X_2, \cdots, X_{j-1})$ $(j \geq 2)$ be one-step linear predictors. Furthermore, let $\nu_n = \mathsf{E}(X_{n+1} - \hat{X}_{n+1})^2, n \geq 0$ and clearly $\nu_0 = \kappa(1, 1)$. The one-step prediction errors $\{X_n - \hat{X}_n\}$ are called the *innovations*. It can be proved that there exist coefficients $\theta_{nj}$ such that the one-step-ahead linear predictors can be computed recursively:

$$\hat{X}_1 = 0, \text{ and } \hat{X}_{n+1} = \sum_{j=1}^{n} \theta_{nj} \left( X_{n+1-j} - \hat{X}_{n+1-j} \right) \text{ for } n = 1, 2, \cdots$$

The following algorithm that recursively generates these coefficients and the mean squared errors $\nu_n$ (remember $\nu_0 = \kappa(1, 1)$ ) is known as the *innovations algorithm*:

$$\begin{cases} \theta_{n,n-k} = \nu_k^{-1} \left( \kappa(n+1, k+1) - \sum_{j=0}^{k-1} \theta_{k,k-j} \theta_{n,n-j} \nu_j \right), & 0 \leq k < n \\ \nu_n = \kappa(n+1, n+1) - \sum_{j=0}^{n-1} \theta_{n,n-j}^2 \nu_j. \end{cases}$$

Remarks on the innovations algorithm are as follows:

- Here interesting is that it can be shown that the innovation $\{X_n - \hat{X}_n\}$ is an uncorrelated time series. Transforming the original series $\{X_t\}$ to the uncorrelated series $\{X_n - \hat{X}_n\}$ is called *prewhitening*.
- The h-step-ahead predictor $\hat{X}_n(h)$ may also be calculated by

$$\hat{X}_n(h) = \sum_{j=h}^{n+h-1} \theta_{n+h-1,j} \left( X_{n+h-j} - \hat{X}_{n+h-j} \right) \text{ for } n = 1, 2, \cdots$$

- If $\{X_t\}$ is a stationary time series with mean zero, then the innovations algorithm is reduced to

$$\begin{cases} \theta_{n,n-k} = \nu_k^{-1}\left(\gamma(n-k) - \sum_{j=0}^{k-1}\theta_{k,k-j}\theta_{n,n-j}\nu_j\right), \ 0 \le k < n \\ \nu_n = \gamma(0) - \sum_{j=0}^{n-1}\theta_{n,n-j}^2\nu_j. \end{cases}$$

Brockwell and Davis (1991, 2016) excellently elaborate on the innovations algorithm. More details can be found there.

### *4.2.2 Method of Moments*

The method of moments was introduced by P. Chebyshev in 1887 and K. Pearson in 1894. The key idea is to firstly express the population (theoretic) moments as functions of the parameters of interest. Then those expressions are set equal to the corresponding sample moments, and the number of such equations should be the same as the number of parameters to be estimated. Solving those equations obtains so-called *moment estimates (estimators)* of those parameters. /COMPSet "moment estimates (estimators)" to Roman Generally speaking, moment estimates are regarded as preliminary estimates which are used as start values for other estimating procedures.

Recall Definition 1.7 in Sect. 1.3 that given a sample $\{X_t; 1 \le t \le n\}$ from a stationary time series $\{X_t\}$, the sample autocovariance and autocorrelation functions are, respectively, defined by

$$\hat{\gamma}_k = c_k = \frac{1}{n}\sum_{t=1}^{n-k}\left(X_{t+k} - \bar{X}\right)\left(X_t - \bar{X}\right) \text{ and } \hat{\rho}_k = r_k = \frac{c_k}{c_0}$$

where $\bar{X} = \sum_{t=1}^{n} X_t/n$ is the sample mean. Both $\bar{X}$ and $\hat{\gamma}_k$ are sample moments, and $\hat{\rho}_k$ are a function of the sample moments. In addition, $\bar{X}$, $\hat{\gamma}_k$, and $\hat{\rho}_k$ are, respectively, moment estimates of the mean, autocovariance, and autocorrelation of the time series. Specifically, $\hat{\gamma}_0$ is the estimate of the series variance $\gamma_0 = \text{Var}(X_t)$.

In Sect. 3.3.3, for an AR($p$) model, we have derived the Yule-Walker equations:

$$\varphi_1\rho_{k-1} + \varphi_2\rho_{k-2} + \cdots + \varphi_p\rho_{k-p} = \rho_k, \ 1 \le k \le p$$

and

$$\sigma_\epsilon^2 = \gamma_0(1 - \varphi_1\rho_1 - \varphi_2\rho_2 - \cdots - \varphi_p\rho_p).$$

Substituting $\rho_k$ in the equations with $\hat{\rho}_k = r_k$ for $1 \leq k \leq p$, we can theoretically solve the Yule-Walker equations to obtain the moment estimators $\{\hat{\varphi}_{1:p}\}$ for $\{\varphi_{1:p}\}$, which are called the *Yule-Walker estimates (estimators)*. In matrix notation, the Yule-Walker equations can be expressed by

$$\boldsymbol{\rho}\varphi = \rho \text{ and } \sigma_\epsilon^2 = \gamma_0(1 - \varphi'\rho)$$

where

$$\boldsymbol{\rho} = \begin{pmatrix} \rho_0 & \rho_1 & \cdots & \rho_{p-1} \\ \rho_1 & \rho_0 & \cdots & \rho_{p-2} \\ \vdots & \vdots & \vdots & \vdots \\ \rho_{p-1} & \rho_{p-2} & \cdots & \rho_0 \end{pmatrix}, \ \varphi = \begin{pmatrix} \varphi_1 \\ \varphi_2 \\ \vdots \\ \varphi_p \end{pmatrix}, \ \rho = \begin{pmatrix} \rho_1 \\ \rho_2 \\ \vdots \\ \rho_p \end{pmatrix}.$$

The Yule-Walker estimators are

$$\hat{\varphi} = \hat{\boldsymbol{\rho}}^{-1}\hat{\rho} = \boldsymbol{R}^{-1}r \text{ and } \hat{\sigma}_\epsilon^2 = \hat{\gamma}_0(1 - r'\boldsymbol{R}^{-1}r)$$

where

$$\boldsymbol{R} = \begin{pmatrix} 1 & r_1 & \cdots & r_{p-1} \\ r_1 & 1 & \cdots & r_{p-2} \\ \vdots & \vdots & \vdots & \vdots \\ r_{p-1} & r_{p-2} & \cdots & 1 \end{pmatrix}, \ r = \begin{pmatrix} r_1 \\ r_2 \\ \vdots \\ r_p \end{pmatrix}.$$

and possess the following large-sample property:

**Theorem 4.1** *If $\hat{\varphi}$ is the Yule-Walker estimator of $\varphi$, then the distribution of $\sqrt{n}(\hat{\varphi} - \varphi)$ asymptotically converges $N(\boldsymbol{0}, \ \sigma_\epsilon^2\gamma_0^{-1}\boldsymbol{\rho}^{-1})$.*

In practice, we tend to use the Durbin-Levinson recursion algorithm to compute the Yule-Walker estimate of $\varphi$ without expensively inverting $\boldsymbol{R}$. In addition to the Yule-Walker procedure, the Burg algorithm also applies to preliminary estimation of AR models. As to MA($q$) as well as ARMA($p, q$)($q > 0$) models, both innovations algorithm and Hannan-Rissanen algorithm are used to obtain the preliminary estimates of their parameters. About all those details, you can find in Brockwell and Davis (1991, 2016).

### 4.2.3 *Method of Conditional Least Squares*

The method of least squares (LS) is probably the most popular estimating technique in statistics. In his memoir, C. F. Gauss mentioned that he had previously discovered LS and used it as early as 1795 in estimating the orbit of an asteroid. On the other hand, A.M. Legendre (1805) also contributed to the method of least squares. For details, you can refer to Stigler (1981).

Still consider an AR($p$) model:

$$X_t = \varphi_1 X_{t-1} + \cdots + \varphi_p X_{t-p} + \varepsilon_t = \varphi' \boldsymbol{X}_{t-1} + \varepsilon_t$$

where

$$\varphi' = \left(\varphi_1\ \varphi_2\ \cdots\ \varphi_p\right) \text{ and } \boldsymbol{X}_{t-1} = \left(X_{t-1}\ X_{t-2}\ \cdots\ X_{t-p}\right)'$$

Letting $\boldsymbol{X}_0 = (X_0\ X_{-1}\ \cdots\ X_{1-p})' = \boldsymbol{0}$ as an added condition, then the conditional sum of squared (css) errors is

$$S_c(\varphi) = \sum_{t=1}^{n} \varepsilon_t^2 = \sum_{t=1}^{n} (X_t - \varphi' \boldsymbol{X}_{t-1})^2.$$

According to standard regression analysis procedure, the (conditional) least squares estimator of $\varphi$ is

$$\hat{\varphi} = \left(\sum_{t=1}^{n} \boldsymbol{X}_{t-1} X_t\right) \left(\sum_{t=1}^{n} \boldsymbol{X}_{t-1} \boldsymbol{X}_{t-1}'\right)^{-1}.$$

For example, in the simple case that $p = 1$, we have

$$\hat{\varphi}_1 = \left(\sum_{t=1}^{n} \boldsymbol{X}_{t-1} X_t\right) \left(\sum_{t=1}^{n} \boldsymbol{X}_{t-1} \boldsymbol{X}_{t-1}'\right)^{-1} = \left(\sum_{t=2}^{n} X_{t-1} X_t\right) \left(\sum_{t=2}^{n} X_{t-1}^2\right)^{-1}.$$

For a general ARMA($p, q$) model, similarly, we need a condition that

$$X_0 = X_{-1} = \cdots = X_{1-p} = \varepsilon_0 = \varepsilon_{-1} = \cdots = \varepsilon_{1-q}$$

in order to estimate its parameters using the method of least squares. Under this condition, we can represent unobservable error $\varepsilon_t$ in terms of the sample $\{X_{1:n}\}$ and coefficients. In the simple case that $p = q = 1$, for example, we have

$$X_t = \varphi_1 X_{t-1} + \varepsilon_t + \theta_1 \varepsilon_{t-1} \text{ and } X_0 = \varepsilon_0 = 0.$$

Then

$$\varepsilon_1 = X_1, \varepsilon_2 = X_2 - \varphi_1 X_1 - \theta_1 X_1, \cdots, \varepsilon_n = X_n - \varphi_1 X_{n-1} - \theta_1 \varepsilon_{n-1}.$$

Now the conditional sum of squared errors for the ARMA($p, q$) model is

$$S_c(\varphi, \theta) = \sum_{t=1}^{n} \varepsilon_t^2 = \sum_{t=1}^{n} (X_t - \varphi_1 X_{t-1} - \cdots - \varphi_p X_{t-p} - \theta_1 \varepsilon_{t-1} - \cdots - \theta_q \varepsilon_{t-q})^2$$

where $\theta = (\theta_1 \ \cdots \ \theta_q)'$. Minimizing $S_c(\varphi, \theta)$ with respect to $(\varphi, \theta)$ yields the conditional least squares estimators. The difficulty is that if $q > 0$ the problem is nonlinear and has no closed form solution. Thus we need to rely on numerical optimization technique, for example, the Gauss-Newton method. More details can be found in Shumway and Stoffer (2017).

### 4.2.4 Method of Maximum Likelihood

The method of maximum likelihood (ML) was proposed by R. A. Fisher in 1922. It is a computing procedure with the most statistical flavor. Nowadays, there are various kinds of maximum likelihood procedures.

Assume that $\{X_t\}$ is a normal (Gaussian) time series with mean zero and autocovariance function $\kappa(i, j) = \mathsf{E}(X_i X_j)$. Let $\boldsymbol{X}_n = (X_1 \ X_2 \ \cdots \ X_n)'$ and $\hat{\boldsymbol{X}}_n = (\hat{X}_1 \ \hat{X}_2 \ \cdots \ \hat{X}_n)'$ where $\hat{X}_1 = 0$ and $\hat{X}_j = \mathsf{E}(X_j | X_1, X_2, \cdots, X_{j-1})$ $(j \geq 2)$. Furthermore, let $\Gamma_n$ be the autocovariance matrix $\Gamma_n = \mathsf{E}(\boldsymbol{X}_n \boldsymbol{X}_n')$. Then the likelihood function of $\boldsymbol{X}_n$ is

$$L(\Gamma_n) = (2\pi)^{-\frac{n}{2}} (\det \Gamma_n)^{-\frac{1}{2}} \exp(-\frac{1}{2} \boldsymbol{X}_n' \Gamma_n^{-1} \boldsymbol{X}_n).$$

Applying the innovations algorithm, the likelihood can be reduced to

$$L(\Gamma_n) = (2\pi)^{-\frac{n}{2}} (\nu_0 \nu_1 \cdots \nu_{n-1})^{-\frac{1}{2}} \exp\left[ -\frac{1}{2} \sum_{j=1}^{n} (X_j - \hat{X}_j)^2 / \nu_{j-1} \right]. \tag{4.3}$$

Now suppose that $\{X_t\}$ is a causal and invertible Gaussian (normal) time series and satisfies the ARMA($p, q$) model:

$$\varphi(B) X_t = \theta(B) \varepsilon_t, \{\varepsilon_t\} \sim \text{WN}(0, \sigma_\epsilon^2)$$

where $\varphi(z) = 1 - \varphi_1 z - \cdots - \varphi_p z^p$, $\theta(z) = 1 + \theta_1 z + \cdots + \theta_q z^q$. Let $\varphi = (\varphi_1 \ \varphi_2 \ \cdots \ \varphi_p)'$, $\theta = (\theta_1 \ \theta_2 \ \cdots \ \theta_q)'$ and $\beta = (\varphi, \theta)'$. At this point, the likelihood function (4.3) can be rewritten as

$$L(\beta, \sigma_\epsilon^2) = (2\pi)^{-\frac{n}{2}} \sigma_\epsilon^{-n} [r_0(\beta) r_1(\beta) \cdots r_{n-1}(\beta)]^{-\frac{1}{2}} \exp\left[ -\frac{S(\beta)}{2\sigma_\epsilon^2} \right] \tag{4.4}$$

where

$$S(\beta) = \sum_{j=1}^{n} [(X_j - \hat{X}_j(\beta))^2 / r_{j-1}(\beta)] \tag{4.5}$$

and $r_j(\beta) = \nu_j/\sigma_\epsilon^2$, $\hat{X}_j(\beta) = \hat{X}_j$. Both $r_j = \nu_j/\sigma_\epsilon^2$ and $\hat{X}_j$ are functions of $\beta$ alone, and we make this fact explicit in (4.4) and (4.5). Differentiating $\log L(\beta, \sigma_\epsilon^2)$ partially with respect to $\sigma_\epsilon^2$ and noting that $\hat{X}_j(\beta)$ and $r_j(\beta)$ are independent of $\sigma_\epsilon^2$ ($\nu_j$ can be represented by $\sigma_\epsilon^2 f_j(\beta)$), we find that the *maximum likelihood estimators* (MLE) of $\beta$ and $\sigma_\epsilon^2$ are

$$\hat{\beta} \text{ and } \hat{\sigma}_\epsilon^2 = n^{-1} S(\hat{\beta})$$

where $\hat{\beta}$ minimizes

$$\ell(\beta) = \log[n^{-1} S(\beta)] + n^{-1} \sum_{j=1}^{n} \log r_{j-1}(\beta).$$

In addition, Eq. (4.5) may be regarded as a transformed form of the sum of squared errors, and the value $\tilde{\beta}$ of $\beta$ that minimizes $S(\beta)$ in (4.5) is called the *(unconditional) least squares estimator*.

There are a few remarks as follows:

- Whether maximum likelihood estimation or least squares estimation of $\beta$, they must generally be done numerically.
- Even if $\{X_t\}$ is not Gaussian, we may still regard (4.4) as a measure of goodness of fit to the data and choose the parameters $(\beta, \sigma_\epsilon^2)$ that maximize this measure. In fact, as long as $\{\varepsilon_t\} \sim \text{iid}(0, \sigma_\epsilon^2)$, regardless of the joint distributions of the time series, the large-sample distribution of the estimator $\hat{\beta}$ is the same (asymptotically normal), and at the same time the estimator $\hat{\sigma}_\epsilon^2$ converges $\sigma_\epsilon^2$ in probability (see Brockwell and Davis 1991, Section 10.8 as well as Fan and Yao 2003, Page 97).
- Due to the reason above, we shall always refer to (4.4) as the "likelihood" function (or "Gaussian likelihood" function) of the sample $\boldsymbol{X}_n = (X_1\ X_2\ \cdots\ X_n)'$ and the estimators derived from maximizing the (Gaussian) likelihood as "maximum likelihood" estimators, regardless of the underlying distribution.

At the end of this section, we give a conclusion about the large-sample properties of all the estimators above: under appropriate conditions, for causal and invertible ARMA models, the maximum likelihood, the unconditional least squares, and the conditional least squares estimators, each initialized by the method of moments estimator, all provide optimal estimators of $\sigma_\epsilon^2$ and $\beta = (\varphi, \theta)'$ in the sense that $\hat{\sigma}_\epsilon^2$ is consistent, and the asymptotic distribution of $\hat{\beta}$ is the best asymptotic normal distribution. In other words, they are asymptotically equivalent. You can find more details in Brockwell and Davis (1991, 2016), Fan and Yao (2003) as well as Shumway and Stoffer (2017).

## 4.3 Order Determination

So far we have only a procedure for selecting appropriate values for the order $(p, q)$ of an ARMA$(p, q)$ model. That is, according to Table 3.1, if the sample ACF cuts off after lag $q$, we choose order $(0, q)$ as the order of our desired model; if the sample PACF cuts off after lag $p$, we select order $(p, 0)$. But if both the sample ACF and PACF simultaneously tail off, then we cannot determine the order using Table 3.1. Fortunately, based on the Kullback-Leibler information index (see Kullback and Leibler 1951), an information criterion is introduced by Akaike (1973), which can be used to select appropriate order $(p, q)$ of an ARMA model (so-called model specification or identification). Nowadays, this criterion is known as *Akaike Information Criterion* (AIC). Its general expression is

$$\text{AIC} = -2(\text{maximized log likelihood}) + 2(\text{No. of estimated parameters}).$$

In the context of ARMA$(p, q)$ modeling, specifically we have

$$\text{AIC} = -2\log[L(\hat{\beta}, n^{-1}S(\hat{\beta}))] + 2(p + q + 1)$$

where $\hat{\beta}$ is the MLE of $\beta$ and $S(\cdot)$ defined in (4.5).[1] If order $(\hat{p}, \hat{q})$ minimizes AIC, then ARMA$(\hat{p}, \hat{q})$ is preliminarily selected as our desired model.

Since AIC has a tendency to overestimate the model, various procedures have been proposed to correct the overfitting nature of AIC, two of which are as follows. One is *Bayesian Information Criterion* (BIC) defined as

$$\text{BIC} = -2(\text{maximized log likelihood}) + \log(n)(\text{No. of estimated parameters}).$$

The name BIC is due to the fact that the BIC criterion is derived from diverse Bayesian arguments. See, for example, Akaike (1977) and Schwarz (1978). Another is *Hannan-Quinn Information Criterion* (HQIC) defined as

$$\text{HQIC} = -2(\text{maximized log likelihood}) + \log\log(n)(\text{No. of estimated parameters}).$$

We also choose order $(\hat{p}, \hat{q})$ such that it minimizes BIC or HQIC. The HQIC criterion is firstly proposed by Hannan and Quinn (1979), and it is easy to see that HQIC is between AIC and BIC, that is, HQIC increases the penalty for

[1] AIC tends to overestimate the order. Hurvich and Tsai (1989) propose a corrected form of AIC: AICC$= -2\log[L(\hat{\beta}, n^{-1}S(\hat{\beta}))] + 2n(p + q + 1)(n - p - q - 2)^{-1}$.

model complexity to a certain extent but not so big as BIC. As to comprehensive explanations of various information criteria, you are referred to Konishi and Kitagawa (2008). Two remarks on the three criteria are as follows:

- Two different criteria (e.g., BIC and HQIC) are not comparable for selecting order $(p, q)$. Each criterion should compare different orders (viz., models) and then make a decision.
- These information criteria tend not to be good guides to selecting the appropriate order of differencing $(d)$ of an ARIMA$(p, d, q)$ model, but only for selecting the values of $(p, q)$. This is because the differencing changes the data on which the likelihood is estimated, making the AIC (BIC or HQIC) values between models with different orders of differencing not comparable. In other words, we should identify the differencing order $(d)$ before determining order $(p, q)$.

In the implementation of the criteria with Python, there is a function in the module `statsmodels.tsa` that can be used to select order $(p, q)$ of an ARMA model. It is the function `sm.tsa.arma_order_select_ic`. Its usage is demonstrated in examples below. What is more, the function `choose_arma` in the Python package `PythonTsa` is also used to determine order $(p, q)$. Notice that there is a parameter (argument) called the *coefficient of control* in this function and denoted by `ctrl`. Its value range is $1.0 \leq ctrl \leq 1.1$ and in general $1.0 \leq ctrl \leq 1.05$. Now we explain why we introduce the control coefficient. In practice, we find that for some fitted ARMA models which are identified according to the AIC (BIC or HQIC), there are such roots of their AR or/and MA polynomials that their abstract values (moduli) are almost equal to 1. In other words, such models can hardly be stationary or invertible and are not thought appropriate. Hence, for a fitted ARMA model, if there exits such a root of its AR or MA polynomials that the modulus of the root is not greater than the control coefficient `ctrl`, we abandon the model although the AIC (BIC or HQIC) for it might be the minimum. Examples of using the function `choose_arma` are shown below. At last, you are kindly reminded that there is no such a once-in-a-lifetime or fully automatic selection procedure that can always select an appropriate model at one time from numerous models.

## 4.4 Diagnosis of Models

It is well known that any statistical model is only an approximation to reality. Thus it does matter that we conduct a post-estimating diagnostic check to see whether the estimated model fits well to the data and to what extent it explains the background of the process from which the data comes.

First of all, for the fitted ARMA$(p, q)$ model, the estimators $\hat{\varphi}_p$ and $\hat{\theta}_q$ of coefficients $\varphi_p$ and $\theta_q$ should be significantly different from zero. Otherwise, the model is not of order $(p, q)$.

If we have fitted an ARMA$(p, q)$ model to the data $\{X_{1:n}\}$ and $\{\hat{X}_j(\hat{\beta}); 1 \leq j \leq n\}$ are the predicted (fitted) values obtained from the fitted model, then

$$\{\hat{\varepsilon}_j\} = \{X_j - \hat{X}_j(\hat{\beta})\} \text{ and } \{[X_j - \hat{X}_j(\hat{\beta})]/[\hat{\sigma}_\epsilon^2 r_{j-1}(\hat{\beta})]^{1/2}\}$$

are, respectively, called the *residuals* and *standardized residuals*. It is easily understandable that the properties of the residuals have a decisive influence on the goodness of fit of the statistical model to a dataset. Thus we need conduct a full and explicit analysis of the residuals. On the other hand, if the estimated model is appropriate (adequate), then the residuals should have nearly the properties of white noise. In order to check whether the residuals resemble a white noise, we analyze the residuals from four aspects:

- *ACF plotting* If the fitted model is appropriate, the autocorrelation function (ACF) of the residuals should look like ACF of white noise.
- *Ljung-Box (portmanteau) testing* we have introduced the Ljung-Box (Q or portmanteau) statistic in Sect. 2.2, namely,

$$Q_{LB}(m) = n(n+2) \sum_{k=1}^{m} \frac{r_k^2}{n-k} \tag{4.6}$$

  to test if a stationary time series is a white noise. Now we utilize the statistic to test whether the residuals look like a white noise (i.e., whether the fitted model is adequate). Under the null hypothesis of model adequacy, $Q_{LB}(m)$ asymptotically ($n \to \infty$) follows the chi-squared distribution $\chi^2(m-p-q)$. Note that building the model results in a loss of $p+q$ degrees of freedom. Details can be found in Box and Pierce (1970) as well as Ljung and Box (1978).
- *QQ plotting* or other *normality testing* In practice, generally speaking, we do not know if $\{\varepsilon_t\}$ is normally distributed. Hence to test the normality of the residuals is actually to test that of $\{\varepsilon_t\}$. At this point, QQ plotting is most used and other normality test procedures may be used too.
- *ACF plotting of the squared residuals* This action is not to test if the residuals are a white noise. It is to check whether the squared residuals are autocorrelated (if yes, we say that the residuals have ARCH effect that is discussed in detail in Chap. 6).

There are two functions in package `PythonTsa` that can be used to analyze the residuals by visualizing. One is the function

```
plot_LB_pvalue(x,noestimatedcoef,nolags)
```

where `x` is a series to be analyzed or residuals to be diagnosed when modeling; `noestimatedcoef` is the number of estimated coefficients ($p+q$ for ARMA$(p, q)$ model) when modeling; and `nolags` is maximum number of added terms in LB statistic (4.6), noting that `nolags` must be greater than `noestimatedcoef`. This function plots the $p$-values for Ljung-Box test statistic against $m \leq$ `nolags`. The other is the function

```
plot_ResidDiag(x,noestimatedcoef,nolags,lag)
```

where the three former parameters are the same as the first function's and `lag` is the number of lags for ACF of `x` and ACF of the squared `x`. It simultaneously plots the $p$-values for Ljung-Box statistic, QQ plot, ACF of the residuals, and ACF of the squared residuals. So you need not use the first function if you utilize the second. Note that the degrees of freedom of $Q_{LB}(m)$ remain unchanged whether we estimate the mean of the time series $\{X_t\}$ or not. See, for example, Li (2004, Section 2.2).

## 4.5 Forecasting

Forecasting is one of key tasks in time series analysis. Let $\{X_t\}$ be a stationary time series with mean zero, and the sample (data) $\{X_{1:n}\}$ is given. There are two types of forecast, which need to distinguish one another. One is in-sample prediction or ex post prediction, and actually there has been an observation value at the predicting time point $t_0$ ( $1 \leq t_0 \leq n$ ). In this case, predicted values are called *fitted values* or *in-sample predictors*. It is used as a means to check against the given data so that the predicting model can be evaluated. The other is out-of-sample forecasting or ex ante forecasting. At this point, we utilize the known data $\{X_{1:n}\}$ to forecast future values of the time series beyond the present $n$ ($n$ is called the *forecast origin*). We denote the forecast of $X_{n+h}$ as $\hat{X}_n(h)$, $P_n X_{n+h}$ or $X^n_{n+h}$ ($h$ is called the *lead time*) and call it the *h-step-ahead predictor(forecast)*. In addition, we denote the forecast error as $e_n(h) = X_{n+h} - \hat{X}_n(h)$. In order to compare various forecasts for the same $X_{n+h}$, we need a certain criterion, and the *minimum mean squared error criterion* is most used. That is, our forecast $\hat{X}_n(h)$ for $X_{n+h}$ must satisfy

$$\mathsf{E}[e_n(h)]^2 = \mathsf{E}[X_{n+h} - \hat{X}_n(h)]^2 = \min_g \left\{\mathsf{E}[X_{n+h} - g(X_{1;n})]^2\right\}$$

where $g(X_{1;n})$ is any measurable function of the observations $\{X_{1:n}\}$. It can be proved that

(1) $\hat{X}_{n+1} = \hat{X}_n(1) = \mathsf{E}(X_{n+1}|X_1, X_2, \cdots, X_n)$ $(n \geq 1)$. Thus $\hat{X}_n(h)$ can be recursively computed with the innovations algorithm (see Sect. 4.2.1 and (4.7) below).
(2) There exist coefficients $\{\phi_{nj}; 1 \leq j \leq n\}$ such that $\hat{X}_{n+1} = \phi_{n1}X_n + \cdots + \phi_{nn}X_1$, which is known as the *best linear predictor* for $X_{n+1}$.

(3) For a causal ARMA$(p, q)$ model with mean zero, the one-step-ahead predictors are recursively obtained by

$$\hat{X}_{n+1} = \begin{cases} 0, & n = 0 \\ \sum_{j=1}^{n} \theta_{nj}(X_{n+1-j} - \hat{X}_{n+1-j}), & 1 \leq n < \max(p, q) \\ \sum_{i=1}^{p} \varphi_i X_{n+1-i} + \sum_{j=1}^{q} \theta_{nj}(X_{n+1-j} - \hat{X}_{n+1-j}), & n \geq \max(p, q) \end{cases}$$

where $\theta_{nj}$ are found from the innovations' algorithm. If $n > \max(p, q)$ (in practice, this inequality is almost always true), then for all $h \geq 1$, the h-step-ahead predictor is

$$\hat{X}_n(h) = \sum_{i=1}^{p} \varphi_i \hat{X}_n(h - i) + \sum_{j=h}^{q} \theta_{n+h-1,j}(X_{n+h-j} - \hat{X}_{n+h-j}), \tag{4.7}$$

noting that $\hat{X}_{n+h-j}$ is a one-step-ahead predictor.

For more details on forecasting by the innovations algorithm, see, for example, Brockwell and Davis (1991, 2016).

## 4.6 Examples

*Example 4.1 (The NAO Index Since January 1950)* The time series dataset "nao" in the folder Ptsadata is the monthly mean north Atlantic oscillation (NAO) index since January 1950. You can also download it from the web page

"https://www.cpc.ncep.noaa.gov/products/precip/CWlink/pna/nao.shtml".

The NAO index is important for weather research and is based on the surface sea level pressure difference between the subtropical (Azores) high and the subpolar low. From its time series plot shown in Fig. 4.1 and ACF plot shown in the top panel of Fig. 4.2, we could say that the time series appears stationary. What is more, we use the KPSS stationarity test on it and obtain the $p$-value 0.096, which validates that it has stationarity.

Now imagine that the NAO series is from an ARMA process and we want to build an ARMA$(p, q)$ model for the process using the NAO series sample. First of all, we need to determine order $(p, q)$. Observing the PACF plot shown in the bottom panel of Fig. 4.2, we clearly see that it cuts off after lag 1. Thus we should choose an AR(1) model for the process according to Table 3.1. One may argue that the ACF also cuts off after lag 1, and so an MA(1) model could be chosen for the process. At this point, however, the AIC, BIC, and HQIC for the MA(1) model are 2358.75, 2368.19, and 2362.37, respectively. They are, respectively, greater than the AIC (2356.02), BIC (2365.47), and HQIC (2359.64) for the AR(1) model. Thus we finally select the AR(1) model for the process.

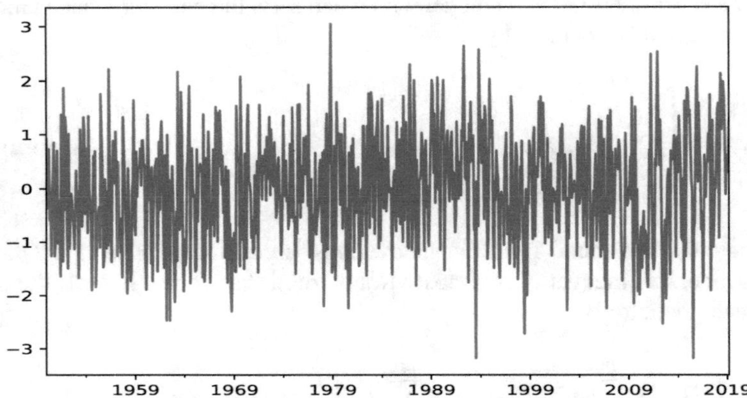


**Fig. 4.1** Time series plot of the NAO index series

 pyTSA_NaoARMA

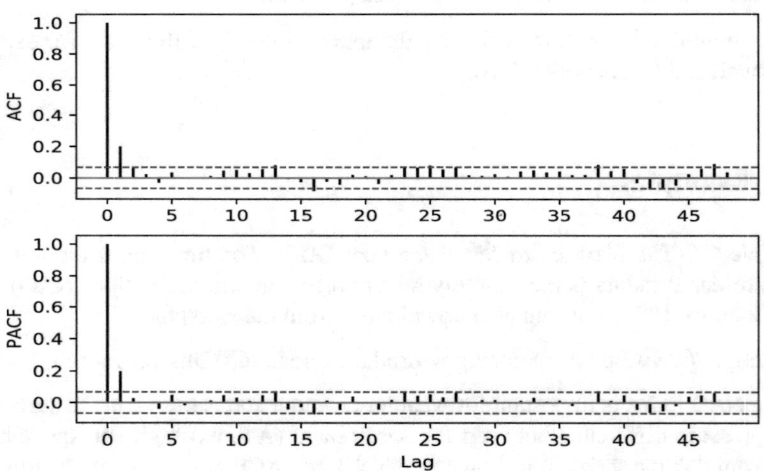


**Fig. 4.2** ACF and PACF plots of the NAO index series

pyTSA_NaoARMA

In building the model, we find that the const (intercept) term is not significantly different from zero and so let it be zero. The ACF and PACF plots of the modelling residuals are shown in Fig. 4.3 and behave like ones of a white noise. Moreover, using the function `plot_LB_pvalue` in the Python package `PythonTsa`, the $p$-values for Ljung-Box statistic of the residuals are plotted in Fig. 4.4, and all are greater than 0.1, which suggests that the residuals should be a white noise. In other words, the resulting model $X_t = 0.1996X_{t-1} + \varepsilon_t$ is sufficient for the NAO sample. Note that the lag on the horizontal axis in Fig. 4.4 is actually the number of the added terms in Ljung-Box statistic and greater than or equal to 2. At last,

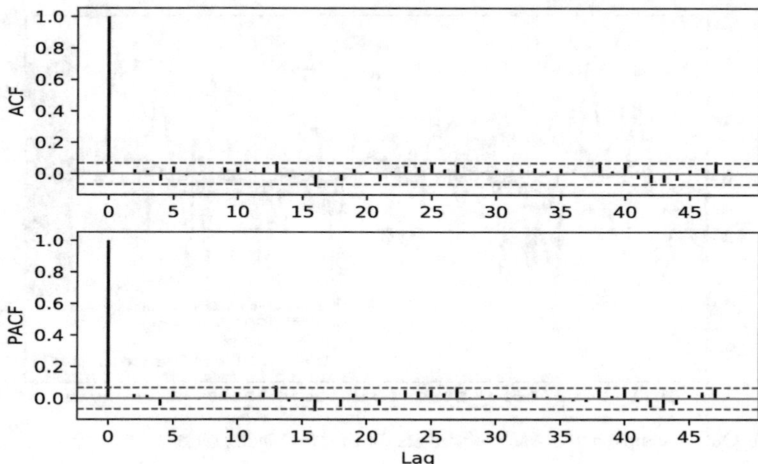


Fig. 4.3 ACF and PACF plots of the NAO index AR(1) model residuals

pyTSA_NaoARMA

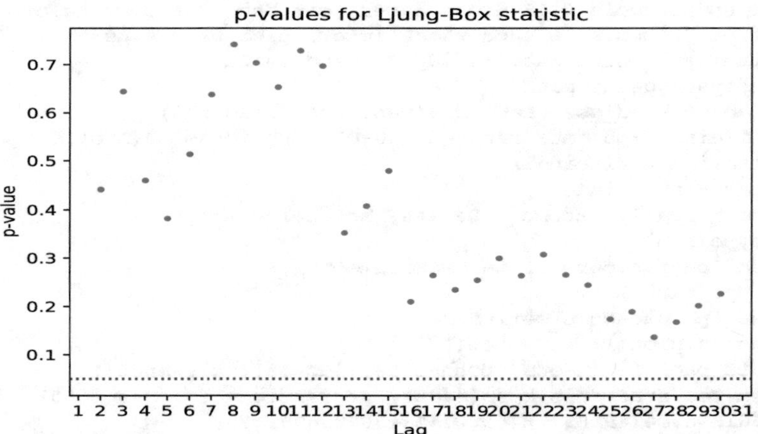


Fig. 4.4 *p*-values for Ljung-Box statistic of the NAO index AR(1) model

pyTSA_NaoARMA

using the resulting model, we can plot the out-of-sample and in-sample predicts shown in Fig. 4.5. The pity is that from Fig. 4.5 we see that the forecasts are not so good although the resulting model fits well. The Python code for this example is as follows.

```
>>> import pandas as pd
>>> import matplotlib.pyplot as plt
>>> import statsmodels.api as sm
>>> from PythonTsa.plot_acf_pacf import acf_pacf_fig
```

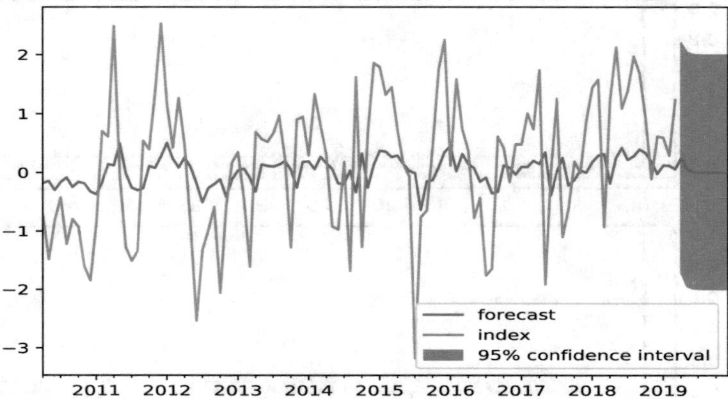


**Fig. 4.5** Out-of-sample and in-sample predicts for the NAO index series

pyTSA_NaoARMA

```
>>> from statsmodels.tsa.arima_model import ARMA
# for statsmodels 0.13.0 and later, see the last code below
>>> from PythonTsa.LjungBoxtest import plot_LB_pvalue
>>> from PythonTsa.datadir import getdtapath
>>> dtapath=getdtapath()
>>> nao=pd.read_csv(dtapath +'nao.csv', header=0)
>>> timeindex=pd.date_range('1950-01', periods=len(nao),freq='M')
>>> nao.index=timeindex
>>> naots=nao['index']
# automatically become a Series, see below
>>> type(nao)
<class 'pandas.core.frame.DataFrame'>
>>> type(naots)
<class 'pandas.core.series.Series'>
>>> naots.plot(); plt.show()
>>> acf_pacf_fig(naots, both=True, lag=48); plt.show()
>>> sm.tsa.stattools.kpss(naots, regression="c", nlags=50)
(0.3561836394749189, 0.09604153470908669, 50,
{'10%': 0.347, '5%': 0.463, '2.5%': 0.574, '1%': 0.739})
>>> ar1=ARMA(naots, order=(1,0)).fit(trend='c', disp=False)
>>> print(ar1.summary())
                              ARMA Model Results
==============================================================================
Dep. Variable:                  index   No. Observations:                  831
Model:                     ARMA(1, 0)   Log Likelihood               -1176.007
Method:                       css-mle   S.D. of innovations              0.996
Date:                Wed, 17 Jul 2019   AIC                           2358.013
Time:                        20:37:38   BIC                           2372.181
Sample:                    01-31-1950   HQIC                          2363.446
                         - 03-31-2019
==============================================================================
                 coef    std err          z      P>|z|      [0.025      0.975]
------------------------------------------------------------------------------
```

```
const          0.0040      0.043      0.092      0.927     -0.081      0.089
ar.L1.index    0.1996      0.034      5.867      0.000      0.133      0.266
                                    Roots
=============================================================================
                  Real          Imaginary           Modulus         Frequency
-----------------------------------------------------------------------------
AR.1            5.0109           +0.0000j            5.0109            0.0000
-----------------------------------------------------------------------------
# the const is not significantly different from zero.
>>> ar1=ARMA(naots, order=(1,0)).fit(trend='nc', disp=False)
>>> print(ar1.summary())
                              ARMA Model Results
==============================================================================
Dep. Variable:                  index   No. Observations:                  831
Model:                     ARMA(1, 0)   Log Likelihood               -1176.011
Method:                       css-mle   S.D. of innovations              0.996
Date:                Wed, 01 May 2019   AIC                           2356.022
Time:                        15:49:57   BIC                           2365.467
Sample:                    01-31-1950   HQIC                          2359.644
                         - 03-31-2019
==============================================================================
                 coef    std err          z      P>|z|      [0.025      0.975]
------------------------------------------------------------------------------
ar.L1.index    0.1996      0.034      5.867      0.000       0.133      0.266
                                    Roots
=============================================================================
                  Real          Imaginary           Modulus         Frequency
-----------------------------------------------------------------------------
AR.1            5.0108           +0.0000j            5.0108            0.0000
-----------------------------------------------------------------------------
>>> resid1 = ar1.resid
>>> acf_pacf_fig(resid1, both=True, lag=48); plt.show()
>>> plot_LB_pvalue(resid1, noestimatedcoef=1, nolags=30)
>>> plt.show()
# noestimatedcoef = number of estimated coefficients
# nolags = max number of added terms in LB statistic.
>>> ar1.plot_predict(start='2010-04', end='2019-12')
>>> plt.show()
>>> ma1=ARMA(naots, order=(0,1)).fit(trend='nc', disp=False)
>>> ma1.aic; ma1.bic; ma1.hqic
2358.7456128850167
2368.1908724747277
2362.367530551026
# so the AR(1) is better than MA(1)
# the following is for statsmodels of v. 0.13.0 and later
>>> from statsmodels.tsa.arima.model import ARIMA
>>> ar1=ARIMA(naots, order=(1,0,0),trend='c').fit()
>>> ar1=ARIMA(naots, order=(1,0,0),trend='n').fit()
>>> pred=ar1.get_prediction(start='2010-04', end='2019-12')
>>> predicts=pred.predicted_mean
>>> predconf=pred.conf_int()
>>> predframe=pd.concat([naots['2010-04-30':], predicts,
                predconf['2019-04-30':]], axis=1)
>>> predframe.plot(); plt.show()
```

*Example 4.2 (Simulating and Building an ARMA(2, 2) Model)* Suppose that an ARMA(2, 2) model is given below:

$$X_t = 0.8X_{t-1} - 0.6X_{t-2} + \varepsilon_t + 0.7\varepsilon_{t-1} + 0.4\varepsilon_{t-2}$$

or

$$X_t - 0.8X_{t-1} + 0.6X_{t-2} = \varepsilon_t + 0.7\varepsilon_{t-1} + 0.4\varepsilon_{t-2}$$

where $\varepsilon_t \sim \text{iidN}(0, 1)$ is a standard normal white noise.

First of all, with Python, it is easy to know that the given model is both stationary and invertible. Then we generate a sample of size 500 from the given ARMA(2, 2) model. The time series plot of the sample is shown in Fig. 4.6, and its ACF and PACF plots are shown in Fig. 4.7. It is without saying that the time series sample is stationary. Now we pretend that we have forgotten where the sample is from and are going to build an ARMA($p, q$) model for it. The difficulty is to determine order $(p, q)$. We use the function `choose_arma( )` in the Python package `PythonTsa` to select $(p, q)$ and let the control coefficient `ctrl`=1.02. It turns out that the three criteria(AIC, BIC, HQIC) derive the same result $(p, q)$=(2, 2). This result is desired since the sample actually comes from the given ARMA(2, 2) model. If we use the function `sm.tsa.arma_order_select_ic` to choose $(p, q)$, then its BIC and HQIC also derive the same $(p, q)$=(2, 2), but its AIC derives $(p, q)$=(6, 7) that is obviously inappropriate. Taking $(p, q)$=(2, 2), the estimated model is

$$X_t = 0.8108X_{t-1} - 0.6313X_{t-2} + \varepsilon_t + 0.7086\varepsilon_{t-1} + 0.4677\varepsilon_{t-2},$$

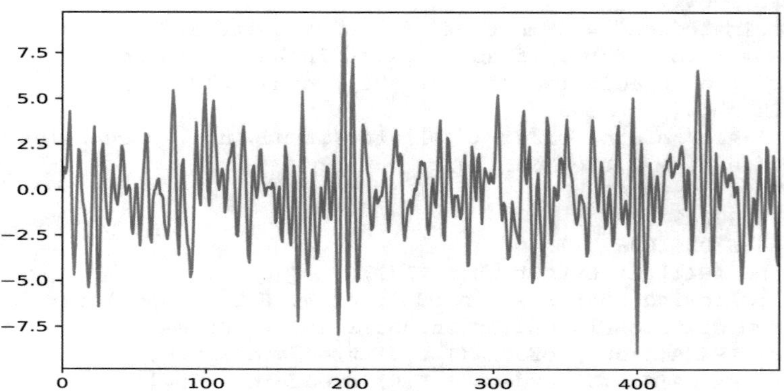


**Fig. 4.6** Time series plot of the simulated time series sample in Example 4.2

pyTSA_ARMA2

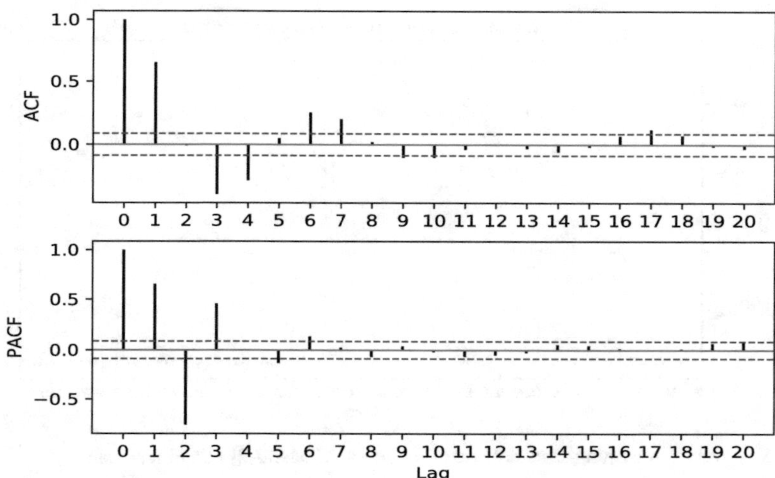


**Fig. 4.7** ACF and PACF plots of the simulated time series sample in Example 4.2

pyTSA_ARMA2

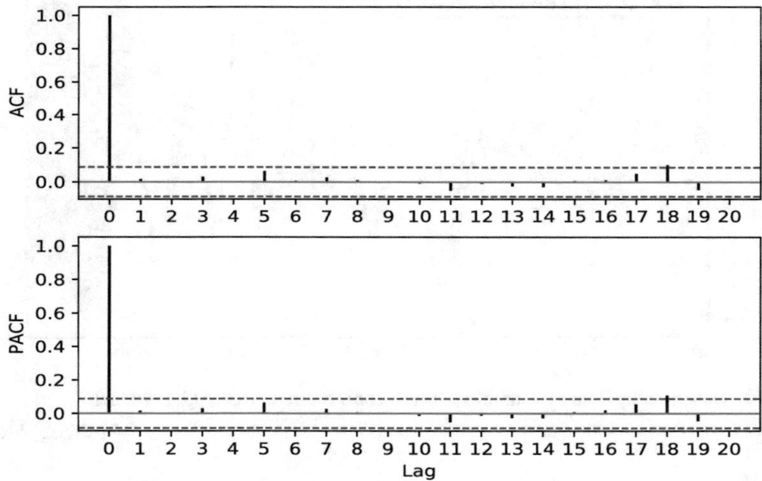


**Fig. 4.8** ACF and PACF plots of the fitted ARMA(2,2) model residuals

pyTSA_ARMA2

which is almost the same as the given model and all its coefficients are significantly different from zero. What is more, the modeling residual series passes the normality test, and from Figs. 4.8 and 4.9, we can conclude that the estimated model fits well to the simulated sample. Besides, both its out-of-sample forecasts and in-sample predicts appear satisfactory in light of Fig. 4.10. The following is the Python code for this example.

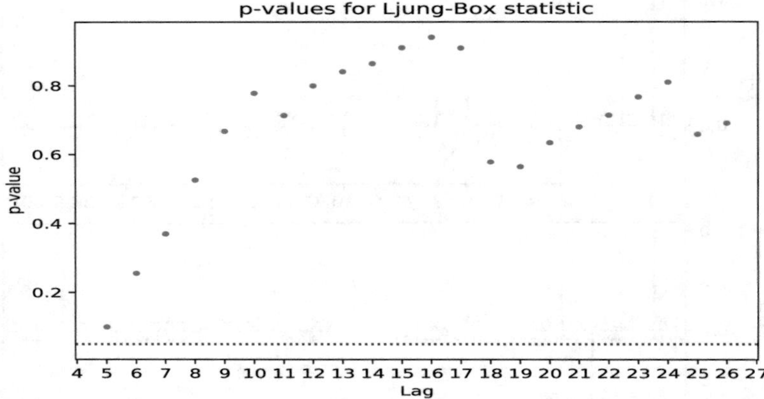


**Fig. 4.9** *p*-values for Ljung-Box statistic of the fitted ARMA(2,2) model residuals

pyTSA_ARMA2

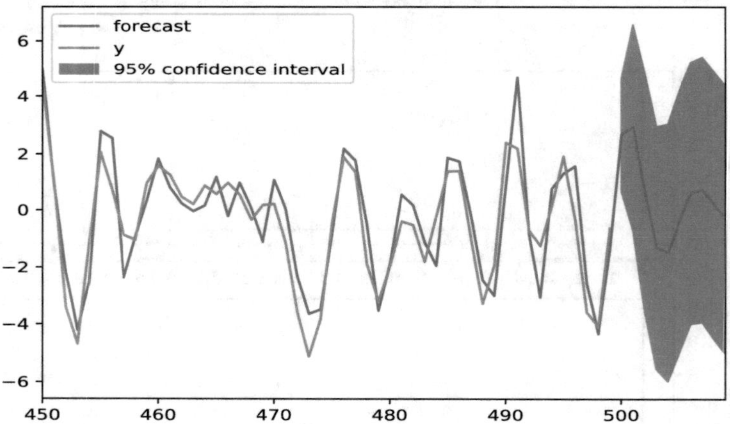


**Fig. 4.10** Out-of-sample and in-sample predicts for the fitted ARMA(2,2) model

pyTSA_ARMA2

```
>>> import numpy as np
>>> import pandas as pd
>>> import matplotlib.pyplot as plt
>>> import statsmodels.api as sm
>>> from statsmodels.tsa.arima_process import arma_generate_sample
>>> from PythonTsa.plot_acf_pacf import acf_pacf_fig
>>> from PythonTsa.Selecting_arma import choose_arma
# for statsmodels 0.13.0 and later, see the last code below
>>> from statsmodels.tsa.arima_model import ARMA
# for statsmodels 0.13.0 and later, see the last code below
>>> from PythonTsa.LjungBoxtest import plot_LB_pvalue
```

```
>>> from scipy import stats
>>> ar=np.array([1, -0.8, 0.6])
>>> ma=np.array([1, 0.7, 0.4])
>>> arma_process = sm.tsa.ArmaProcess(ar, ma)
>>> arma_process.isstationary  # check stationarity
True
>>> arma_process.isinvertible  # check invertibility
True
>>> np.random.seed(12357)
>>> y=arma_generate_sample(ar=ar, ma=ma, nsample=500)
>>> y=pd.Series(y, name='y')
# It is always a good idea to make the data a Series!
>>> y.plot(); plt.show()
>>> acf_pacf_fig(y, both=True, lag=20)
>>> plt.show()
>>> choose_arma(y, max_p=6, max_q=7, ctrl=1.02)
AIC:
         0        1        2        3 ...       6        7
0      NaN  1859.36      NaN      NaN ...     NaN      NaN
1  2030.95      NaN  1582.53      NaN ... 1479.85      NaN
2  1594.04  1511.55  1456.82  1457.63 ... 1462.64  1462.78
3  1475.13  1477.12  1457.68  1459.22 ...     NaN  1464.76
4  1477.11  1475.17  1459.67  1461.22 ... 1465.89      NaN
5  1469.46  1467.28  1460.79  1462.30 ...     NaN      NaN
6  1462.48  1464.14  1462.64  1464.29 ...     NaN      NaN
AIC minimum is 1456.82
(p, q)= (array([2], dtype=int64), array([2], dtype=int64))
BIC:
         0        1        2        3 ...       6        7
0      NaN  1872.01      NaN      NaN ...     NaN      NaN
1  2043.59      NaN  1603.60      NaN ... 1517.78      NaN
2  1610.90  1532.63  1482.11  1487.13 ... 1504.79  1509.14
3  1496.20  1502.41  1487.18  1492.94 ...     NaN  1515.33
4  1502.40  1504.67  1493.39  1499.15 ... 1516.47      NaN
5  1498.96  1501.00  1498.73  1504.44 ...     NaN      NaN
6  1496.19  1502.08  1504.78  1510.65 ...     NaN      NaN
BIC minimum is 1482.11
(p, q)= (array([2], dtype=int64), array([2], dtype=int64))
HQIC:
         0        1        2        3 ...       6        7
0      NaN  1864.32      NaN      NaN ...     NaN      NaN
1  2035.91      NaN  1590.80      NaN ... 1494.73      NaN
2  1600.66  1519.82  1466.75  1469.21 ... 1479.18  1480.97
3  1483.40  1487.04  1469.25  1472.45 ...     NaN  1484.60
4  1487.04  1486.75  1472.91  1476.10 ... 1485.74      NaN
5  1481.03  1480.51  1475.68  1478.83 ...     NaN      NaN
6  1475.71  1479.03  1479.18  1482.48 ...     NaN      NaN
HQIC minimum is 1466.75
(p, q)= (array([2], dtype=int64), array([2], dtype=int64))
>>> inf=sm.tsa.arma_order_select_ic(y, max_ar=6, max_ma=7,
                       ic=['aic', 'bic', 'hqic'], trend='nc')
# for statsmodels 0.13.0 and later, trend='n'
```

```
>>> inf.aic_min_order
(6, 7)
>>> inf.bic_min_order
(2, 2)
>>> inf.hqic_min_order
(2, 2)
>>> arma22=ARMA(y, order=(2,2)).fit(trend="nc")
>>> print(arma22.summary())
                              ARMA Model Results
==============================================================================
Dep. Variable:                      y   No. Observations:                  500
Model:                     ARMA(2, 2)   Log Likelihood                -722.842
Method:                       css-mle   S.D. of innovations              1.024
Date:                Mon, 29 Apr 2019   AIC                           1455.684
Time:                        21:59:45   BIC                           1476.757
Sample:                             0   HQIC                          1463.953

==============================================================================
                 coef    std err          z      P>|z|      [0.025      0.975]
------------------------------------------------------------------------------
ar.L1.y      0.8108      0.047     17.203      0.000       0.718       0.903
ar.L2.y     -0.6313      0.042    -15.147      0.000      -0.713      -0.550
ma.L1.y      0.7086      0.052     13.732      0.000       0.607       0.810
ma.L2.y      0.4677      0.050      9.444      0.000       0.371       0.565
                                    Roots
==============================================================================
                 Real          Imaginary           Modulus         Frequency
------------------------------------------------------------------------------
AR.1          0.6422          -1.0824j            1.2586           -0.1648
AR.2          0.6422          +1.0824j            1.2586            0.1648
MA.1         -0.7576          -1.2508j            1.4623           -0.3367
MA.2         -0.7576          +1.2508j            1.4623            0.3367
------------------------------------------------------------------------------
>>> resid22 = arma22.resid
>>> acf_pacf_fig(resid22, both=True, lag=20)
>>> plt.show()
>>> plot_LB_pvalue(resid22, noestimatedcoef=4, nolags=26)
>>> plt.show()
>>> stats.normaltest(resid22)
NormaltestResult(statistic=0.48049991217901883,
                 pvalue=0.786431263213941)
>>> arma22.plot_predict(start=450, end=509)
>>> plt.show()
# the following is for statsmodels of v. 0.13.0 and later
>>> from PythonTsa.Selecting_arma2 import choose_arma2
>>> choose_arma2(y, max_p=6, max_q=7, ctrl=1.02)
>>> from statsmodels.tsa.arima.model import ARIMA
>>> arma22=ARIMA(y, order=(2,0,2),trend='n').fit()
>>> pred=arma22.get_prediction(start=450, end=509)
>>> predicts=pred.predicted_mean
>>> predconf=pred.conf_int()
>>> predframe=pd.concat([y[450:], predicts,
```

```
                   predconf.iloc[-10:]], axis=1)
>>> predframe.plot(); plt.show()
```

*Example 4.3 (Global Annual Mean Surface Air Temperature Changes Series from 1880 to 1985)* The time series dataset "Global mean surface air temp changes 1880–1985" in the folder Ptsadata (now denoted as GMSATC) is from Hansen and Lebedeff (1987) that investigates the global warming issue. From its time series plot shown in Fig. 4.11, a global warming trend is clearly displayed in the period from 1880 to 1985, and the time series is not stationary. So we take a first-order difference on it to make it stationary. Observing the time series plot as well as ACF and PACF plots of the differenced series (see Figs. 4.12 and 4.13) and noticing that the $p$-value

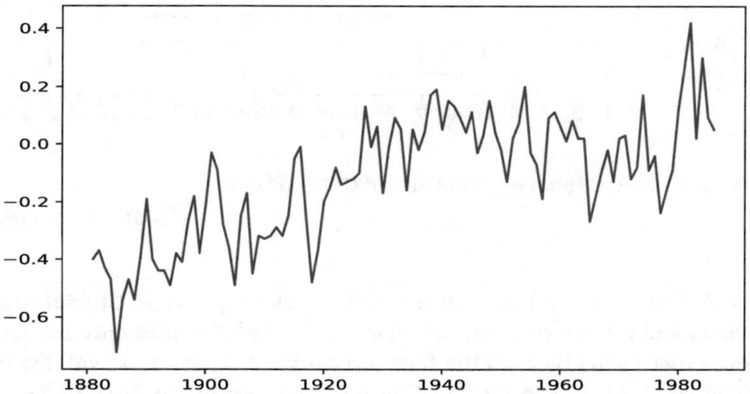


**Fig. 4.11** Time series plot of the GMSATC series

pyTSA_AirTempChange

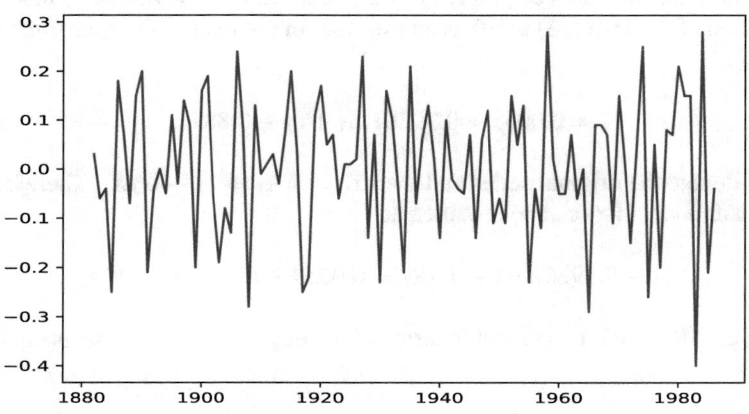


**Fig. 4.12** Time series plot of the differenced GMSATC series

pyTSA_AirTempChange

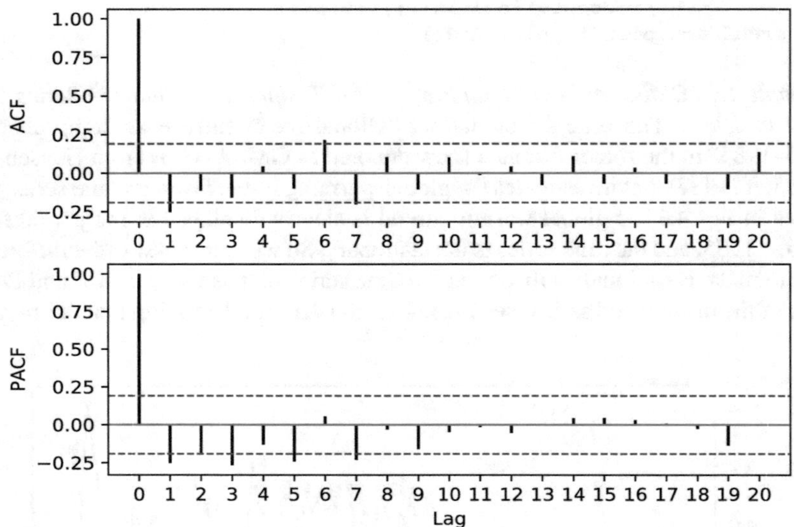


**Fig. 4.13** ACF and PACF plots of the differenced GMSATC series

pyTSA_AirTempChange

for the KPSS test statistic is greater than 0.1, we can say that the differenced series is stationary and can build an ARMA($p, q$) model with the differenced series. Now we select order ($p, q$) by both the function `choose_arma( )` with the control coefficient `ctrl=1.03` and the function `sm.tsa.arma_order_select_ic`. Executing the Python commands, we see that the two functions obtain the same result: AIC and HQIC derive ($p, q$)=(1, 3) and BIC derives ($p, q$)=(1, 1). Here we parsimoniously take ($p, q$)=(1, 1) as the order of our ARMA($p, q$) model, and estimation of ARMA(1, 3) is left as an exercise for the reader. Thus the fitted model is

$$Y_t = 0.0053 + 0.3926Y_{t-1} + \varepsilon_t - 0.8876\varepsilon_{t-1} \tag{4.8}$$

Let $X_t$ denote the original series and then $Y_t = \nabla X_t = (1 - B)X_t$. Therefore, the estimated model (4.8) can be rewritten as

$$(1 - 0.3926B)(1 - B)X_t = 0.0053 + (1 - 0.8876B)\varepsilon_t$$

This is an ARIMA(1,1,1) model in terms of the original series $X_t$ that is equivalent to (4.8).

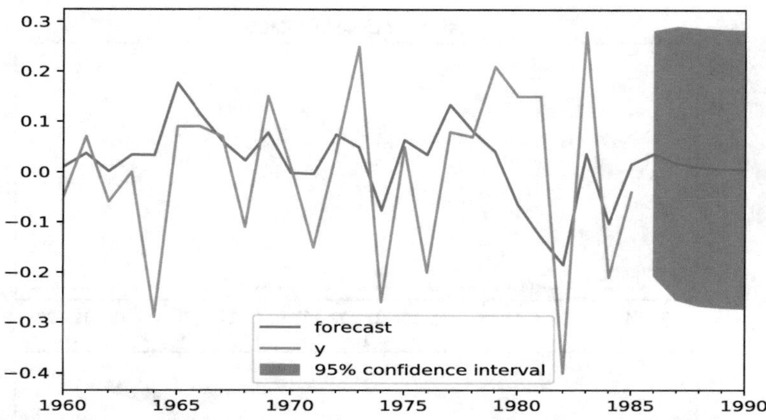


**Fig. 4.14** Out-of-sample and in-sample forecasts of the fitted ARMA(1,1)

pyTSA_AirTempChange

Now we diagnose this fitted model. First, we see that the modeling residuals pass the normality test with the $p-$ value $= 0.8997$. Then, we use the function `plot_ResidDiag` in the Python package PythonTsa to further analyze the residuals. Running the function `plot_ResidDiag`, we arrive at Fig. 4.15 where in the top panel is a plot of the $p$-values for Ljung-Box statistic of the residuals; the second panel normal Q-Q plot; the third panel ACF plot of the residuals; and the bottom panel ACF plot of the squared residuals. Inspecting the $p$-value plot, Q-Q plot, and ACF plot, respectively, shown in the top, second, and third panels of Fig. 4.15 concludes that the residual series behaves like a normal white noise, and so the estimated model (4.1) fits very well to the differenced series data. Besides, ACF plot of the squared residuals in the bottom panel of Fig. 4.15 suggests that there is no ARCH effect in the residuals. Unfortunately, from Fig. 4.14, we see that both out-of-sample and in-sample forecasts of the fitted ARMA(1,1) model could not be satisfactory. We have left an alternative ARMA(1, 3) model in the previous paragraph. How about it? That is an interesting question to be answered. The Python code for this example is below (Fig. 4.15).

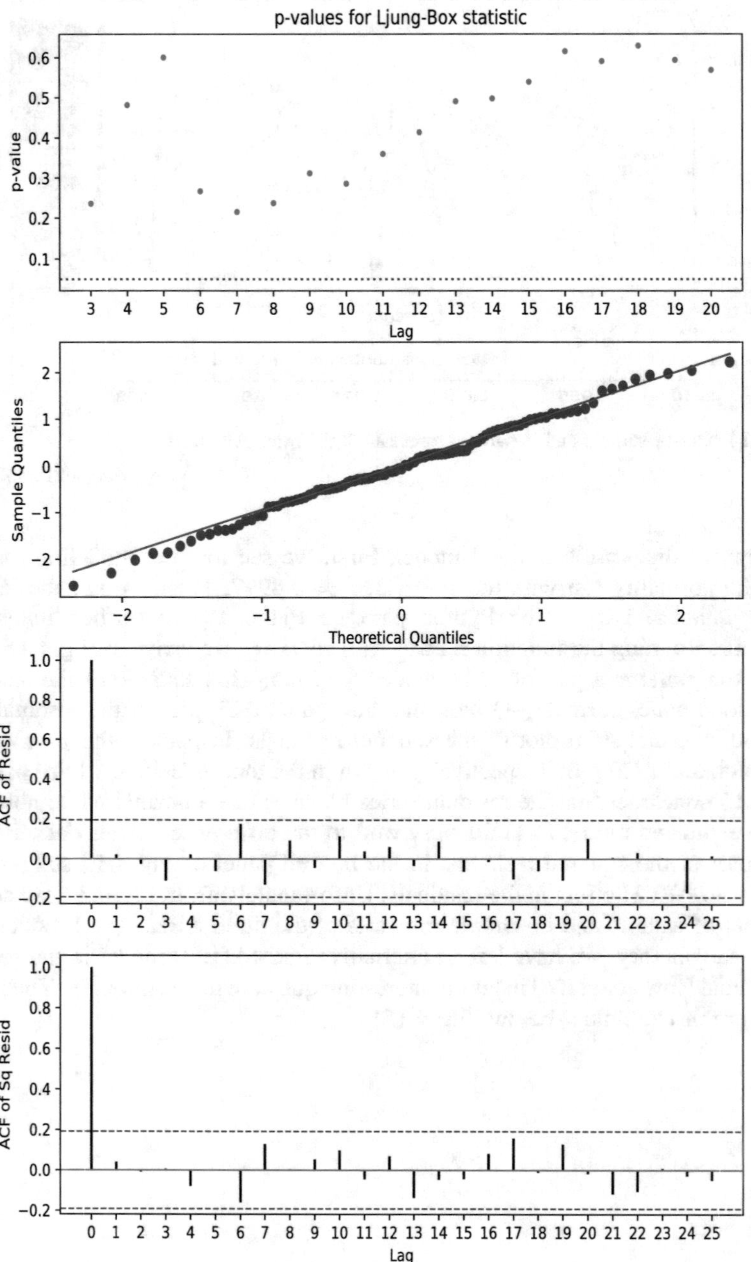


**Fig. 4.15** Diagnostic plots of ARMA(1,1) for the differenced GMSATC series

pyTSA_AirTempChange

```
>>> import pandas as pd
>>> import numpy as np
>>> import matplotlib.pyplot as plt
>>> import statsmodels.api as sm
>>> from PythonTsa.plot_acf_pacf import acf_pacf_fig
>>> from PythonTsa.Selecting_arma import choose_arma
>>> from statsmodels.tsa.arima_model import ARMA
# for statsmodels 0.13.0 and later, see the last code below
>>> from scipy import stats
>>> from PythonTsa.ModResidDiag import plot_ResidDiag
>>> from PythonTsa.datadir import getdtapath
>>> dtapath=getdtapath()
>>> tep=pd.read_csv(dtapath + 'Global mean surface air
                    temp changes 1880-1985.csv', header=None)
>>> dates=pd.date_range('1880-12',periods=len(tep),freq='A-DEC')
>>> tep.index=dates
>>> tepts=pd.Series(tep[0], name='tep')
>>> plt.plot(tepts, color='b'); plt.show()
>>> dtepts=tepts.diff(1)
>>> dtepts=dtepts.dropna()
>>> dtepts.name='dtep'
>>> plt.plot(dtepts, color='b'); plt.show()
>>> sm.tsa.kpss(dtepts, regression='c', nlags='auto')
InterpolationWarning:
         p-value is greater than the indicated p-value
(0.08259101825264815, 0.1, 12,
{'10%': 0.347, '5%': 0.463, '2.5%': 0.574, '1%': 0.739})
>>> acf_pacf_fig(dtepts, both=True, lag=20)
>>> plt.show()
>>> choose_arma(dtepts, max_p=7, max_q=7, ctrl=1.03)
AIC:
        0       1       2       3      4    5    6    7
0     NaN -122.23 -129.96 -129.74    NaN  NaN  NaN  NaN
1 -113.35 -130.71 -129.04 -135.56 -133.7  NaN  NaN  NaN
2 -115.28 -129.25     NaN     NaN    NaN  NaN  NaN  NaN
3 -120.97 -128.65     NaN     NaN    NaN  NaN  NaN  NaN
4 -121.02 -126.65     NaN     NaN    NaN  NaN  NaN  NaN
5 -126.98 -125.61 -130.38 -130.29    NaN  NaN  NaN  NaN
6 -125.47     NaN -129.46 -128.63    NaN  NaN  NaN  NaN
7 -131.25 -129.47 -129.32 -128.75    NaN  NaN  NaN  NaN
AIC minimum is -135.56
(p, q)= (array([1], dtype=int64), array([3], dtype=int64))
BIC:
        0       1       2       3       4    5    6    7
0     NaN -114.27 -119.34 -116.47     NaN  NaN  NaN  NaN
1 -105.39 -120.10 -115.77 -119.63 -115.12  NaN  NaN  NaN
2 -104.67 -115.98     NaN     NaN     NaN  NaN  NaN  NaN
3 -107.70 -112.73     NaN     NaN     NaN  NaN  NaN  NaN
4 -105.10 -108.08     NaN     NaN     NaN  NaN  NaN  NaN
5 -108.40 -104.38 -106.49 -103.75     NaN  NaN  NaN  NaN
6 -104.24     NaN -102.92  -99.43     NaN  NaN  NaN  NaN
7 -107.37 -102.93 -100.13  -96.91     NaN  NaN  NaN  NaN
BIC minimum is -120.1
(p, q)= (array([1], dtype=int64), array([1], dtype=int64))
```

```
HQIC:
         0       1       2       3       4    5    6    7
0      NaN -119.01 -125.65 -124.36     NaN  NaN  NaN  NaN
1 -110.12 -126.41 -123.66 -129.11 -126.17  NaN  NaN  NaN
2 -110.98 -123.87     NaN     NaN     NaN  NaN  NaN  NaN
3 -115.59 -122.20     NaN     NaN     NaN  NaN  NaN  NaN
4 -114.57 -119.13     NaN     NaN     NaN  NaN  NaN  NaN
5 -119.45 -117.00 -120.70 -119.53     NaN  NaN  NaN  NaN
6 -116.87     NaN -118.70 -116.80     NaN  NaN  NaN  NaN
7 -121.57 -118.71 -117.49 -115.85     NaN  NaN  NaN  NaN
HQIC minimum is -129.11
(p, q)= (array([1], dtype=int64), array([3], dtype=int64))
>>> inf= sm.tsa.arma_order_select_ic(dtepts, max_ar=7,
          max_ma=7, ic=['aic', 'bic', 'hqic'], trend='c')
>>> inf.aic_min_order
(1, 3)
>>> inf.bic_min_order
(1, 1)
>>> inf.hqic_min_order
(1, 3)
>>> arma11= ARMA(dtepts, order=(1,1)).fit(trend='c', disp=False)
# here ARMA()=sm.tsa.ARMA()
>>> print(arma11.summary())
                              ARMA Model Results
==============================================================================
Dep. Variable:                   dtep   No. Observations:                  105
Model:                     ARMA(1, 1)   Log Likelihood                  69.356
Method:                       css-mle   S.D. of innovations              0.124
Date:                Wed, 17 Jul 2019   AIC                           -130.713
Time:                        13:33:56   BIC                           -120.097
Sample:                    12-31-1881   HQIC                          -126.411
                         - 12-31-1985
==============================================================================
                 coef    std err          z      P>|z|      [0.025      0.975]
------------------------------------------------------------------------------
const          0.0053      0.002      2.196      0.030       0.001       0.010
ar.L1.y        0.3926      0.118      3.329      0.001       0.161       0.624
ma.L1.y       -0.8876      0.060    -14.738      0.000      -1.006      -0.770
                                    Roots
==============================================================================
                 Real          Imaginary           Modulus         Frequency
------------------------------------------------------------------------------
AR.1           2.5473          +0.0000j           2.5473            0.0000
MA.1           1.1266          +0.0000j           1.1266            0.0000
------------------------------------------------------------------------------
>>> resid11 = arma11.resid
>>> stats.normaltest(resid11)
NormaltestResult(statistic=0.21129120460097417,
                 pvalue=0.8997434585916072)
>>> plot_ResidDiag(resid11,noestimatedcoef=2,nolags=20,lag=25)
>>> plt.show()
# noestimatedcoef = number of estimated coefficients
# nolags = max number of added terms in LB statistic.
# lag = number of lags for ACF
```

```
>>> arma11.plot_predict(start='1960-12', end='1990-12')
>>> plt.show()
# the following is for statsmodels of v. 0.13.0 and later
>>> from PythonTsa.Selecting_arma2 import choose_arma2
>>> choose_arma2(dtepts, max_p=7, max_q=7, ctrl=1.03)
>>> from statsmodels.tsa.arima.model import ARIMA
>>> arma11=ARIMA(dtepts, order=(1,0,1),trend='c').fit()
>>> pred=arma11.get_prediction(start='1960-12', end='1990-12')
>>> predicts=pred.predicted_mean
>>> predconf=pred.conf_int()
>>> predframe=pd.concat([dtepts['1960-12-31':], predicts,
            predconf['1986-01-31':]], axis=1)
>>> predframe.plot(); plt.show()
```

*Example 4.4 (US Government Treasury Bills from Jan. 1950 to June 1988)* The time series dataset "USbill" in the folder Ptsadata consists of the monthly interest rate on 3-month US government treasury bills from January 1950 to June 1988. We leave the last six items for forecast comparison before using the time series data. Its time series plot is shown in Fig. 4.16, and it is easy to see that the time series is nonstationary. In order to make it stationary, take a logarithm on it and then first-difference the logarithmed series. Observing Figs. 4.17 and 4.18 and noting that the $p$-value for the KPSS test is greater than 0.1, it follows that the differenced and logarithmed series is stationary.

Now let us select feasible order for our model. Checking Fig. 4.18 and according to Table 3.1, we might choose (6,0) or (0,6) as the order of our model. On the other side, using the function `choose_arma`, we get order (6, 0) and (0, 1), while using the function `sm.tsa.arma_order_select_ic`, we obtain order (6,7), (0, 1) and (6,0). We notice that the three approaches all derive order (6,0). Hence we take (6,0) as the order of our model. Models with order (0, 1), (0,6), or (6,7) are left to the reader as an exercise. For predict convenience, we adopt the function `ARIMA( )` in Python package `statsmodels` to estimate our model. The distinction between

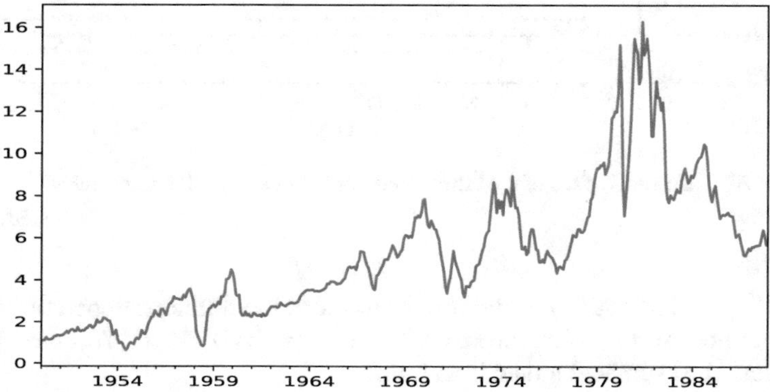


**Fig. 4.16** Time series plot of the US treasury bill monthly interest rates

pyTSA_USbill

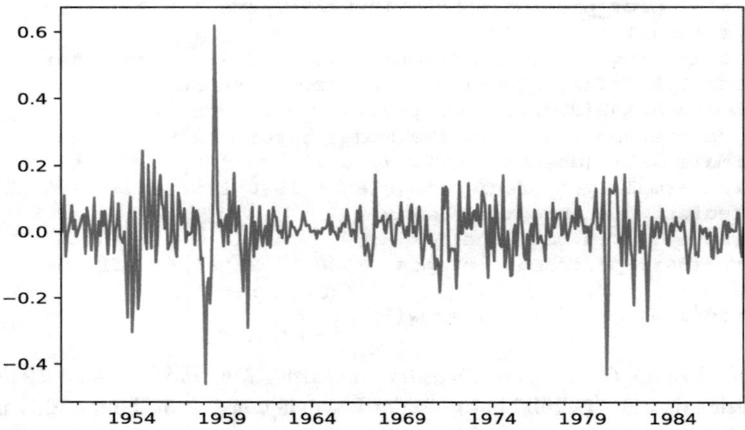


**Fig. 4.17** Time series plot of the differenced log US treasury bill interest rates

pyTSA_USbill

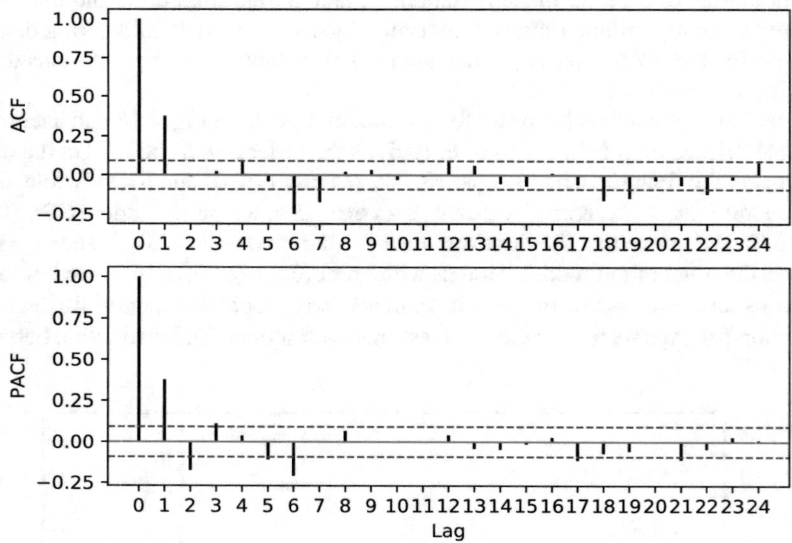


**Fig. 4.18** ACF and PACF plots of the differenced log US treasury bill interest rates

pyTSA_USbill

`ARIMA( )` and `ARMA( )` is the former includes the difference operation and the latter does not. At this point, the series `ly=log(y)` (viz., $\log(Y_t)$) is used to build the model. The estimated model is as follows:

$$X_t = 0.439X_{t-1} - 0.186X_{t-2} + 0.088X_{t-3} + 0.045X_{t-4} - 0.010X_{t-5} - 0.207X_{t-6} + \varepsilon_t$$

**Table 4.1** Comparison of forecasted and actual interest rates of US bills in 1988

| Time | Jan. | Feb. | Mar. | Apr. | May | June |
|---|---|---|---|---|---|---|
| Actual rates | 5.81 | 5.66 | 5.70 | 5.91 | 6.26 | 6.46 |
| Forecasted rates | 5.87 | 5.78 | 5.64 | 5.66 | 5.78 | 5.80 |

where the estimated coefficients at lag 4 and 5 are not significant, which does not matter, but the estimated coefficient at lag 6 is significant, which does matter since the order $p = 6$. From the $p$-values plot and ACF plot, respectively, shown in the top and third panels of Fig. 4.19, it turns out that the estimated model AR(6) fits very well to the series $X_t = \log(Y_t) - \log(Y_{t-1})$ where $Y_t$ is the original series. The Q-Q plot in the second panel of Fig. 4.19 as well as the normal test result indicates that the residuals do not follow a normal distribution. An interesting phenomenon is that ACF plot of the residuals and ACF plot of the squared residuals, respectively, shown in the third and bottom panels of Fig. 4.19 demonstrate that there exists no autocorrelation in the residual series while there is autocorrelation in the squared residuals. This phenomenon suggests that the residuals have ARCH effect.

At the beginning of this example, we have left a sub-sample of size 6 for comparing with forecasts. Now let us observe Table 4.1 by comparing. We gladly see that the forecasted values in the first 3 months are very close to the actual ones. The pity is that the forecasted values in the latter 3 months are not so close to the actual ones. However, this is common in that no matter what method you use, the farther away from the present, the harder to predict. Besides, from the out-of-sample and in-sample predicts shown in Fig. 4.20, we can say again that the forecasting is excellent.

At last, you can use the following Python code to reproduce all results and figures in this example.

```
>>> import numpy as np
>>> import pandas as pd
>>> import matplotlib.pyplot as plt
>>> from PythonTsa.plot_acf_pacf import acf_pacf_fig
>>> import statsmodels.api as sm
>>> from PythonTsa.Selecting_arma import choose_arma
>>> from statsmodels.tsa.arima_model import ARIMA
# for statsmodels 0.13.0 and later, see the last code below
>>> from PythonTsa.ModResidDiag import plot_ResidDiag
>>> from scipy import stats
>>> from PythonTsa.datadir import getdtapath
>>> dtapath=getdtapath()
>>> rat=pd.read_csv(dtapath + 'USbill.csv',header=None)
>>> rat.tail(6)
            0     1
456   1988/1/31  5.81
457   1988/2/29  5.66
458   1988/3/31  5.70
459   1988/4/30  5.91
460   1988/5/31  6.26
461   1988/6/30  6.46
>>> y=rat[:456]
```

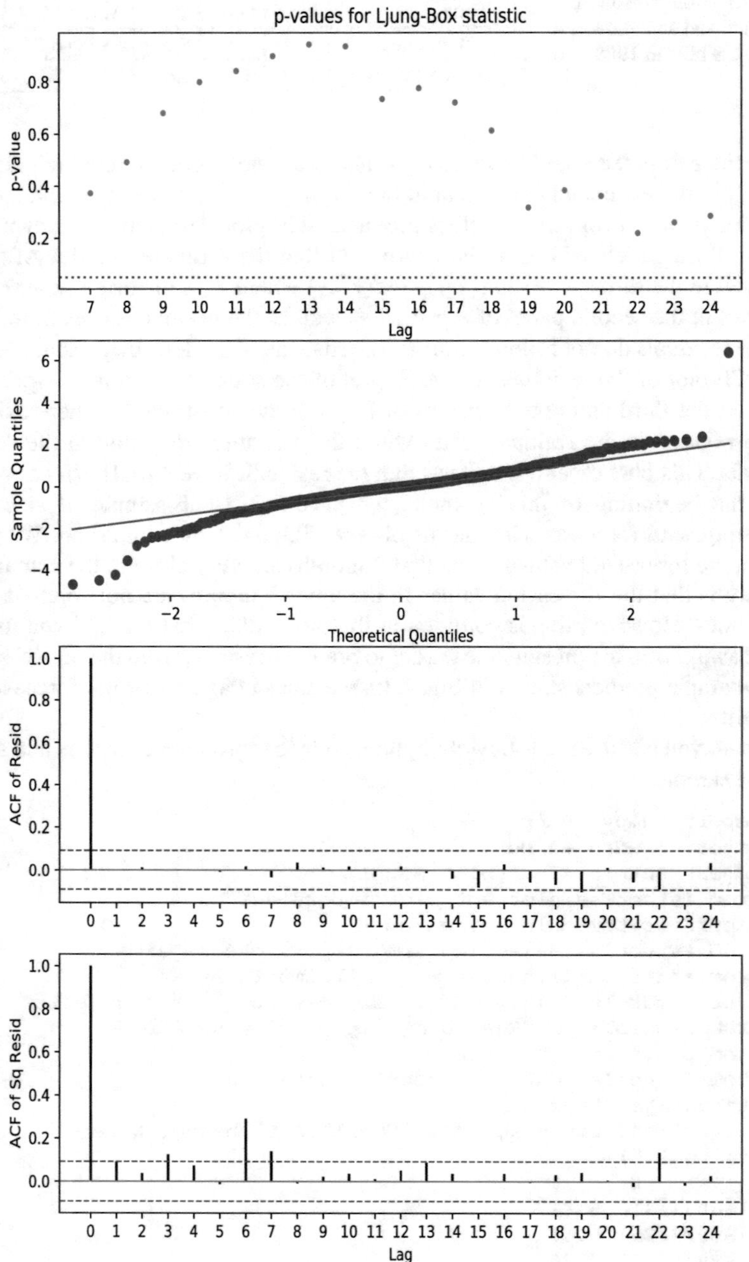


**Fig. 4.19** Diagnostic plots of ARIMA(6,1,0) for the log of the US bill series

pyTSA_USbill

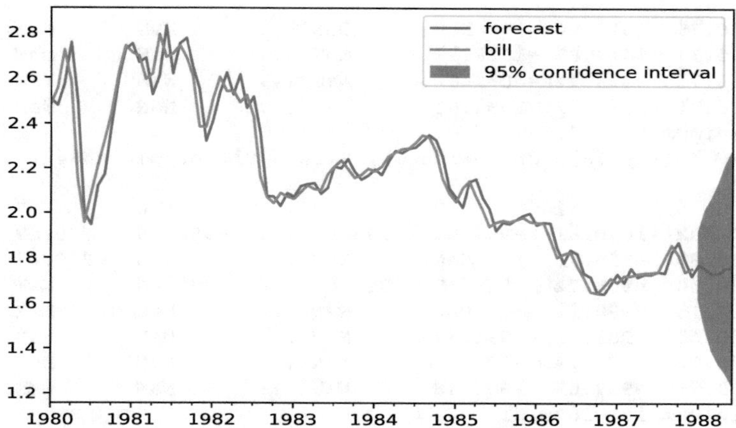


**Fig. 4.20** Out-of-sample and in-sample predicts for the log of US bill interest rates

pyTSA_USbill

```
# leave the last 6 items for forecast comparison
>>> y.tail(6)
              0     1
450   1987/7/31  5.69
451   1987/8/31  6.04
452   1987/9/30  6.40
453  1987/10/31  6.13
454  1987/11/30  5.69
455  1987/12/31  5.77
>>> y.rename(columns={0:'time', 1:'bill'},inplace=True)
# ARIMA requires 'strings' for column names.
>>> dates=pd.date_range('1950-1',periods=len(y),freq='M')
>>> y.index=dates
>>> y=y['bill']
>>> y.plot(); plt.show()
>>> ly=np.log(y)
>>> dly=ly.diff(1)
>>> dly=dly.dropna()
>>> dly.plot(); plt.show()
>>> acf_pacf_fig(dly, both=True, lag=24)
>>> plt.show()
>>> sm.tsa.stattools.kpss(dly, regression="c", nlags='auto')
InterpolationWarning:
           p-value is greater than the indicated p-value
(0.06937042978189786, 0.1, 5,
{'10%': 0.347, '5%': 0.463, '2.5%': 0.574, '1%': 0.739})
>>> choose_arma(dly, max_p=6, max_q=7, ctrl=1.05)
AIC:
         0        1        2        3 ...        6        7
0      NaN -1017.97 -1015.99 -1014.41 ... -1018.82 -1032.96
1 -1000.99 -1015.99      NaN      NaN ...      NaN -1031.27
2 -1012.98 -1014.34 -1012.87 -1019.94 ... -1028.06      NaN
```

```
3 -1016.75 -1014.94       NaN       NaN ...      NaN      NaN
4 -1015.37 -1013.88 -1028.33        NaN ...      NaN      NaN
5 -1018.71 -1025.74 -1031.12        NaN ...      NaN      NaN
6 -1036.71 -1034.72 -1034.68        NaN ...      NaN      NaN
AIC minimum is -1036.71
(p, q)= (array([6], dtype=int64), array([0], dtype=int64))
BIC:
        0        1        2        3 ...        6        7
0     NaN -1005.61  -999.51  -993.81 ...  -985.86  -995.87
1  -988.63  -999.51      NaN      NaN ...      NaN  -990.07
2  -996.50  -993.74  -988.15  -991.09 ...  -986.86      NaN
3  -996.15  -990.22      NaN      NaN ...      NaN      NaN
4  -990.65  -985.04 - 995.37      NaN ...      NaN      NaN
5  -989.86  -992.78  -994.04      NaN ...      NaN      NaN
6 -1003.75  -997.63  -993.48      NaN ...      NaN      NaN
BIC minimum is -1005.61
(p, q)= (array([0], dtype=int64), array([1], dtype=int64))
HQIC:
        0        1        2        3 ...        6        7
0     NaN -1013.10 -1009.50 -1006.30 ... -1005.84 -1018.35
1  -996.12 -1009.50      NaN      NaN ...      NaN -1015.04
2 -1006.48 -1006.23 -1003.13 -1008.57 ... -1011.83      NaN
3 -1008.63 -1005.20      NaN      NaN ...      NaN      NaN
4 -1005.63 -1002.52 -1015.34      NaN ...      NaN      NaN
5 -1007.34 -1012.75 -1016.51      NaN ...      NaN      NaN
6 -1023.73 -1020.11 -1018.45      NaN ...      NaN      NaN
HQIC minimum is -1023.73
(p, q)= (array([6], dtype=int64), array([0], dtype=int64))
>>> res = sm.tsa.arma_order_select_ic(dly, max_ar=6, max_ma=7,
         ic=['aic', 'bic', 'hqic'], trend='nc')
# for statsmodels 0.13.0 and later, trend='n'
>>> res.aic_min_order
(6, 7)
>>> res.bic_min_order
(0, 1)
>>> res.hqic_min_order
(6, 0)
>>> arima610=ARIMA(ly, order=(6,1,0)).fit(trend='nc')
# the const is significantly zero.
>>> print(arima610.summary())
                    ARIMA Model Results
==============================================================
Dep. Variable:          D.bill  No. Observations:          455
Model:          ARIMA(6, 1, 0)  Log Likelihood         525.973
Method:                css-mle  S.D. of innovations      0.076
Date:         Sun, 12 May 2019  AIC                  -1037.946
Time:                 20:24:03  BIC                  -1009.104
Sample:             02-28-1950  HQIC                 -1026.583
                  - 12-31-1987
==============================================================
                coef  std err       z  P>|z|  [0.025   0.975]
--------------------------------------------------------------
ar.L1.D.bill  0.4385  0.046     9.561  0.000   0.349    0.528
ar.L2.D.bill -0.1860  0.050    -3.697  0.000  -0.285   -0.087
```

```
ar.L3.D.bill  0.0878  0.051   1.712  0.088  -0.013    0.188
ar.L4.D.bill  0.0445  0.051   0.866  0.387  -0.056    0.145
ar.L5.D.bill -0.0101  0.051  -0.200  0.841  -0.109    0.089
ar.L6.D.bill -0.2065  0.046  -4.497  0.000  -0.296   -0.116
                              Roots
=============================================================
              Real          Imaginary      Modulus   Frequency
-------------------------------------------------------------
AR.1         1.0999          -0.5491j        1.2293    -0.0737
AR.2         1.0999          +0.5491j        1.2293     0.0737
AR.3         0.0861          -1.2211j        1.2242    -0.2388
AR.4         0.0861          +1.2211j        1.2242     0.2388
AR.5        -1.2105          -0.8205j        1.4624    -0.4052
AR.6        -1.2105          +0.8205j        1.4624     0.4052
-------------------------------------------------------------
>>> resid610=arima610.resid
>>> plot_ResidDiag(resid610,noestimatedcoef=6,nolags=24,lag=24)
>>> plt.show()
>>> stats.normaltest(resid610)
NormaltestResult(statistic=67.59229104473481,
                 pvalue=2.101456913555842e-15)
>>>fo=arima610.predict(start='1988-01',end='1988-06',typ='levels')
#typ='levels' means: predict the levels of the variable 'ly'.
#typ='linear' to predict 'dly'
>>> np.exp(fo)
1988-01-31    5.871458
1988-02-29    5.776394
1988-03-31    5.639406
1988-04-30    5.663803
1988-05-31    5.783622
1988-06-30    5.798004
Freq: M, dtype: float64
>>> arima610.plot_predict(start='1980-01', end='1988-06')
>>> plt.show()
# the following is for statsmodels of v. 0.13.0 and later
>>> from PythonTsa.Selecting_arma2 import choose_arma2
>>> choose_arma2(dly, max_p=6, max_q=7, ctrl=1.05)
>>> from statsmodels.tsa.arima.model import ARIMA
>>> arima610=ARIMA(ly, order=(6,1,0),trend='n').fit()
>>> pred=arima610.get_prediction(start='1980-01',end='1988-06')
>>> predicts=pred.predicted_mean
>>> predconf=pred.conf_int()
>>> np.exp(predicts.tail(6))
>>> predframe=pd.concat([ly['1980-01-31':], predicts,
             predconf['1988-01-31':]], axis=1)
>>> predframe.plot(); plt.show()
```

## Problems

**4.1** Use the function `sm.tsa.arma_order_select_ic` to determine order $(p, q)$ to build the model ARMA$(p, q)$ for the time series data NAO in Example 4.1. Then estimate the model ARMA$(p, q)$ where $(p, q) \neq (1, 0)$ and explain it.

**4.2** For the simulated time series sample in Example 4.2, we observe that its PACF shown in Fig. 4.7 seems to cut off after lag 3. As a result, an ARMA(3,0) model could be used to fit the sample. Please build this model and compare it with the fitted model in Example 4.2.

**4.3** We get the time series `dtepts` in Example 4.3. Now use an ARMA(1,3) model to fit it, and then compare the fitted model with the estimated model ARMA(1,1) in Example 4.3.

**4.4** Use models ARIMA(6,1,7), ARIMA(0,1,1), and ARIMA(0,1,6), respectively, to fit the data `ly` in Example 4.4. Are they appropriate models? Why?

**4.5** Suppose that time series $X_t$ follows the MA(1) model $X_t = \varepsilon_t + \theta\varepsilon_{t-1}, \varepsilon_t \sim \mathrm{WN}(0, \sigma_\epsilon^2)$. Reduce the innovations' algorithm for this series.

**4.6** For the AR(1) model $X_t = \varphi X_{t-1} + \varepsilon_t$, find the moment estimates for the coefficient $\varphi$ and $\sigma_\epsilon^2$.

**4.7** For the MA(1) model $X_t = \varepsilon_t + \theta\varepsilon_{t-1}$, find the moment estimate for the coefficient $\theta$, and then give your comment.

**4.8** For a causal and invertible ARMA(1, 1) model $X_t - \varphi X_{t-1} = \varepsilon_t + \theta\varepsilon_{t-1}, \varepsilon_t \sim \mathrm{WN}(0, \sigma_\epsilon^2)$, reduce the innovations algorithm, and find the one-step predictor $\hat{X}_{n+1}$.

**4.9** Let $X_t$ be a causal and invertible time series with mean $\mu$ and satisfy the ARMA(1,1) model:

$$(1 - \varphi B)(X_t - \mu) = \varepsilon_t - \theta\varepsilon_{t-1}$$

(1) Derive a formula for forecast function $\hat{X}_n(h)$ in terms of $\varphi, \theta$, and $\varepsilon_t$.
(2) What would $\hat{X}_n(h)$ tend to as $h \to \infty$? Interpret this limit.
(3) Find the formula for $\mathrm{Var}(e_n(h))$. What is $\lim_{h\to\infty} \mathrm{Var}(e_n(h))$?

# Chapter 5
# Nonstationary Time Series Models

This chapter focuses on the Box-Jenkins approach to building models for nonstationary time series. It contains ARIMA modeling for nonseasonal time series presented in Chap. 4 and SARIMA modeling for seasonal time series to be considered in this chapter. Through case study, we demonstrate how to use Python to implement the Box-Jenkins method. In addition, we also discuss REGARMA models.

## 5.1 The Box-Jenkins Method

In Chap. 4, we considered the problem of modeling nonseasonal time series. We adopted a procedure like this: if a time series is nonstationary, we make it stationary by differencing (sometimes plus transforming) and then use an appropriate ARMA model to fit the differenced data. This is a special case of the Box-Jenkins modeling approach (see Box et al. 2016). For seasonal time series, a similar way is also proposed to model them.

### *5.1.1 Seasonal Differencing*

Suppose that we have a nonstationary time series $\{X_t\}$ with trend and seasonal period $s$. If the differenced series

$$Y_t = \nabla^d \nabla_s^D X_t = (1 - B)^d (1 - B^s)^D X_t \tag{5.1}$$

is stationary, then we can be going to consider the problem of finding an appropriate model for $\{Y_t\}$. In Eq. (5.1), $B$ is the backshift operator; $\nabla^d = (1 - B)^d$ is the

 
C. Huang, A. Petukhina, *Applied Time Series Analysis and Forecasting with Python*, Statistics and Computing,
https://doi.org/10.1007/978-3-031-13584-2_5

(ordinary) differencing of order $d$; and $\nabla_s^D = (1 - B^s)^D$ is the seasonal differencing of order $D$. In most cases, $D \leq 2$. Note that differencing in Eq. (5.1) drops $d + Ds$ data points, that is, $\{Y_t\}$ is $d + Ds$ fewer data points than $\{X_t\}$. Now let us look at a real example.

*Example 5.1 (America Monthly Employment Figures for Females Aged 20 Years and Over)* The time series dataset "USFemalesAged20+Job1948-81" in the folder Ptsadata is from the Andrews and Herzberg Datasets. It is US monthly employment figures for females aged 20 years and over from 1948 to 1981. Its seasonal plots are shown in Fig. 5.1 and they display almost the same pattern. From its time series plot shown in Fig. 5.2 as well as the time series plot of its subseries (1963–

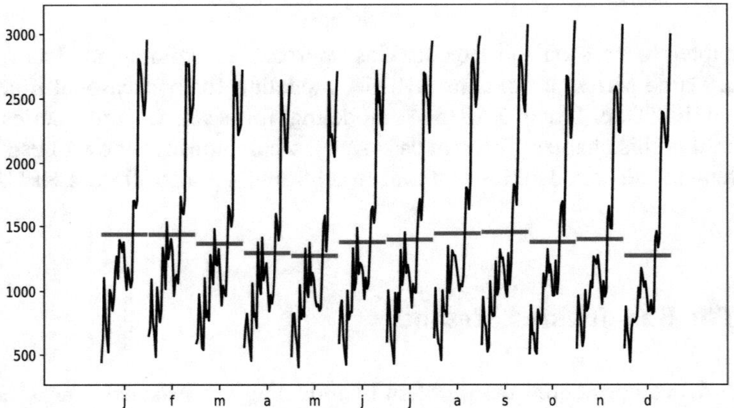


**Fig. 5.1** Seasonal plots of the female aged 20 and over job data in Example 5.1
pyTSA_FemaleLabor

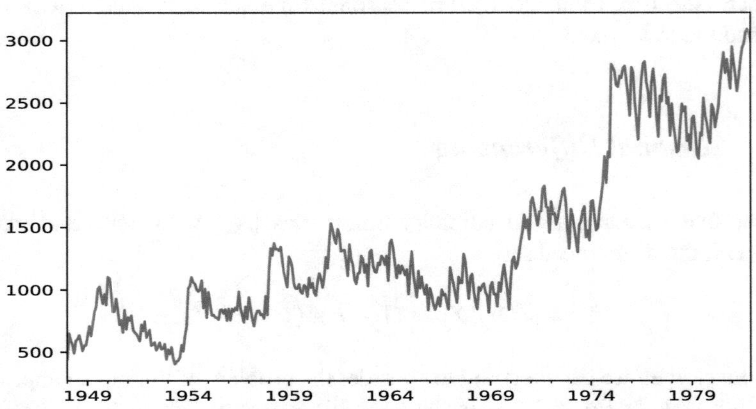


**Fig. 5.2** Time series plot of the female aged 20 and over job data in Example 5.1
pyTSA_FemaleLabor

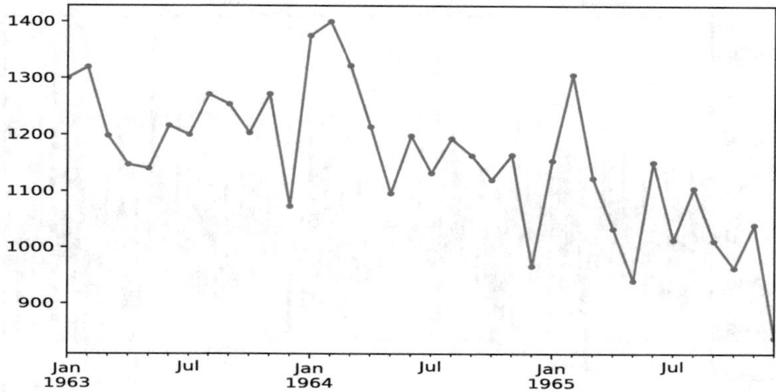


**Fig. 5.3** Time series plot of the subseries (1963–1965) in Example 5.1

pyTSA_FemaleLabor

1965) shown in Fig. 5.3 , we can observe that the time series has both trend and seasonality although the seasonality is affected by noise and seems not be so strong. Therefore, (1) we take a seasonal difference and then a first-order difference on the data; (2) we check if the differenced series is stationary. The Python function `sm.tsa.statespace.tools.diff` can be used to simultaneously implement the seasonal and ordinary differencing. The usage for it is demonstrated in the Python code below. In the present example, $D = 1$; $s = 12$; and $d = 1$. After double differencing, the differenced series `dDx` is from Feb. 1949 and loses $Ds + d = 13$ data points. It is easy to see that the differenced series is stationary from whether its time series plot shown in Fig. 5.4 or $p$-value $> 0.1$ for its KPSS stationarity test. What is more, we find that the differenced series has seasonal correlation since the ACF value at seasonal lag 12 shown in Fig. 5.5 is significantly different from zero. Thus, a question comes out: how to model the seasonal correlation?

```
>>> import pandas as pd
>>> import numpy as np
>>> import statsmodels.api as sm
>>> import matplotlib.pyplot as plt
>>> from statsmodels.graphics.tsaplots import month_plot
>>> from PythonTsa.plot_acf_pacf import acf_pacf_fig
>>> from PythonTsa.datadir import getdtapath
>>> dtapath=getdtapath()
>>> x=pd.read_csv(dtapath + 'USFemalesAged20
                   +Job1948-81.csv', header=None)
>>> dates = pd.date_range(start='1948-01',periods=len(x),freq='M')
>>> x.index=dates
>>> x=pd.Series(x[0])
>>> month_plot(x); plt.show()
>>> x.plot(); plt.show()
>>> x['1963-01' : '1965-12'].plot(marker='.'); plt.show()
>>> dDx=sm.tsa.statespace.tools.diff(x,k_diff=1,
```

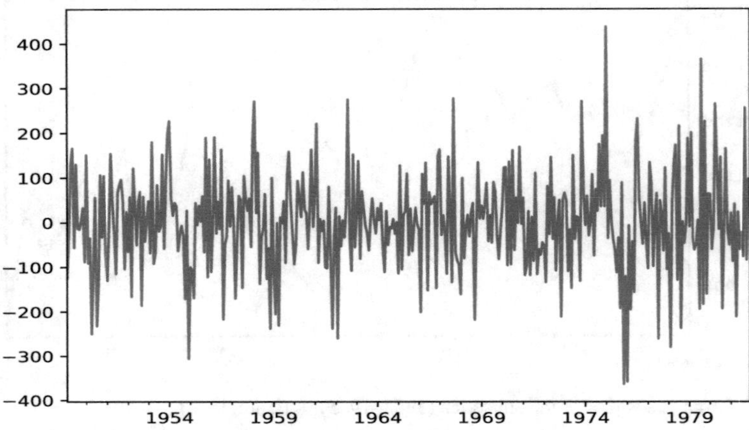


**Fig. 5.4** Time series plot of the seasonally firstly differenced data in Example 5.1

pyTSA_FemaleLabor

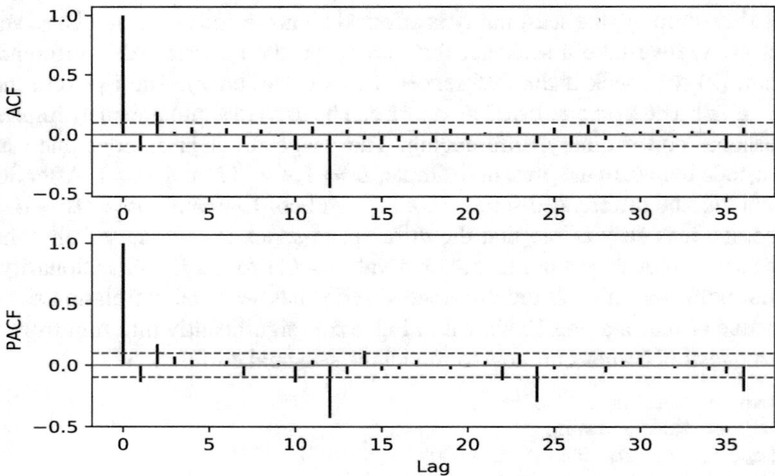


**Fig. 5.5** ACF and PACF plots of the seasonal first differences in Example 5.1

pyTSA_FemaleLabor

```
                        k_seasonal_diff=1,seasonal_periods=12)
# k_diff is the order number of the ordinary difference.
# k_seasonal_diff is the order number of the seasonal difference.
# seasonal_periods is the seasonal period.
>>> dDx.head()
1949-02-28   -124.0
1949-03-31    -21.0
1949-04-30    126.0
1949-05-31    166.0
```

```
1949-06-30    -56.0
Freq: M, Name: 0, dtype: float64
>>> dDx.plot(); plt.show()
>>> acf_pacf_fig(dDx, both=True, lag=36)
>>> plt.show()
>>> sm.tsa.kpss(dDx, regression='c', nlags='auto')
InterpolationWarning: p-value is greater than the indicated p-value
(0.022502332513746816, 0.1, 5,
{'10%': 0.347, '5%': 0.463, '2.5%': 0.574, '1%': 0.739})
```

### *5.1.2 SARIMA Models*

As we know, we use an ARMA model to characterize the ordinary correlation of a time series, and ARMA models are of $\varphi(B)Y_t = \theta(B)\varepsilon_t$, which reminds us that we could use such a form of $\Phi(B^s)Y_t = \Theta(B^s)\varepsilon_t$ to model seasonal correlation. If a time series has both ordinary correlation and seasonal correlation, for example, the differenced series in Example 5.1 has this two kinds of correlation, which can be observed from Fig. 5.5, then we should combine $\varphi(B)Y_t = \theta(B)\varepsilon_t$ and $\Phi(B^s)Y_t = \Theta(B^s)\varepsilon_t$ together in order to model these two types of correlation. This leads to the following definition.

**Definition 5.1** If $d$ and $D$ are nonnegative integers and $B$ is the backshift operator, then time series $\{X_t\}$ is a SARIMA$(p, d, q)(P, D, Q)_s$ process with *seasonal period* $s$ if the differenced series $Y_t = (1 - B)^d(1 - B^s)^D X_t$ is stationary and $\{X_t\}$ satisfies the following model

$$\varphi(B)\Phi(B^s)(1 - B)^d(1 - B^s)^D X_t = \theta(B)\Theta(B^s)\varepsilon_t, \quad \varepsilon_t \sim \mathrm{WN}(0, \sigma_\epsilon^2) \tag{5.2}$$

where $\varphi(z) = 1 - \varphi_1 z - \cdots - \varphi_p z^p$ ($\varphi_p \neq 0$) and $\theta(z) = 1 + \theta_1 z + \cdots + \theta_q z^q$ ($\theta_q \neq 0$), as we know, are, respectively, the AR polynomial and MA polynomial and $\Phi(z) = 1 - \Phi_1 z - \cdots - \Phi_P z^P$ ($\Phi_P \neq 0$) and $\Theta(z) = 1 + \Theta_1 z + \cdots + \Theta_Q z^Q$ ($\Theta_Q \neq 0$) are, respectively, called the *seasonal AR polynomial* and *seasonal MA polynomial*. Besides, Eq. (5.2) itself is known as a SARIMA$(p, d, q)(P, D, Q)_s$ model where $Y_t = (1 - B)^d(1 - B^s)^D X_t$ is still required to be stationary.

Five remarks on Definition 5.1 are as follows:

- Just as both $\varphi(B)$ and $\theta(B)$ are for modeling ordinary correlation, both $\Phi(B^s)$ and $\Theta(B^s)$ are to model seasonal correlation.
- As in ARMA modeling, we require that the resulting model SARMA$(p, q)$ $(P, Q)_s$ for $Y_t = (1 - B)^d(1 - B^s)^D X_t$ is both causal and invertible. It can be proved that the resulting model is causal if and only if $\varphi(z) \neq 0$ and $\Phi(z) \neq 0$ for any $|z| \leq 1$; it is invertible if and only if $\theta(z) \neq 0$ and $\Theta(z) \neq 0$ for any $|z| \leq 1$. Note that $\Phi(z)$ and $\Theta(z)$ are, respectively, of degrees $P$ and $Q$, not of degrees $Ps$ and $Qs$ (see Remark 1 on page 177, Brockwell and Davis 2016).

- If $X_t$ is stationary, namely, $d = D = 0$, then model (5.2) is reduced to $\varphi(B)\Phi(B^s)X_t = \theta(B)\Theta(B^s)\varepsilon_t$, which is SARMA$(p, q)(P, Q)_s$ models.
- If $p = q = d = 0$, then model (5.2) is reduced to $\Phi(B^s)(1 - B^s)^D X_t = \Theta(B^s)\varepsilon_t$, which is called (pure) SARIMA$(P, D, Q)_s$ models.
- If $X_t$ is stationary, namely, $d = D = 0$ and $p = q = 0$, then model (5.2) is reduced to $\Phi(B^s)X_t = \Theta(B^s)\varepsilon_t$, which is known as (pure) SARMA$(P, Q)_s$ models. Furthermore, if $Q = 0$, SARMA$(P, Q)_s$ reduces to (pure) SAR$(P)_s$ and if $P = 0$, SARMA$(P, Q)_s$ reduces to (pure) SMA$(Q)_s$.

*Example 5.2 (Pure Seasonal AR, MA, and ARMA Models)* Suppose time series $\{X_t\}$ is stationary and of seasonal period 4, satisfying $f(X_t, \varepsilon_t) = 0$, then

(1) If the equation $f(X_t, \varepsilon_t) = 0$ can be expressed by

$$(1 - 0.36B^4)X_t = \varepsilon_t \text{ or } X_t = 0.36X_{t-4} + \varepsilon_t,$$

then it is a seasonal AR(1) model with seasonal period 4, namely, a pure SAR$(1)_4$ model.

(2) If the equation $f(X_t, \varepsilon_t) = 0$ can be expressed by

$$X_t = (1 + 0.46B^4)\varepsilon_t \text{ or } X_t = \varepsilon_t + 0.46\varepsilon_{t-4},$$

then it is a seasonal MA(1) model with seasonal period 4, namely, a pure SMA$(1)_4$ model.

(3) If the equation $f(X_t, \varepsilon_t) = 0$ can be expressed by

$$(1 - 0.36B^4)X_t = (1 + 0.46B^4)\varepsilon_t \text{ or } X_t - 0.36X_{t-4} = \varepsilon_t + 0.46\varepsilon_{t-4},$$

then it is a seasonal ARMA(1,1) model with seasonal period 4, namely, a pure SARMA$(1, 1)_4$ model.

Now let us plot their true ACF and PACF and then observe these plots. The following is the Python code that we run.

```
>>> import numpy as np
>>> import matplotlib.pyplot as plt
>>> from PythonTsa.True_acf import Tacf_pacf_fig
>>> ar1=np.array([1,0,0,0,-0.36])
>>> Tacf_pacf_fig(ar=ar1,ma=[1], both=True, lag=20)
>>> plt.show()
>>> ma1=np.array([1,0,0,0,0.46])
>>> Tacf_pacf_fig(ar=[1], ma=ma1, both=True, lag=20)
>>> plt.show()
>>> Tacf_pacf_fig(ar=ar1, ma=ma1, both=True, lag=20)
>>> plt.show()
```

From the true ACF and PACF plots of the pure SAR$(1)_4$ model shown in Fig. 5.6, we see that the true PACF cuts off after seasonal lag 4 and is all zero at lags 1, 2,

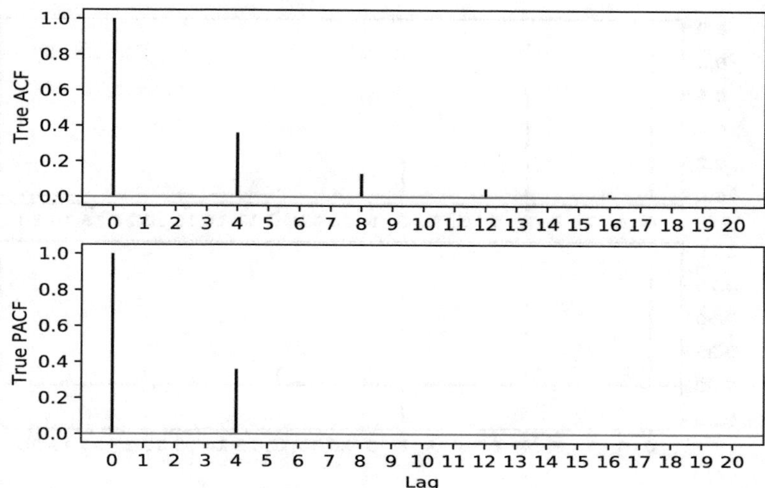


**Fig. 5.6** True ACF and PACF plots of the pure SAR$(1)_4$ model in Example 5.2

pyTSA_SARMA

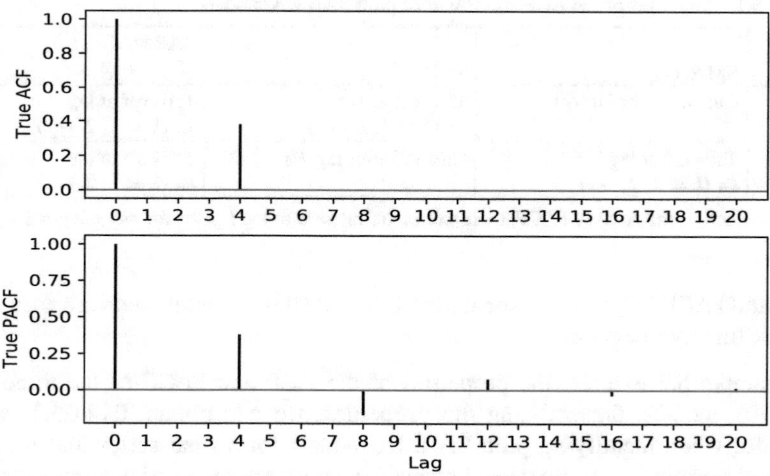


**Fig. 5.7** True ACF and PACF plots of the pure SMA$(1)_4$ model in Example 5.2

pyTSA_SARMA

and 3, while the true ACF tails off at seasonal lags 4, 8, and so on. From the true ACF and PACF plots of the pure SMA$(1)_4$ model shown in Fig. 5.7, we see that the true ACF cuts off after seasonal lag 4 and is all zero at lags 1, 2, and 3 while the true PACF tails off at seasonal lags 4, 8, and so forth. From the true ACF and PACF plots of the pure SARMA$(1, 1)_4$ model shown in Fig. 5.8, we see that both the true

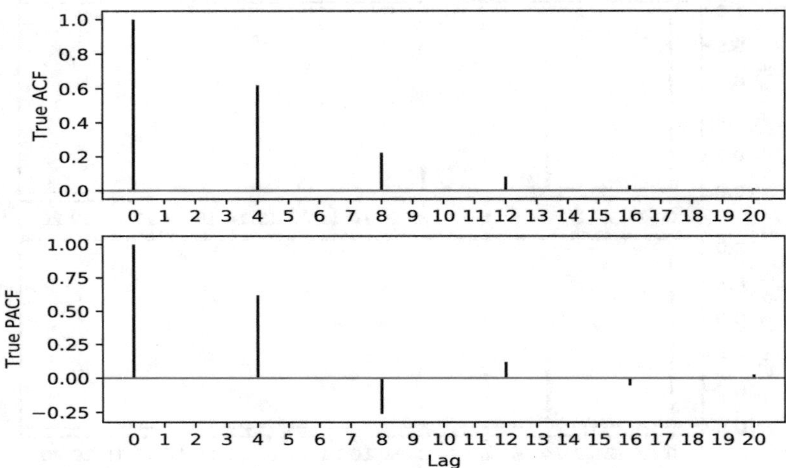


**Fig. 5.8** True ACF and PACF plots of the pure SARMA$(1, 1)_4$ in Example 5.2

pyTSA_SARMA

**Table 5.1** Properties of the ACF and PACF of pure SARMA models

| | SMA$(Q)_s$ | SAR$(P)_s$ | SARMA$(P, Q)_s$ $(P > 0, Q > 0)$ |
|---|---|---|---|
| ACF | Cuts off after lag $Qs$ | Tails off at lag $ks$ $(k = 1, 2, \cdots)$ | Tails off at lag $ks$ $(k = 1, 2, \cdots)$ |
| PACF | Tails off at lag $ks$ $(k = 1, 2, \cdots)$ | Cuts off after lag $Ps$ | Tails off at lag $ks$ $(k = 1, 2, \cdots)$ |

Note: whether the ACF or PACF, their values are all zero at lag $l \neq ks$ for any integer $k$.

ACF and PACF tail off at seasonal lags 4, 8, and so forth, while both are zero at lag $h \neq ks$ for any integer $k$.

Example 5.2 exhibits the properties of the ACF and PACF of the three pure SARMA models. Generalizing the properties, we can obtain Table 5.1, which is useful when identifying pure SARMA models for a time series that has only seasonal correlation. In practice, however, a stationary time series with a seasonal period tends to have both seasonal correlation and ordinary correlation. In addition, this type of time series tends to come from a SARMA$(p, q)(P, Q)_s$ process. For example, from the ACF plot (shown in Fig. 5.5) of the differenced series in Example 5.1, we see that the ACF values only at lags 2 and 3 and seasonal lag 12 are significantly nonzero. In other words, the differenced series has both seasonal correlation and ordinary correlation, and we could think that the differenced series was generated from a SARMA$(0, 2)(0, 1)_{12}$ process. In order to better understand behavior of the ACF and PACF of a general SARMA$(p, q)(P, Q)_s$ model, let us look at the following example.

*Example 5.3 (Simulating SARMA Processes)* Now we have three SARMA processes with seasonal period 4 as follows:

(1) It is a $\mathrm{SAR}(1)(1)_4 = \mathrm{SARMA}(1, 0)(1, 0)_4$ process

$$(1 - 0.2B)(1 - 0.3B^4)X_t = \varepsilon_t, \ \varepsilon_t \sim \mathrm{iidN}(0, 2^2),$$

which is a special case of $(1 - \varphi_1 B)(1 - \Phi_1 B^4)X_t = \varepsilon_t, \ \varepsilon_t \sim \mathrm{iidN}(0, \sigma_\epsilon^2)$. We first simulate a sample of size 1000 for this process and then plot the sample ACF and PACF shown in Fig. 5.9. From the PACF plot, we see that only at lag 1 and seasonal lag 4 the PACF values are significantly nonzero. From the ACF plot, we find that the ACF tails off at seasonal lag 4, 8, and so on. In other words, if we want to use the simulated sample as a dataset to build a time series model, we could use a $\mathrm{SAR}(1)(1)_4$ model to fit the sample. By executing the corresponding Python commands, we obtain that the estimators of $\varphi_1$, $\Phi_1$, and $\sigma_\epsilon^2$ are, respectively, 0.1748, 0.3334, and 4.1549. And they are all significantly nonzero and, respectively, close to the true values 0.2, 0.3, and 4. Details on building SARMA models with Python can be found in Sect. 5.2.

(2) It is a $\mathrm{SMA}(1)(1)_4 = \mathrm{SARMA}(0, 1)(0, 1)_4$ process

$$X_t = (1 + 0.5B)(1 + 0.4B^4)\varepsilon_t, \ \varepsilon_t \sim \mathrm{iidN}(0, 2^2),$$

which is a special case of $X_t = (1 + \theta_1 B)(1 + \Theta_1 B^4)\varepsilon_t, \ \varepsilon_t \sim \mathrm{iidN}(0, \sigma_\epsilon^2)$. The ACF and PACF plots of the simulated sample of this process are shown in Fig. 5.10. From the ACF plot, we observe that at lag 1 and seasonal lag 4, the ACF values are significantly nonzero and the seasonal ACF cuts off after seasonal lag $1s = 4$. Besides, the ACF values at lags 3 and 5 appear also to be

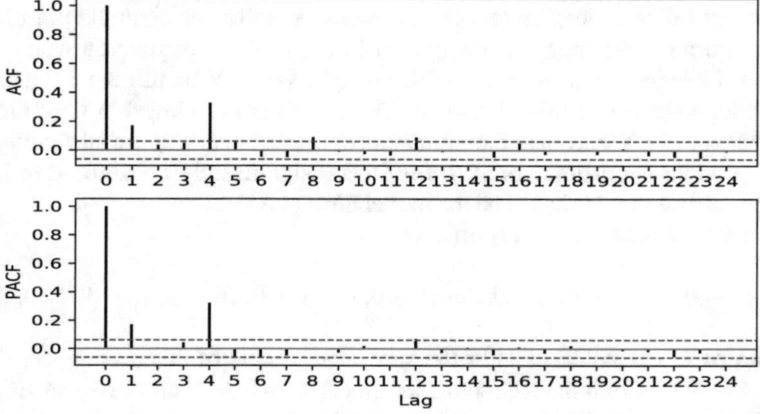


**Fig. 5.9** Sample ACF and PACF plots of the process $\mathrm{SAR}(1)(1)_4$ in Example 5.3

pyTSA_SimSARMA

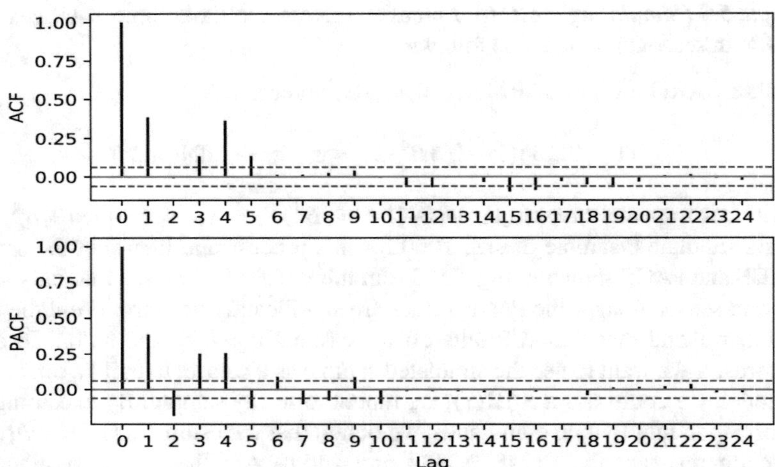


**Fig. 5.10** Sample ACF and PACF plots of the process SMA(1)(1)$_4$ in Example 5.3

pyTSA_SimSARMA

significantly nonzero. In fact, for SMA$(1)(1)_4$ models, it is easy to show that

$$\rho_1 = \frac{\theta_1}{1+\theta_1^2},\ \rho_4 = \frac{\Theta_1}{1+\Theta_1^2},\ \rho_3 = \rho_5 = \frac{\theta_1\Theta_1}{(1+\theta_1^2)(1+\Theta_1^2)},\ \rho_k = 0\ (k \neq 1, 3, 4, 5).$$

Hence, as long as $\theta_1 \neq 0$ and $\Theta_1 \neq 0$, then $\rho_3 = \rho_5 \neq 0$. In addition, we have that $\rho_k = 0$ for all $k > 5$. Then, might we select $q = 5$ as an order of the desired model SMA$(q)(Q)_4$? Note that lags 3 and 5 are neighbor to seasonal lag 4 and $\rho_3$ and $\rho_5$ are influenced by $\rho_4$, namely, $\Theta_1$ (this phenomenon is typical for seasonal time series). In this case, the sample value for $\rho_5$ might not be used to determine the order of MA part in SMA$(q)(Q)_4$. Maybe you would think that the evidence is not reasonable enough. Well, let us use the SMA$(5)(1)_4$ model to fit the simulated sample. The results are displayed in the following Python code. We see that the estimators of $\theta_2, \theta_3, \theta_4, and \theta_5$ are all significantly zero while the estimators of $\theta_1 and \Theta_1$ are significantly nonzero. That is, the appropriate model should be the model SMA$(1)(1)_4$.

(3) It is a SARMA$(1, 1)(1, 1)_4$ process

$$(1 - 0.2B)(1 - 0.3B^4)X_t = (1 + 0.5B)(1 + 0.4B^4)\varepsilon_t,\ \varepsilon_t \sim \text{iidN}(0, 2^2).$$

The ACF and PACF plots of the simulated sample of this process are shown in Fig. 5.11. From the ACF plot, we find that both ACF and seasonal ACF tail off. From the PACF plot, we see that the PACF tails off and the seasonal PACF seems to cut off after lag $1s = 4$. Therefore, in this case, we can not determine orders of a SARMA$(p, q)(P, G)_s$ model by its ACF or PACF.

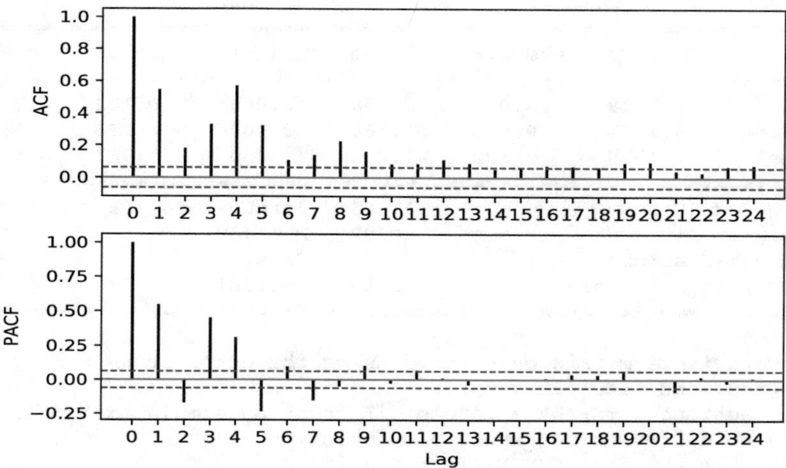


**Fig. 5.11** Sample ACF and PACF of the model SARMA(1, 1)(1, 1)$_4$ in Example 5.3

pyTSA_SimSARMA

```
# Below is the code for Example 5.3 (1).
>>> import pandas as pd
>>> import numpy as np
>>> import matplotlib.pyplot as plt
>>> import statsmodels.api as sm
>>> from PythonTsa.plot_acf_pacf import acf_pacf_fig
>>> phi = np.r_[0.2]
>>> theta = np.r_[0.0]
>>> Phi = np.r_[0.3]
>>> Theta = np.r_[0.0]
>>> sigma2=4.0
>>> params=np.r_[phi, theta, Phi, Theta, sigma2]
>>> sarsim=sm.tsa.SARIMAX([0], order=(1,0,1), seasonal_order
      =(1,0,1,4)).simulate(params=params,nsimulations=1000)
>>> simts=pd.Series(sarsim)
>>> acf_pacf_fig(simts, both=True, lag=24)
>>> plt.show()
>>> sarima1010 = sm.tsa.SARIMAX(simts, order=(1,0,0),
                                seasonal_order=(1,0,0,4))
>>> sarimaRes1010=sarima1010.fit(disp=False)
>>> print(sarimaRes1010.summary())
                     SARIMAX Results
==================================================================
Dep. Variable:                     y No. Observations:       1000
Model:           SARIMAX(1,0,0)x(1,0,0,4) Log Likelihood   -2131.336
Date:                Sun, 29 Sep 2019 AIC                  4268.672
Time:                        05:29:11 BIC                  4283.395
Sample:                            0 HQIC                 4274.268
                              - 1000
Covariance Type:                 opg
```

```
==============================================================================
                 coef    std err          z      P>|z|      [0.025      0.975]
------------------------------------------------------------------------------
ar.L1          0.1748      0.032      5.490      0.000       0.112       0.237
ar.S.L4        0.3334      0.030     11.160      0.000       0.275       0.392
sigma2         4.1549      0.190     21.822      0.000       3.782       4.528
==============================================================================
Ljung-Box (Q):                      31.63   Jarque-Bera (JB):              0.83
Prob(Q):                             0.82   Prob(JB):                      0.66
Heteroskedasticity (H):              0.95   Skew:                         -0.05
Prob(H) (two-sided):                 0.65   Kurtosis:                      2.90
==============================================================================
Warnings:
[1] Covariance matrix calculated using the outer product of
    gradients (complex-step).
# the estimates may be a little different as simulated sample
# may be different for every simulation.
# Below is the code for Example 5.3 (2).
>>> phi = np.r_[0.0]
>>> theta = np.r_[0.5]
>>> Phi = np.r_[0.0]
>>> Theta = np.r_[0.4]
>>> sigma2=4.0
>>> params=np.r_[phi, theta, Phi, Theta, sigma2]
>>> smasim=sm.tsa.SARIMAX([0], order=(1,0,1), seasonal_order
       =(1,0,1,4)).simulate(params=params,nsimulations=1000)
>>> simts=pd.Series(smasim)
>>> acf_pacf_fig(simts, both=True, lag=24)
>>> plt.show()
>>> sarima0501 = sm.tsa.SARIMAX(simts, order=(0,0,5),
    seasonal_order=(0,0,1,4))  # now order=(0,0,[1,1,1,0,1])
>>> sarimaRes0501=sarima0501.fit(disp=False)
>>> print(sarimaRes0501.summary())
                            SARIMAX Results
==============================================================================
Dep. Variable:                      y   No. Observations:              1000
Model:          SARIMAX(0,0,5)x(0,0,1,4)   Log Likelihood         -2134.646
Date:                Mon, 30 Sep 2019   AIC                        4283.293
Time:                        20:47:59   BIC                        4317.647
Sample:                             0   HQIC                       4296.350
                               - 1000
Covariance Type:                  opg
==============================================================================
                coef    std err          z      P>|z|      [0.025      0.975]
------------------------------------------------------------------------------
ma.L1         0.4811      0.032     14.988      0.000       0.418       0.544
ma.L2        -0.0035      0.036     -0.097      0.923      -0.075       0.068
ma.L3         0.0271      0.037      0.741      0.459      -0.045       0.099
ma.L4         0.0612      0.099      0.616      0.538      -0.133       0.256
ma.L5         0.0292      0.059      0.499      0.618      -0.085       0.144
ma.S.L4       0.3667      0.091      4.026      0.000       0.188       0.545
sigma2        4.1805      0.185     22.577      0.000       3.818       4.543
==============================================================================
Ljung-Box (Q):                     25.80   Jarque-Bera (JB):               2.68
```

```
Prob(Q):                        0.96   Prob(JB):                  0.26
Heteroskedasticity (H):         1.09   Skew:                     -0.12
Prob(H) (two-sided):            0.42   Kurtosis:                  3.08
===================================================================
Warnings:
[1] Covariance matrix calculated using the outer product of
    gradients (complex-step).
# Below is the code for Example 5.3 (3).
>>> phi = np.r_[0.2]
>>> theta = np.r_[0.5]
>>> Phi = np.r_[0.3]
>>> Theta = np.r_[0.4]
>>> sigma2=4.0
>>> params=np.r_[phi, theta, Phi, Theta, sigma2]
>>> smasim=sm.tsa.SARIMAX([0], order=(1,0,1), seasonal_order
       =(1,0,1,4)).simulate(params=params,nsimulations=1000)
>>> simts=pd.Series(smasim)
>>> acf_pacf_fig(simts, both=True, lag=24)
>>> plt.show()
```

## 5.2 SARIMA Model Building

In this section, we consider how to build an appropriate SARIMA model for a given time series data with seasonality as well as how to implement this with Python.

### 5.2.1 General Idea

For a given time series dataset $\{X_t\}$, a tough task is selecting an appropriate model to fit it. If the time series is nonstationary, whether seasonal or nonseasonal, according to the Box-Jenkins method, first, we should make it stationary through differencing (sometimes plus transforming). Second, we should select order $(p, q)(P, G)$ for a SARMA$(p, q)(P, G)_s$ model, which is used to fit the differenced series$\{Y_t = (1 - B)^d(1 - B^s)^D X_t\}$. Third, estimate the model. Fourth, diagnose the model, and if the estimated model is unsatisfactory, we should restart from step 1. Finally, use the resulting model to forecast development of the time series in the future. When selecting order $(p, q)(P, G)$, as a good rule of thumb, we can inspect the seasonal ACF and PACF by Table 5.1 and determine orders $P$ or $Q$; meanwhile, we also check the ordinary ACF and PACF by Table 3.1 and determine orders $p$ or $q$. If both of the seasonal ACF and PACF tail off or both of the ordinary ACF and PACF tail off, then we need the information criteria (such as AICC, BIC, and HQIC) to identify models.

As we know, differencing (5.1) drops $d + Ds$ frontmost data points. Since the $d + Ds$ residuals at the head are estimated without using the resulting model, when

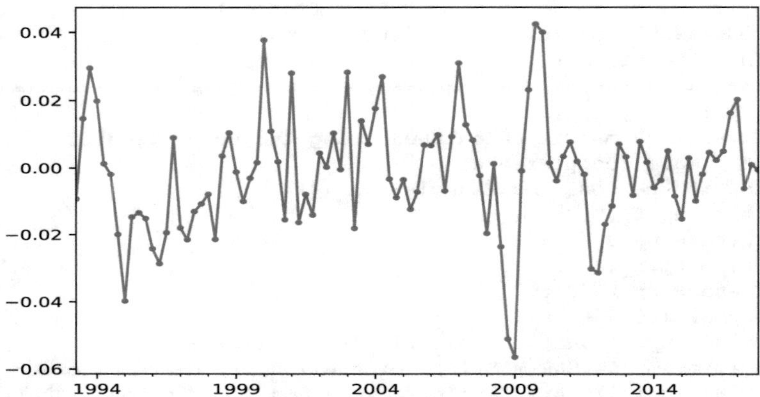


**Fig. 5.12** Time series plot of dDlx of Chinese quarterly GDP in Example 5.4
pyTSA_GDPChinaSARIMA

analyzing residuals, we should abandon these $d + Ds$ residuals in order to obtain right results. Besides, for seasonal time series, as we have seen in Example 3.3, we should first conduct seasonal differencing, and if the seasonally differenced series is stationary, then the ordinary differencing should not be taken.

### 5.2.2 Case Studies

*Example 5.4 (Revisiting the Chinese Quarterly GDP)* We have analyzed the Chinese quarterly GDP series several times in the previous chapters. From its time series plot shown in Fig. 1.7, we observe that its variance becomes bigger and bigger over time. So we first log the series `x` and get the logged series `lx`. Then we seasonally and firstly difference the logged series and obtain the differenced logged series `dDlx`. It is easy to see that the series `dDlx` is stationary (Fig. 5.12). From the ACF and PACF plots shown in Fig. 5.13, both of the ordinary ACF and PACF appear to tail off, and the ACF value as well as PACF value at lag 1 is very significant. Thus, orders $(p, q) = (1, 1)$, $(p, q) = (2, 0)$, and $(p, q) = (0, 2)$ are alternative. Here we select $(p, q) = (0, 2)$ and the other two are left as an exercise. On the other hand, the seasonal PACF values are not significantly nonzero at lags 4, 8, 12, 16, and so on; the seasonal ACF values at lags 4 and 8 are not significantly nonzero, while the seasonal ACF value at lag 12 appears to be significant. Hence, orders $(P, Q) = (0, 0)$ and $(P, Q) = (0, 3)$ are alternative. We here choose order $(P, Q) = (0, 0)$ and the other is left for the reader. In summary, we select the model SARIMA$(0, 1, 2)(0, 1, 0)_4$ to fit the logged series `lx`.

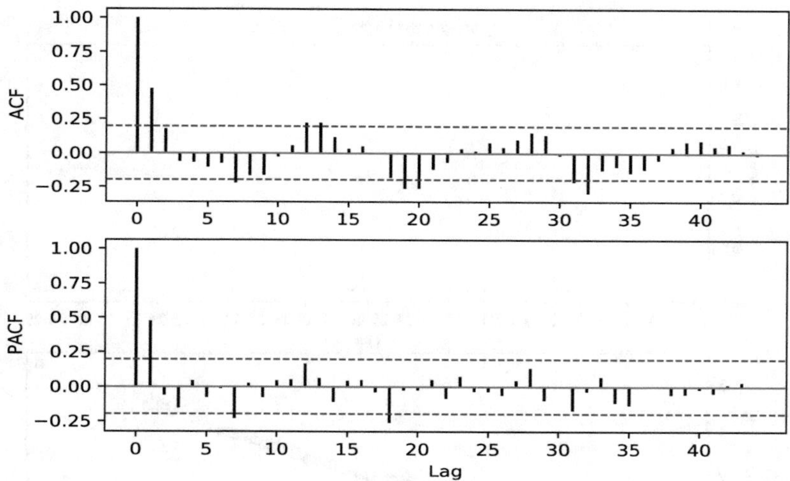


**Fig. 5.13** ACF and PACF of dDlx of Chinese quarterly GDP in Example 5.4

pyTSA_GDPChinaSARIMA

From the following `SARIMAX Results`, we see that the estimators of $\theta_1$, $\theta_2$, and $\sigma_\epsilon^2$ are all significant. From the model diagnostic plots shown in Fig. 5.14, we conclude that the resulting model is appropriate, and it can be written as

$$(1 - B)(1 - B^4)\ln(X_t) = (1 + 0.4940B + 0.2688B^2)\varepsilon_t$$

Moreover, the residual series should have ARCH effect since the ACF value of the squared residuals is significant at lag 1. When predicting, we should first forecast $\ln(X_t)$ and then obtain $X_t$ by $X_t = \exp[\ln(X_t)]$. Through comparing with the actual values (see Example 2.6), the forecast values are barely satisfactory. You can run the following Python code to reproduce the results in this example.

```
>>> import pandas as pd
>>> import numpy as np
>>> import statsmodels.api as sm
>>> import matplotlib.pyplot as plt
>>> from PythonTsa.plot_acf_pacf import acf_pacf_fig
>>> from PythonTsa.ModResidDiag import plot_ResidDiag
>>> from PythonTsa.datadir import getdtapath
>>> dtapath=getdtapath()
>>> x=pd.read_csv(dtapath + 'gdpquarterlychina1992.1-2017.4.csv',
                  header=0)
>>> dates = pd.date_range(start='1992',periods=len(x),freq='Q')
>>> x.index=dates
>>> x=pd.Series(x['GDP'])
>>> lx=np.log(x)
>>> dDlx=sm.tsa.statespace.tools.diff(lx, k_diff=1,
              k_seasonal_diff=1, seasonal_periods=4)
```

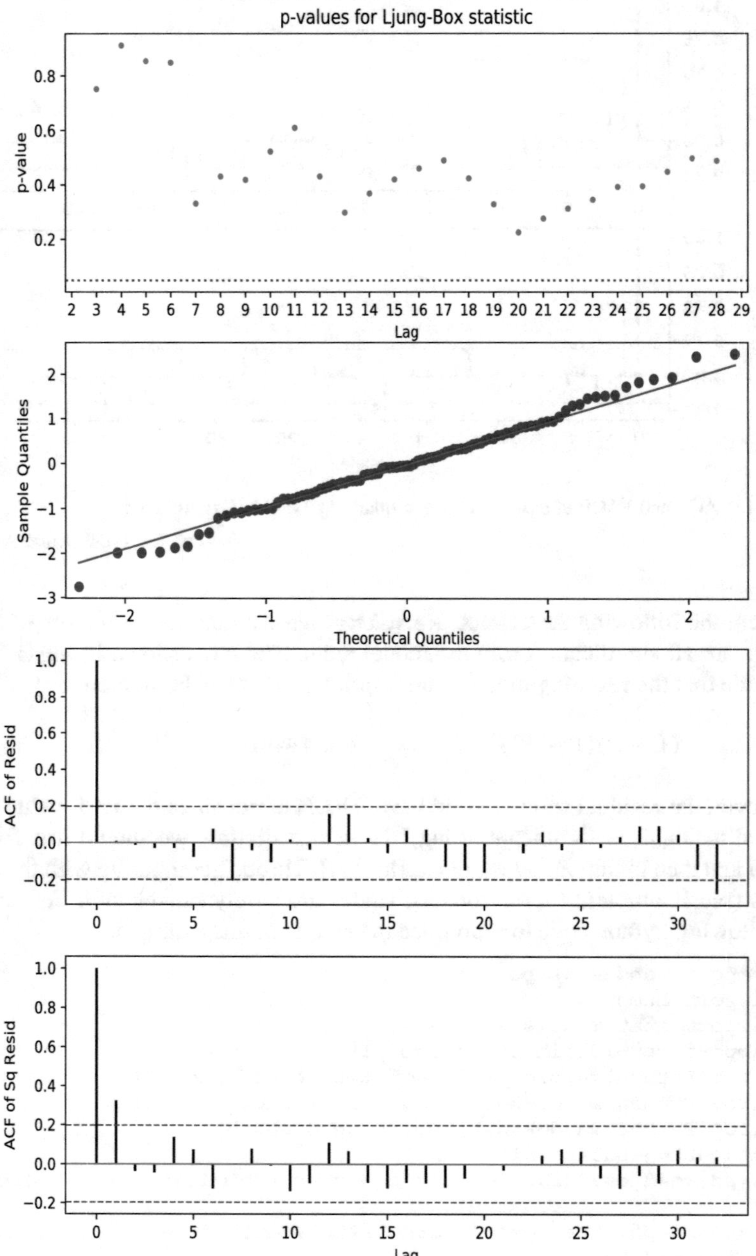


**Fig. 5.14** Model diagnostic plots for log Chinese quarterly GDP series

pyTSA_GDPChinaSARIMA

```
>>> dDlx.plot(marker='o',ms=3); plt.show()
>>> acf_pacf_fig(dDlx, both=True, lag=44)
>>> plt.show()
>>> sm.tsa.kpss(dDlx, regression='c', nlags='auto')
InterpolationWarning:
        p-value is greater than the indicated p-value
(0.10298546572401573, 0.1, 4,
{'10%': 0.347, '5%': 0.463, '2.5%': 0.574, '1%': 0.739})
>>> sarima0200 = sm.tsa.SARIMAX(lx, order=(0,1,2),
                              seasonal_order=(0,1,0,4))
>>> sarimaRes0200=sarima0200.fit(disp=False)
>>> print(sarimaRes0200.summary())
                        SARIMAX Results
================================================================
Dep. Variable:                    GDP   No. Observations:    104
Model:          SARIMAX(0,1,2)x(0,1,0,4)   Log Likelihood   274.805
Date:                Thu, 03 Oct 2019   AIC             -543.610
Time:                        16:18:08   BIC             -535.824
Sample:                    03-31-1992   HQIC            -540.460
                         - 12-31-2017
Covariance Type:                  opg
================================================================
              coef    std err       z     P>|z|    [0.025   0.975]
----------------------------------------------------------------
ma.L1       0.4940      0.080   6.142     0.000     0.336    0.652
ma.L2       0.2688      0.102   2.645     0.008     0.070    0.468
sigma2      0.0002   3.19e-05   7.102     0.000     0.000    0.000
================================================================
Ljung-Box (Q):             45.88   Jarque-Bera (JB):         0.02
Prob(Q):                    0.24   Prob(JB):                 0.99
Heteroskedasticity (H):     0.52   Skew:                    -0.00
Prob(H) (two-sided):        0.07   Kurtosis:                 3.07
================================================================
Warnings:
[1] Covariance matrix calculated using the outer product of
    gradients (complex-step).
>>> resid0200=sarimaRes0200.resid[5:]
# drop 1+4=5 residuals at head.
>>> plot_ResidDiag(resid0200,noestimatedcoef=2,nolags=28,lag=32)
>>> plt.show()
>>> fore0200=sarimaRes0200.predict(start='2018-03-31',
         end='2018-12-31')
>>> np.exp(fore0200)
2018-03-31    200804.519028
2018-06-30    223319.096238
2018-09-30    235608.407056
2018-12-31    261203.751808
Freq: Q-DEC, dtype: float64
```

*Example 5.5 (Corticosteroid Drug Sales in Australia from July 1991 to June 2008)* The time series data set "h02July1991June2008" in the folder Ptsadata is the corticosteroid drug sales in Australia (in millions of scripts per month) from July 1991 to June 2008. You can also find it in Section 8.9, Hyndman and

Athanasopoulos (2018). In what follows, we denote the series by `h02`. From its time series plot shown in Fig. 5.15, we see that the series has strong seasonality. So first we take a seasonal difference on it and then check whether the seasonally differenced series is stationary. Observing the time series plot shown in Fig. 5.16 and the ACF plot shown in Fig. 5.17 as well as the result (the *p*-value is greater than 0.1) for the KPSS test of the seasonally differenced series, we conclude that the differenced series is stationary.

Examining the ACF and PACF plots of the seasonally differenced series shown in Fig. 5.17, we observe that the ordinary ACF cuts off after lag 5 and the ordinary PACF cuts off after lag 3; meanwhile the seasonal ACF cuts off after lag 12 and the seasonal PACF cuts off after lag 24. Thus, we have

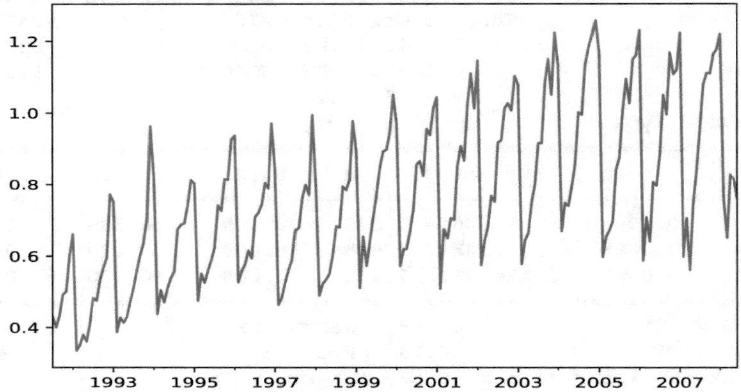


**Fig. 5.15** Time series plot of the corticosteroid drug sales in Australia

pyTSA_CorticoSales

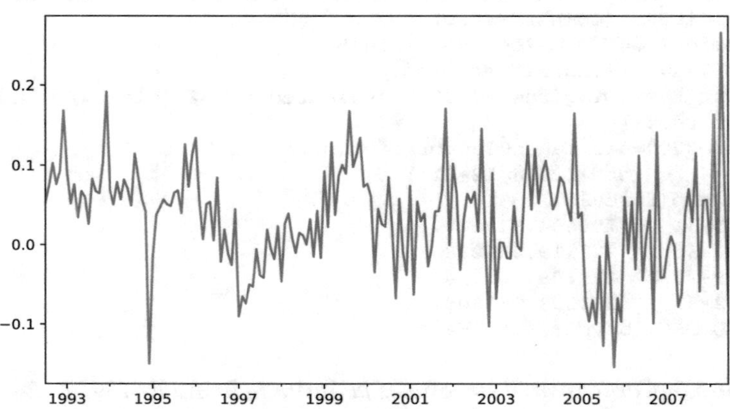


**Fig. 5.16** Time series plot of the seasonal differences Dh02 in Example 5.5

pyTSA_CorticoSales

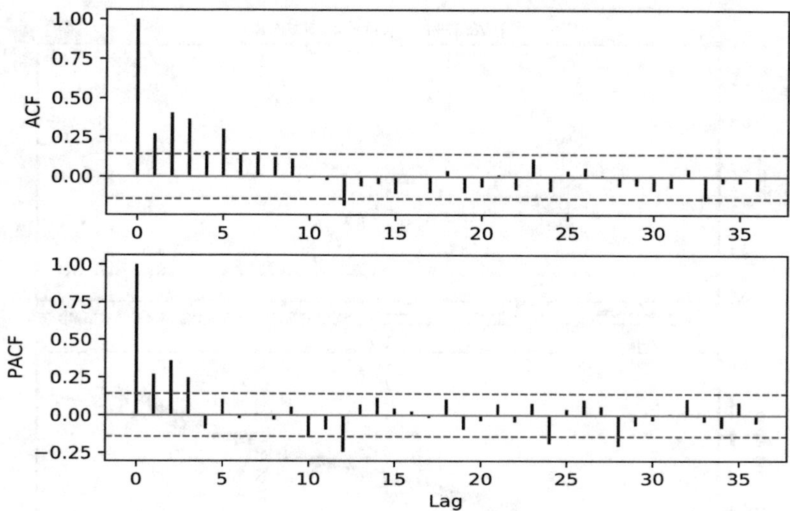


**Fig. 5.17** ACF and PACF plots of the seasonal differences Dh02 in Example 5.5

pyTSA_CorticoSales

four models to choose: SARIMA$(3, 0, 0)(2, 1, 0)_{12}$, SARIMA$(3, 0, 0)(0, 1, 1)_{12}$, SARIMA$(0, 0, 5)(0, 1, 1)_{12}$, and SARIMA$(0, 0, 5)(2, 1, 0)_{12}$. By executing the corresponding Python commands and inspecting estimating results, we find that only model SARIMA$(0, 0, 5)(2, 1, 0)_{12}$ is worth further considering, and its preliminary estimation shown in the following table `SARIMAX Results` looks perfect. However, from the model diagnostic plots shown in Fig. 5.18, we see that the $p$-values for Ljung-Box test of the residuals are smaller than 0.05 at lags 8 and 9. So the preliminary estimated model is not appropriate. Note that the $p$-value for `ma.L4`(viz., the coefficient $\theta_4$) is 0.065, which suggests that $\theta_4 = 0$. Therefore we rebuild the model SARIMA$(0, 0, 5)(2, 1, 0)_{12}$ with $\theta_4 = 0$. At this point, from the model diagnostic plots shown in Fig. 5.19, we gladly see that the $p$-value plot is perfect, and we arrive at an appropriate model for the series `h02`:

$$(1+0.272B^{12}+0.259B^{24})(1-B^{12})X_t = (1+0.205B+0.498B^2+0.510B^3+0.350B^5)\varepsilon_t.$$

In addition, solving the seasonal AR polynomial $\Phi(z) = 1+0.2718z+0.2590z^2$ and the ordinary MA polynomial$\theta(z) = 1+0.2048z+0.4979z^2+0.5102z^3+0.3496z^5$, we find that all the roots of the two polynomials are greater than 1 in modulus. This further validates that the resulting model is satisfactory.

```
>>> import pandas as pd
>>> import numpy as np
>>> import statsmodels.api as sm
>>> import matplotlib.pyplot as plt
>>> from PythonTsa.plot_acf_pacf import acf_pacf_fig
```

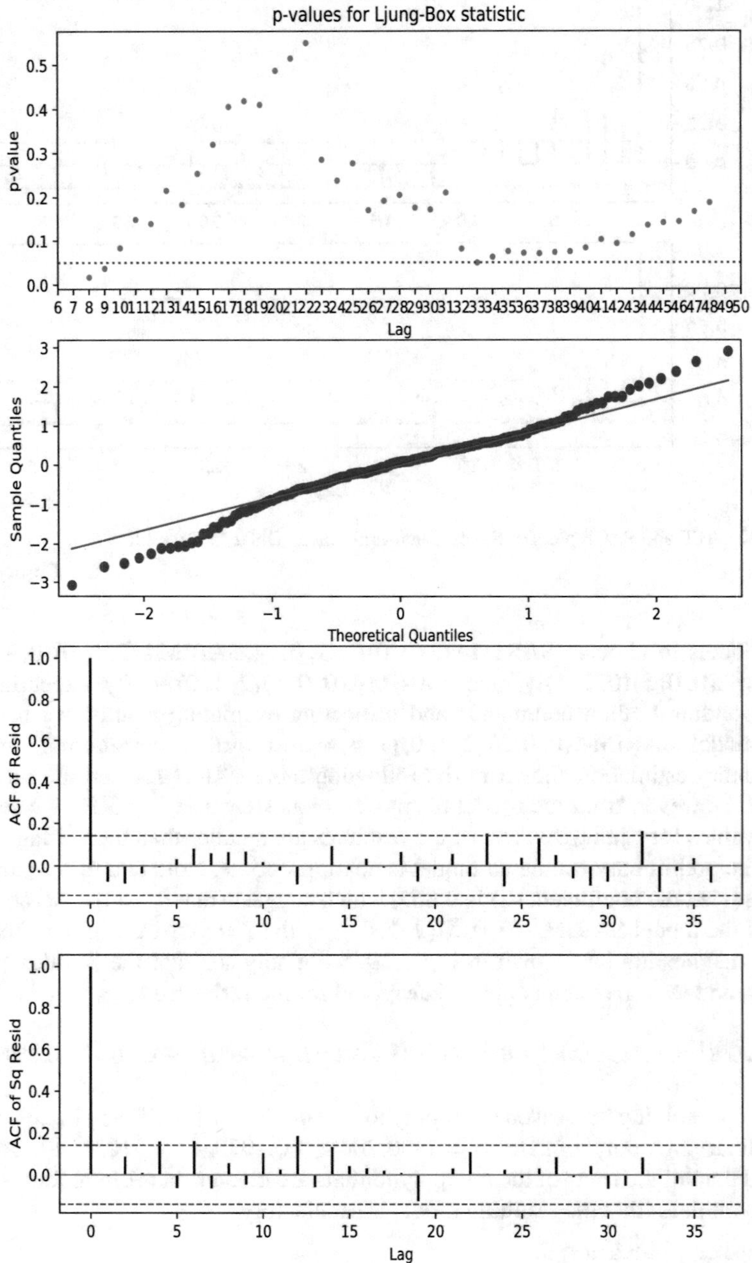


**Fig. 5.18** Model diagnostic plots for the model sarima005210 in Example 5.5

pyTSA_CorticoSales

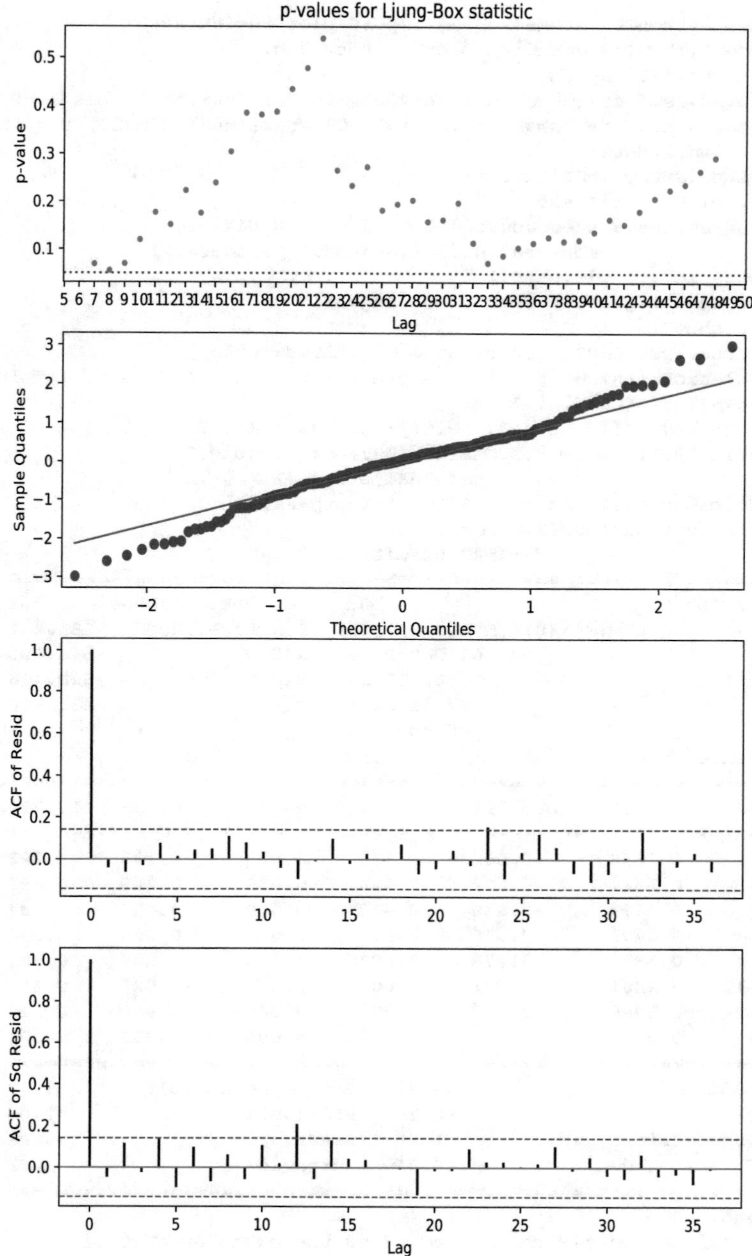


**Fig. 5.19** Model diagnostic plots for the model sarima0054210 in Example 5.5

pyTSA_CorticoSales

```
>>> from PythonTsa.ModResidDiag import plot_ResidDiag
>>> from PythonTsa.datadir import getdtapath
>>> dtapath=getdtapath()
>>> h02=pd.read_csv(dtapath + 'h02July1991June2008.csv', header=0)
>>> dates = pd.date_range(start='1991-07',periods=len(h02),freq='M')
>>> h02.index=dates
>>> h02=pd.Series(h02['h02'])
>>> h02.plot(); plt.show()
>>> Dh02=sm.tsa.statespace.tools.diff(h02, k_diff=0,
              k_seasonal_diff=1,seasonal_periods=12)
>>> Dh02.plot(); plt.show()
>>> acf_pacf_fig(Dh02, both=True, lag=36)
>>> plt.show()
>>> sm.tsa.kpss(Dh02, regression='c', nlags='auto')
InterpolationWarning: p-value is greater than the indicated p-value
(0.30447532679666206, 0.1, 8,
{'10%': 0.347, '5%': 0.463, '2.5%': 0.574, '1%': 0.739})
>>> sarima005210 = sm.tsa.SARIMAX(h02, order=(0,0,5),
                         seasonal_order=(2,1,0,12))
>>> sarimaMod005210=sarima005210.fit(disp=False)
>>> print(sarimaMod005210.summary())
                          SARIMAX Results
=================================================================
Dep. Variable:                     h02   No. Observations:     204
Model:          SARIMAX(0,0,5)x(2,1,0,12)   Log Likelihood   281.231
Date:                 Tue, 08 Oct 2019   AIC                  -546.463
Time:                         21:18:54   BIC                  -520.403
Sample:                     07-31-1991   HQIC                 -535.908
                          - 06-30-2008
Covariance Type:                   opg
=================================================================
                coef    std err          z     P>|z|     [0.025    0.975]
-----------------------------------------------------------------
ma.L1         0.2658      0.067      3.938     0.000      0.133     0.398
ma.L2         0.5511      0.072      7.620     0.000      0.409     0.693
ma.L3         0.5679      0.066      8.632     0.000      0.439     0.697
ma.L4         0.1406      0.076      1.848     0.065     -0.009     0.290
ma.L5         0.3591      0.077      4.693     0.000      0.209     0.509
ar.S.L12     -0.2861      0.071     -4.005     0.000     -0.426    -0.146
ar.S.L24     -0.2968      0.077     -3.858     0.000     -0.448    -0.146
sigma2        0.0031      0.000     10.067     0.000      0.002     0.004
=================================================================
Ljung-Box (Q):                   45.79   Jarque-Bera (JB):           5.17
Prob(Q):                          0.24   Prob(JB):                   0.08
Heteroskedasticity (H):           1.66   Skew:                      -0.21
Prob(H) (two-sided):              0.04   Kurtosis:                   3.69
=================================================================
Warnings:
[1] Covariance matrix calculated using the outer product of
    gradients (complex-step).
>>> resid005210=sarimaMod005210.resid[12:]
>>> plot_ResidDiag(resid005210, noestimatedcoef=7, nolags=48, lag=36)
>>> plt.show()
>>> sarima0054210 = sm.tsa.SARIMAX(h02, order=(0,0,[1,1,1,0,1]),
```

```
                                     seasonal_order=(2,1,0,12))
# [1,1,1,0,1] means ma.L4=0 by default.
>>> sarimaMod0054210=sarima0054210.fit(disp=False)
>>> print(sarimaMod0054210.summary())
                        SARIMAX Results
=================================================================
Dep. Variable:                      h02   No. Observations:    204
Model:SARIMAX(0,0,(1,2,3,5))x(2,1,(),12)  Log Likelihood   279.791
Date:                  Tue, 08 Oct 2019   AIC             -545.583
Time:                          21:47:02   BIC             -522.780
Sample:                      07-31-1991   HQIC            -536.347
                           - 06-30-2008
Covariance Type:                    opg
=================================================================
              coef    std err        z    P>|z|     [0.025   0.975]
-----------------------------------------------------------------
ma.L1       0.2048      0.061    3.346    0.001      0.085    0.325
ma.L2       0.4979      0.064    7.787    0.000      0.373    0.623
ma.L3       0.5102      0.069    7.355    0.000      0.374    0.646
ma.L5       0.3496      0.075    4.684    0.000      0.203    0.496
ar.S.L12   -0.2718      0.068   -3.995    0.000     -0.405   -0.138
ar.S.L24   -0.2590      0.074   -3.477    0.001     -0.405   -0.113
sigma2      0.0031      0.000   10.335    0.000      0.003    0.004
=================================================================
Ljung-Box (Q):                 44.03   Jarque-Bera (JB):      4.04
Prob(Q):                        0.31   Prob(JB):              0.13
Heteroskedasticity (H):         1.54   Skew:                 -0.08
Prob(H) (two-sided):            0.09   Kurtosis:              3.69
=================================================================
Warnings:
[1] Covariance matrix calculated using the outer product of
    gradients (complex-step).
>>> resid0054210=sarimaMod0054210.resid[12:]
>>> plot_ResidDiag(resid0054210, noestimatedcoef=6, nolags=48, lag=36)
>>> plt.show()
>>> arp=[-0.2590,-0.2718, 1]
>>> aroot=np.roots(arp)
>>> abs(aroot)
array([2.55850611, 1.50908526])
>>> map=[0.3496,0,0.5102,0.4979,0.2048, 1]
>>> maroot=np.roots(map)
>>> abs(maroot)
array([1.13613429, 1.28364679, 1.28364679, 1.23610088, 1.23610088])
```

## 5.3 REGARMA Models

As we know, many nonstationary time series tend to have a deterministic trend and/or seasonality, and if we can eliminate the trend and seasonality, they become stationary. In the previous sections, the differencing method is used to eliminate the trend and seasonality, and then the model SARMA$(p, q)(P, Q)_s$ is fitted to the differenced series, which has been stationary. In this section, we consider a special type of nonstationary time series, the deterministic components of which can be represented by a regressive model and the error part can been represented by an ARMA model. Specifically, we introduce a new model: REGARMA model, that is, a regression model with ARMA errors, or ARMAX model. Suppose that the exogenous (viz., independent or explanatory) variables for the endogenous (viz., dependent or response) series $\{Y_t\}$ are $\{(X_{t1}, X_{t2}, \cdots, X_{tk})\}$. Then the REGARMA model is of the form

$$\begin{cases} Y_t = c + \beta_1 X_{t1} + \beta_2 X_{t2} + \cdots + \beta_k X_{tk} + Z_t \\ \varphi(B)Z_t = \theta(B)\varepsilon_t \end{cases}$$

where $c$ is the const term; $\beta_1, \beta_2, \cdots, \beta_k$ are unknown fixed regression coefficients; and $\varepsilon_t \sim \text{WN}(0, \sigma_\epsilon^2)$. In practice, a key point is to find "true" exogenous variables for the endogenous variable.

A time series develops and evolves with time, and then it is evident that the "time" $t$ is the most important exogenous variable. Thus, we could express the deterministic trend and seasonality in terms of the "time" $t$. Now suppose that $\{X_t\}$ is a time series with seasonal period $s$. The so-called harmonic seasonal regression model is of the following form

$$X_t = p_t + \sum_{k=1}^{K} \{a_k \sin(2\pi kt/s)\} + \sum_{j=1}^{J} \{b_j \cos(2\pi jt/s)\} + Z_t \tag{5.3}$$

where $p_t$ is a polynomial of $t$ to be estimated; $a_k$ and $b_j$ are unknown parameters; $K, J \leq s/2$; and $Z_t$ follows an ARMA process. It is well known that Fourier series (viz., series of sine and cosine terms) can approximate any (periodic) function. So we can use sine or cosine terms (viz., Fourier terms) to describe the seasonal behavior of the series. This kind of REGARMA models is very useful for time series with a bigger seasonal period. See, for example, Section 9.5, Hyndman and Athanasopoulos (2018). Now let us look at an example below.

*Example 5.6 (Global Temperature Change Series from 1856 to 2005)* The time series dataset "GlobalTemperature" in the folder Ptsadata is the global temperature series from Jan. 1856 to Dec. 2005, expressed as anomalies from the monthly means over the period 1961–1990. In climate change studies, the series plays a central role. The dataset (now denoted as GMTC) is from Cowpertwait and Metcalfe (2009) and can also be downloaded free of charge from the Internet at: http://www.cru.uea.ac.uk/cru/data/. Its time series plot is shown in Fig. 5.20. And from it, we see that on the whole, the trend ascends with time. In the light of the background for the series, it certainly has strong seasonality although this is masked. Naturally it is the trend that is of most concern; thus we use the `resample` function to remove any seasonal effects within each year and produce a yearly series of temperatures from 1856 to 2005 (denoted as GYTC). The time series plot of the yearly mean temperatures is

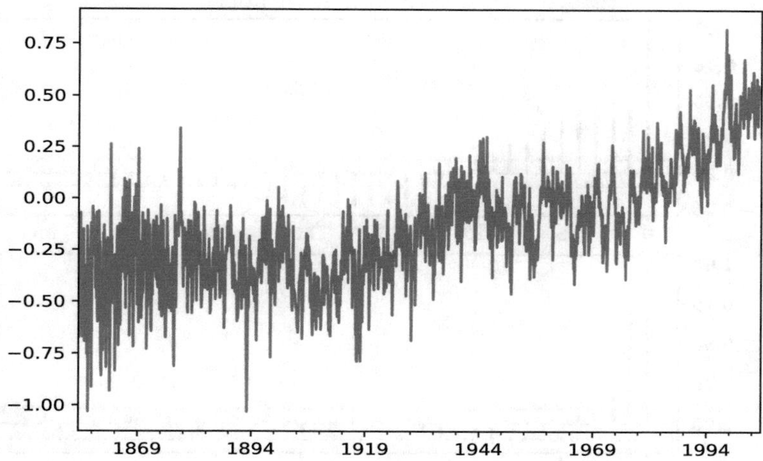


**Fig. 5.20** Time series plot of the GMTC series from 1856 to 2005 in Example 5.6
pyTSA_GlobalTemperature

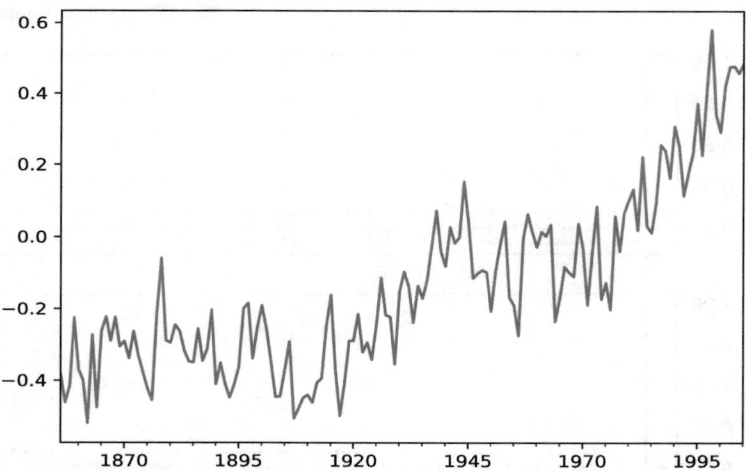


**Fig. 5.21** Time series plot of the GYTC series from 1856 to 2005 in Example 5.6
pyTSA_GlobalTemperature

shown in Fig. 5.21, and clearly the plot is much smoother. Moreover, the ascending trend from about 1970 on has been used as evidence of global warming.

Now we try to fit a harmonic seasonal regression model to the subseries of the GMTC series from 1970 to 2005. First of all, we give the case $K = J = 6$ in (5.3) a try. From the first table in the following Python program, we find that only the coefficients of `const, x1, x3` and `x5` are significant. Therefore, we refit with `const, x1, x3` and `x5` (at this point, $K = 2$ and $J = 0$) and obtain

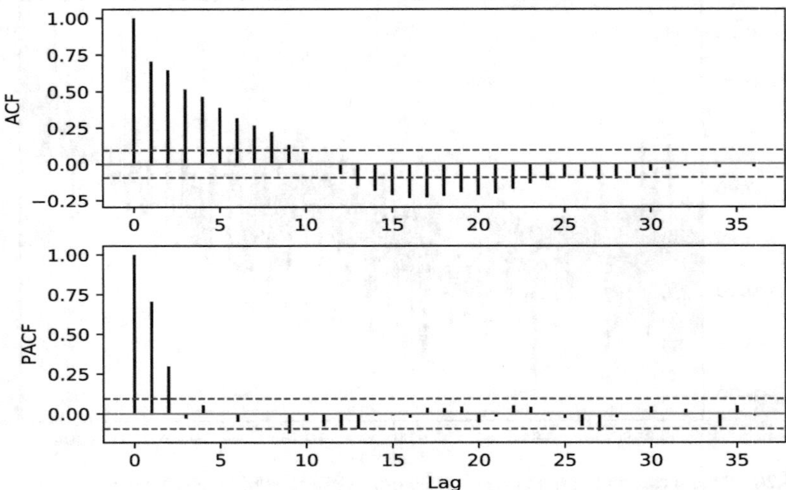


**Fig. 5.22** ACF and PACF of OLS model residuals of the GMTC subseries

pyTSA_GlobalTemperature

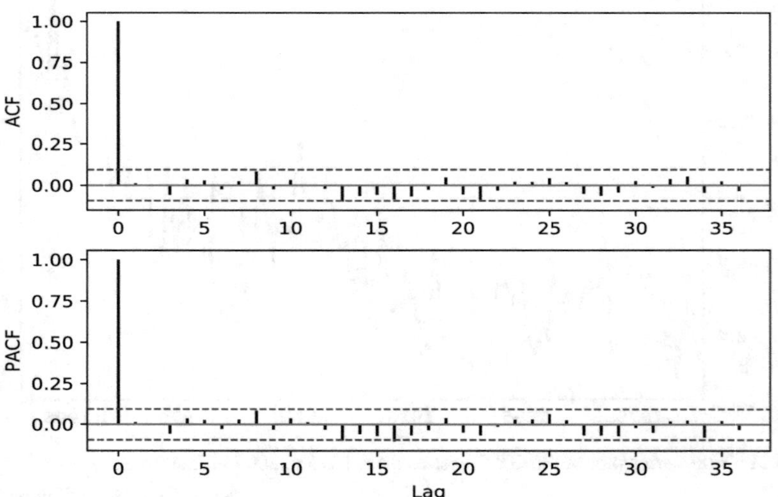


**Fig. 5.23** ACF and PACF of ar2 residuals of OLS residuals of the GMTC subseries

pyTSA_GlobalTemperature

the results shown in the second table in the following program. We see that all the coefficients are significant. Furthermore, the residuals (errors) pass the KPSS stationarity test, and its PACF shown in Fig. 5.22 cuts off after lag 2. This suggests that the errors follows an AR(2) model. Estimating the AR(2) model arrives at the third table in the following program. The ACF and PACF plots of the residuals of

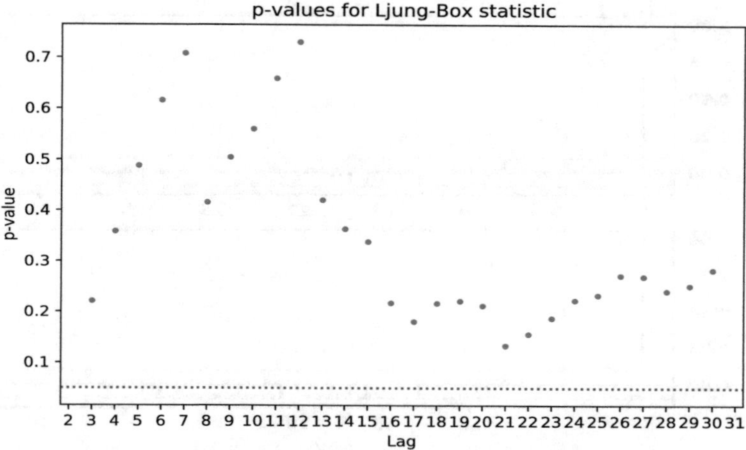


**Fig. 5.24** $p$-values for the L-B statistics of ar2 resid. of OLS resid. of the subseries
pyTSA_GlobalTemperature

the estimated AR(2) model are shown in Fig. 5.23, and furthermore, the estimated model passes the significance test of the coefficients and the Ljung-Box test of the residuals (checking the $p$-value plot shown in Fig. 5.24), which indicates that the estimated model is appropriate. Finally, we have the resulting harmonic seasonal regression model

$$\begin{cases} Y_t = 0.175 + 0.184\frac{m_t - 1987.958}{10.392} + 0.020\sin(2\pi m_t) + 0.016\sin(4\pi m_t) + Z_t \\ Z_t = 0.494Z_{t-1} + 0.307Z_{t-2} + \varepsilon_t \end{cases} \tag{5.4}$$

where $m_t = tim_t$ represents the time variable.

Using the preliminary modeling information above (viz., $K = 2$ and $J = 0$), we can simultaneously estimate the whole model (5.4) applying the function `ARMA` that has just been used above. The results display in the fourth table in the following program. Both the ACF and PACF plots shown in Fig. 5.25 and the $p$-value plot shown in Fig. 5.26 verify that the estimated model in the fourth table is adequate and almost the same as model (5.4). Note that we find that the interval estimates for `const, x1, x2, x3` in the second table and the corresponding interval estimates in the fourth table are a little different. This is because if the error series is autocorrelated, then the standard errors are likely to be underestimated or overestimated. The results in the fourth table are more accurate since the autocorrelation in the error series is considered.

```
>>> import numpy as np
>>> import pandas as pd
>>> import matplotlib.pyplot as plt
>>> import math
```

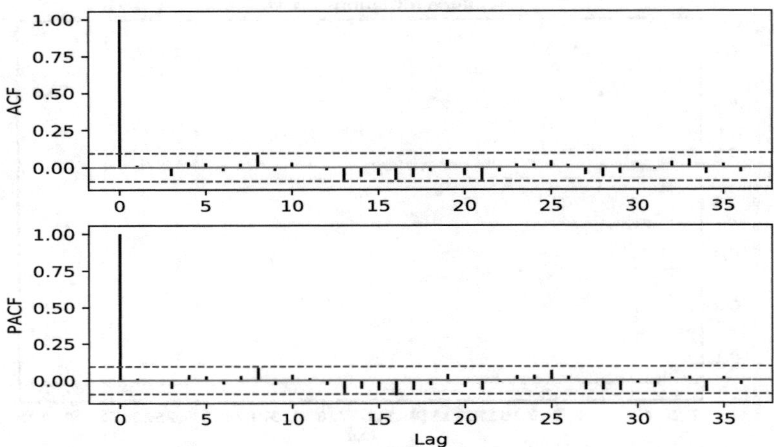


**Fig. 5.25** ACF and PACF of the residuals of the ARMAX model for the subseries

pyTSA_GlobalTemperature

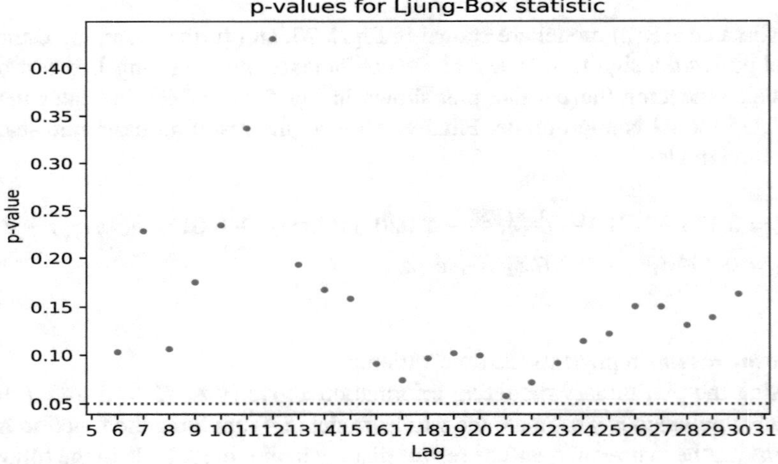


**Fig. 5.26** *p*-values for the L-B statistics of the resid. of the ARMAX model of the subseries

pyTSA_GlobalTemperature

```
>>> import statsmodels.api as sm
>>> from PythonTsa.plot_acf_pacf import acf_pacf_fig
>>> from PythonTsa.LjungBoxtest import plot_LB_pvalue
>>> from statsmodels.tsa.arima_model import ARMA
# for statsmodels of v 0.13.0 and later, see the last code below
>>> from PythonTsa.datadir import getdtapath
>>> dtapath=getdtapath()
>>> gtem=pd.read_csv(dtapath + 'GlobalTemperature.txt',
```

```
                        header=None,sep='\s+')
>>> gtemts=pd.concat([gtem.loc[0],gtem.loc[1]],ignore_index='true')
>>> for i in range(2, len(gtem)):
    gtemts=pd.concat([gtemts,gtem.loc[i]],ignore_index='true')
>>> dates=pd.date_range('1856-01',periods= len(gtemts),freq='M')
>>> gtemts.index=dates
>>> gtemts.plot(); plt.show()
>>> ygtemts=gtemts.resample(rule='12M', kind='period').mean()
# transform gtemts into annual mean data
>>> ydates=pd.date_range('1856',periods=len(ygtemts),freq='A')
>>> ygtemts.index=ydates
>>> ygtemts.plot(); plt.show()
>>> temp=gtemts['1970-01' : '2005-12']
>>> COS=np.zeros((len(temp), 6))
>>> SIN=np.zeros((len(temp), 6))
>>> SIN.shape
(432, 6) # a 432*6 zero matrix
>>> tim=np.zeros((len(temp)))
>>> for i in range(36):
        for j in range(12):
            tim[i*12+j]=1970.0+i+j/12.0
# tim is a time variable similar to the function 'time' in R.
>>> pi=math.pi
>>> for i in range(6):
         COS[:, i]=np.cos(2 * pi * (i+1) * tim)
         SIN[:, i]=np.sin(2 * pi * (i+1) * tim)
>>> TIME = (tim- np.mean(tim))/np.sqrt(np.var(tim))
# tim is standardized for reducing computation error.
>>> np.mean(tim)
1987.9583333333333
>>> np.sqrt(np.var(tim))
10.39227700248104
>>> Z = np.column_stack((TIME, COS[:,0], SIN[:,0], COS[:,1],
        SIN[:,1], COS[:,2], SIN[:,2], COS[:,3], SIN[:,3],
        COS[:,4], SIN[:,4], COS[:,5], SIN[:,5]))
>>> Z= sm.add_constant(Z)
# adding the constant term.
>>> OLSmod= sm.OLS(temp, Z).fit()
>>> print(OLSmod.summary())
                       OLS Regression Results
===============================================================
Dep. Variable:                  y   R-squared:            0.688
Model:                        OLS   Adj. R-squared:       0.678
Method:             Least Squares   F-statistic:          70.94
Date:            Sat, 26 Oct 2019   Prob (F-statistic): 4.93e-97
Time:                    11:09:58   Log-Likelihood:      287.48
No. Observations:             432   AIC:                 -547.0
Df Residuals:                 418   BIC:                 -490.0
Df Model:                      13
Covariance Type:        nonrobust
===============================================================
          coef    std err        t   P>|t|    [0.025    0.975]
---------------------------------------------------------------
```

```
const     0.1750        0.006  28.762  0.000        0.163     0.187
x1        0.1840        0.006  30.214  0.000        0.172     0.196
x2        0.0065        0.009   0.760  0.448       -0.010     0.023
x3        0.0204        0.009   2.373  0.018        0.004     0.037
x4        0.0119        0.009   1.378  0.169       -0.005     0.029
x5        0.0158        0.010   1.605  0.109       -0.004     0.035
x6        0.0055        0.009   0.644  0.520       -0.011     0.022
x7        0.0034        0.009   0.390  0.697       -0.014     0.020
x8        0.0047        0.009   0.545  0.586       -0.012     0.022
x9        0.0015        0.009   0.170  0.865       -0.015     0.018
x10       0.0028        0.009   0.320  0.749       -0.014     0.020
x11       0.0030        0.009   0.354  0.723       -0.014     0.020
x12      -0.0013        0.007  -0.180  0.857       -0.016     0.013
x13   1.197e+08     1.44e+09    0.083  0.934    -2.72e+09  2.96e+09
===============================================================
Omnibus:                 3.694   Durbin-Watson:           0.576
Prob(Omnibus):           0.158   Jarque-Bera (JB):        3.560
Skew:                    0.162   Prob(JB):                0.169
Kurtosis:                3.304   Cond. No.             2.38e+11
===============================================================
Warnings:
[1] Standard Errors assume that the covariance matrix of the
    errors is correctly specified.
[2] The smallest eigenvalue is 7.68e-21. This might indicate
    that there are strong multicollinearity problems or that
    the design matrix is singular.
>>> X = np.column_stack((TIME, SIN[:,0], SIN[:,1]))
>>> X = sm.add_constant(X)
>>> OLSmodel= sm.OLS(temp, X).fit()
>>> print(OLSmodel.summary())
                  OLS Regression Results
=================================================================
Dep. Variable:                 y   R-squared:                0.685
Model:                       OLS   Adj. R-squared:           0.683
Method:            Least Squares   F-statistic:              310.8
Date:           Sun, 20 Oct 2019   Prob (F-statistic):   4.61e-107
Time:                   16:58:16   Log-Likelihood:          285.59
No. Observations:            432   AIC:                     -563.2
Df Residuals:                428   BIC:                     -546.9
Df Model:                      3
Covariance Type:       nonrobust
=================================================================
            coef    std err        t     P>|t|    [0.025    0.975]
-----------------------------------------------------------------
const     0.1750      0.006   28.982    0.000     0.163     0.187
x1        0.1839      0.006   30.443    0.000     0.172     0.196
x2        0.0204      0.009    2.390    0.017     0.004     0.037
x3        0.0162      0.009    1.891    0.059    -0.001     0.033
=================================================================
Omnibus:                5.476   Durbin-Watson:              0.580
Prob(Omnibus):          0.065   Jarque-Bera (JB):           5.412
Skew:                   0.218   Prob(JB):                  0.0668
Kurtosis:               3.332   Cond. No.                    1.42
```

```
==============================================================================
Warnings:
[1] Standard Errors assume that the covariance matrix of
    the errors is correctly specified.
>>> OLSresid=OLSmodel.resid
>>> acf_pacf_fig(OLSresid, both=True, lag=36)
>>> plt.show()
>>> sm.tsa.kpss(OLSresid, regression='c', nlags='auto')
InterpolationWarning: p-value is greater than the indicated p-value
(0.034017819549384036, 0.1, 11,
{'10%': 0.347, '5%': 0.463, '2.5%': 0.574, '1%': 0.739})
>>> ar2=ARMA(OLSresid, order=(2,0)).fit(trend='nc',disp=False)
>>> print(ar2.summary())
                              ARMA Model Results
==============================================================================
Dep. Variable:                      y   No. Observations:                  432
Model:                     ARMA(2, 0)   Log Likelihood                 456.779
Method:                       css-mle   S.D. of innovations              0.084
Date:                Sun, 20 Oct 2019   AIC                           -907.557
Time:                        20:14:48   BIC                           -895.352
Sample:                    01-31-1970   HQIC                          -902.739
                         - 12-31-2005
==============================================================================
                 coef    std err          z      P>|z|      [0.025      0.975]
------------------------------------------------------------------------------
ar.L1.y        0.4938      0.046     10.747      0.000       0.404       0.584
ar.L2.y        0.3072      0.046      6.646      0.000       0.217       0.398
                                    Roots
=============================================================================
                  Real          Imaginary           Modulus         Frequency
-----------------------------------------------------------------------------
AR.1            1.1715          +0.0000j            1.1715            0.0000
AR.2           -2.7791          +0.0000j            2.7791            0.5000
-----------------------------------------------------------------------------
>>> ar2Resid=ar2.resid
>>> acf_pacf_fig(ar2Resid, both=True, lag=36)
>>> plt.show()
>>> plot_LB_pvalue(ar2Resid, noestimatedcoef=2, nolags=30)
>>> plt.show()
>>> Y = np.column_stack((TIME, SIN[:,0], SIN[:,1]))
>>> regar=ARMA(temp,order=(2,0),exog=Y).fit(trend='c',disp=False)
# exog should not include a constant, specifying this in the "fit".
>>> print(regar.summary())
                              ARMA Model Results
==============================================================================
Dep. Variable:                      y   No. Observations:                  432
Model:                     ARMA(2, 0)   Log Likelihood                 456.927
Method:                       css-mle   S.D. of innovations              0.084
Date:                Sun, 20 Oct 2019   AIC                           -899.853
Time:                        21:54:13   BIC                           -871.374
Sample:                    01-31-1970   HQIC                          -888.610
                         - 12-31-2005
==============================================================================
```

```
              coef    std err          z      P>|z|      [0.025   0.975]
---------------------------------------------------------------------------
const       0.1765      0.020      8.790      0.000       0.137    0.216
x1          0.1765      0.020      8.985      0.000       0.138    0.215
x2          0.0179      0.009      2.067      0.039       0.001    0.035
x3          0.0149      0.005      2.969      0.003       0.005    0.025
ar.L1.y     0.4936      0.046     10.747      0.000       0.404    0.584
ar.L2.y     0.3078      0.046      6.660      0.000       0.217    0.398
                              Roots
===========================================================================
              Real        Imaginary         Modulus          Frequency
---------------------------------------------------------------------------
AR.1        1.1709        +0.0000j          1.1709             0.0000
AR.2       -2.7745        +0.0000j          2.7745             0.5000
---------------------------------------------------------------------------
>>> regarResid=regar.resid
>>> acf_pacf_fig(regarResid, both=True, lag=36)
>>> plt.show()
>>> plot_LB_pvalue(regarResid, noestimatedcoef=5, nolags=30)
>>> plt.show()
# the following is for statsmodels of v 0.13.0 and later
>>> from statsmodels.tsa.arima.model import ARIMA
>>> ar2=ARIMA(OLSresid, order=(2,0,0),trend='n').fit()
>>> regar=ARIMA(temp,order=(2,0,0),exog=Y,trend='c').fit()
```

In this chapter, we elaborated on the SARIMA models. This type of models is a powerful tool in the analysis of time series since they are capable of modelling a wide range of time series. On the other hand, naturally, there are time series such as financial time series, which the SARIMA models tend not to be well fitted to. Besides, we also considered the REGARMA models as well as harmonic seasonal regression models. The latter is especially useful for time series with a long seasonal period.

## Problems

**5.1** Build an appropriate SARIMA model for the time series data in Example 5.1, and then diagnose the model.

**5.2** Simulate a sample of size 1000 for the SMA$(1)(1)_4$ process

$$X_t = (1 - 0.5B)(1 + 0.4B^4)\varepsilon_t,\ \varepsilon_t \sim \text{iidN}(0, 2^2).$$

Then use the simulated sample to build an appropriate model and comment on it.

**5.3** Simulate a sample of size 1000 for the SARMA$(0, 1)(1, 0)_6$ process

$$X_t - 0.6X_{t-6} = \varepsilon_t - 0.5\varepsilon_{t-1},\ \varepsilon_t \sim \text{iidN}(0, 2^2).$$

Then use the simulated sample to build an appropriate model and comment on it.

**5.4** Simulate a sample of size 1000 for the SARMA$(1, 0)(0, 1)_6$ process

$$X_t - 0.6X_{t-1} = \varepsilon_t - 0.5\varepsilon_{t-6},\ \varepsilon_t \sim \text{iidN}(0, 2^2).$$

Then use the simulated sample to build an appropriate model and comment on it.

**5.5** Simulate a sample of size 1000 for the SARMA$(1, 1)(1, 1)_4$ process

$$(1 - 0.2B)(1 + 0.3B^4)X_t = (1 - 0.5B)(1 + 0.4B^4)\varepsilon_t,\ \varepsilon_t \sim \text{iidN}(0, 2^2).$$

Then use the simulated sample to build an appropriate model and comment on it.

**5.6** Respectively use models SARIMA$(2, 1, 0)(0, 1, 0)_4$ and SARIMA$(1, 1, 1)$ $(0, 1, 0)_4$ to fit the logged series (viz., `lx`) in Example 5.4, and then compare them with the resulting model in Example 5.4.

**5.7** Use models SARIMA$(0, 1, 2)(0, 1, 3)_4$ to fit the logged series (viz., `lx`) in Example 5.4, and then compare your model with the resulting model in Example 5.4. (Hint: pay attention to the significance of $\Theta_1$ and $\Theta_2$.)

**5.8** Validate that models SARIMA$(3, 0, 0)(2, 1, 0)_{12}$, SARIMA$(3, 0, 0)(0, 1, 1)_{12}$, and
SARIMA$(0, 0, 5)(0, 1, 1)_{12}$ in Example 5.5 are all inappropriate.

**5.9** Take a log on the original series (viz., `h02`) in Example 5.5, and then build an appropriate model for the logged series `lh02=log(h02)`. (Hint: check if `Dlh02` $= (1 - B^{12})\ \ln($`h02`$)$ is stationary.)

**5.10** Extract a subseries from Jan. 1970 to Dec. 2007 from the Southern Hemisphere temperature volatility data series in Example 3.1. And then analyze the subseries using the methods in Example 5.6.

# Chapter 6
# Financial Time Series and Related Models

Financial time series analysis has been one of the hottest research topics in the recent decades. In this chapter, we illustrate the stylized facts of financial time series by real financial data. To characterize these facts, new models different from the Box-Jenkins ones are needed. And for this reason, ARCH models were firstly proposed by R. F. Engle in 1982 and have been extended by a great number of scholars since then. We also demonstrate how to use Python and its libraries to implement ARCH and some extensions modeling.

## 6.1 Stylized Facts of Financial Time Series

As we have seen in Examples 4.4 and 5.4, there are a lot of time series that possess the ARCH effect, that is, although the (modeling residual) series is white noise, its squared series may be autocorrelated. What is more, in practice, a large number of financial time series are found having this property so that the ARCH effect has become one of the stylized facts from financial time series.

### 6.1.1 *Examples of Return Series*

Since the 1970s, the financial industry has been very prosperous with advancement of computer and Internet technology. Trade of financial products (including various derivatives) generates a huge amount of data which form financial time series. For finance, the return on a financial product is most interesting, and so our attention focuses on the return series. As we have seen in Sect. 1.3, if $\{P_t\}$ is the closing price



C. Huang, A. Petukhina, *Applied Time Series Analysis and Forecasting with Python*, Statistics and Computing,
https://doi.org/10.1007/978-3-031-13584-2_6

at time $t$ for a certain financial product, then the return on this product is

$$X_t = (P_t - P_{t-1})/P_{t-1} \approx \log(P_t) - \log(P_{t-1}). \tag{6.1}$$

It is return series $\{X_t\}$ that have been much independently studied. And important stylized features which are common across many instruments, markets, and time periods have been summarized. Note that if you purchase the financial product, then it becomes your asset, and its returns become your asset returns. Now let us look at the following examples.

*Example 6.1 (Monthly Returns of Procter and Gamble Stock from 1961 to 2016)* We have preliminarily discussed this return (PG) series in Example 2.4 and concluded that it is a white noise series. Now we give more evidence to support this conclusion. The $p$-value for the KPSS test of the return series is greater than 0.1, which further illustrates that the series is stationary. And the $p$-value plot for the Ljung-Box test of the return series is shown in Fig. 6.1 that suggests evidently that the series is a white noise. With respect to the squared return series, however, from Figs. 6.2 and 6.3, we clearly see that it is autocorrelated, that is, the return series has an ARCH effect. Furthermore, we can plot the histogram and kernel density estimator (KDE) of the return series data shown in Fig. 6.4. We can also plot a normal density with mean `smean=1.0342` and standard variance `scal=5.5383` where `smean` and `scal` are computed using the return data and show it in Fig. 6.4 as well. Observing Fig. 6.4 carefully, we find that the distribution density of the return data has a fat (heavy) tail on the left compared with the normal density and is asymmetric (the symmetric line is the mean $x = 1.0342$), both of which are stylized facts on financial time series. To reproduce the results and figures in this example, run the following Python code.

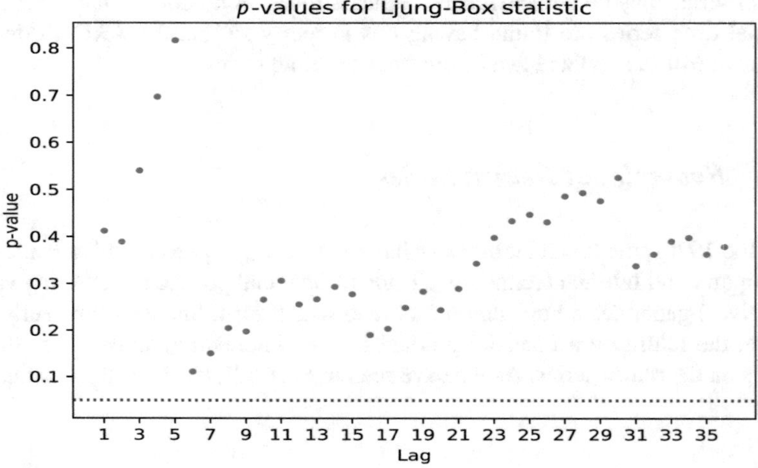


**Fig. 6.1** $p$-Value plot for the Ljung-Box test of the PQ return series in Example 6.1

pyTSA_ReturnsPG2

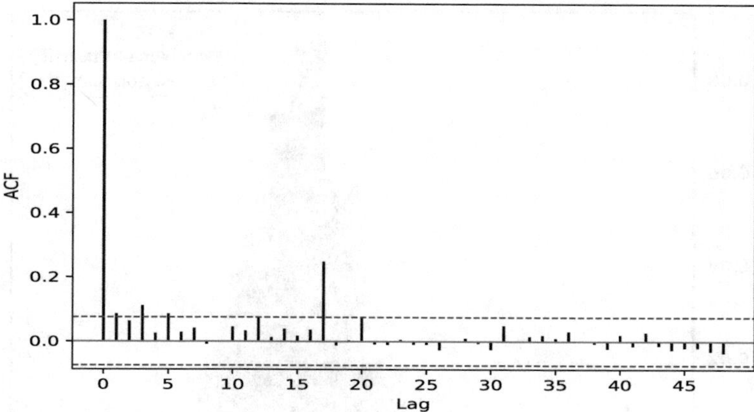


**Fig. 6.2** ACF plot of the squared PQ return series in Example 6.1

pyTSA_ReturnsPG2

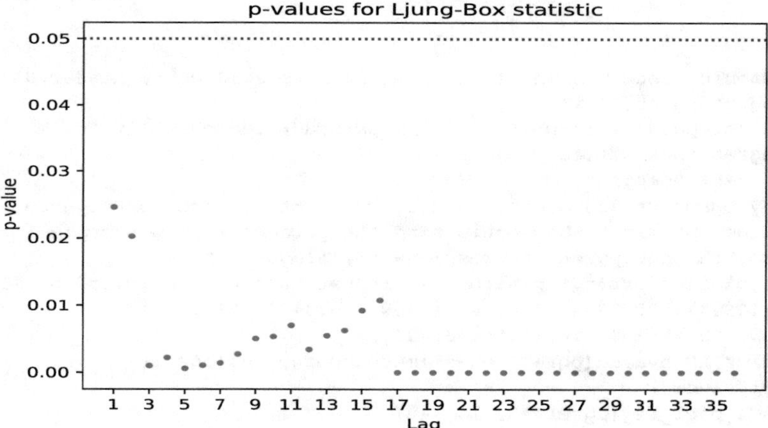


**Fig. 6.3** $p$-Value plot for the Ljung-Box test of the squared PQ return series

pyTSA_ReturnsPG2

```
>>> import pandas as pd
>>> import numpy as np
>>> import matplotlib.pyplot as plt
>>> from PythonTsa.plot_acf_pacf import acf_pacf_fig
>>> import statsmodels.api as sm
>>> from PythonTsa.LjungBoxtest import plot_LB_pvalue
>>> from scipy.stats import norm
>>> from PythonTsa.datadir import getdtapath
>>> dtapath=getdtapath()
>>> pgret=pd.read_csv(dtapath + 'monthly returns of Procter n
```

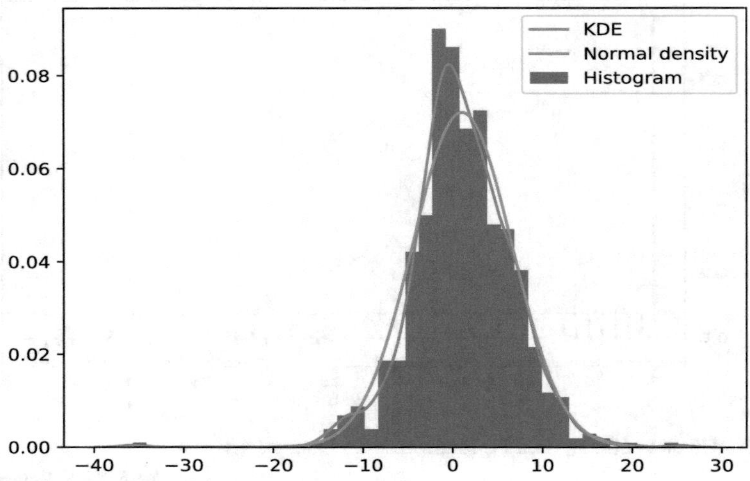


**Fig. 6.4** Histogram, KDE, and normal density for the PQ return series

pyTSA_ReturnsPG2

```
    Gamble stock n 3 market indexes 1961 to 2016.csv', header=0)
>>> pgret=pgret['RET']
>>> dates=pd.date_range('1961-01', periods= len(pgret),freq='M')
>>> pgret.index=dates
>>> pgret=100*pgret
# many pgret values are too small, which may affect convergence
# of the optimizer when estimating the parameters, so rescale it.
>>> sm.tsa.kpss(pgret, regression='c', nlags='auto')
InterpolationWarning: p-value is greater than the indicated p-value
(0.11109642750067264, 0.1, 4, {'10%': 0.347, '5%':
 0.463, '2.5%': 0.574, '1%': 0.739})
>>> plot_LB_pvalue(pgret, noestimatedcoef=0, nolags=36)
>>> plt.show()
>>> acf_pacf_fig(pgret**2, lag=48)
>>> plt.show()
>>> plot_LB_pvalue(pgret**2, noestimatedcoef=0, nolags=36)
>>> plt.show()
>>> fig = plt.figure()
>>> ax = fig.add_subplot(111)
>>> hfig=ax.hist(pgret, bins=40, density=True, label='Histogram')
>>> kde = sm.nonparametric.KDEUnivariate(pgret)
>>> kde.fit()
>>> ax.plot(kde.support, kde.density, label='KDE')
>>> smean=np.mean(pgret)
>>> scal=np.std(pgret, ddof=1)
>>> normden=norm.pdf(kde.support, loc=smean, scale=scal)
>>> ax.plot(kde.support, normden, label='Normal density')
>>> ax.legend(loc='best')
>>> plt.show()
```

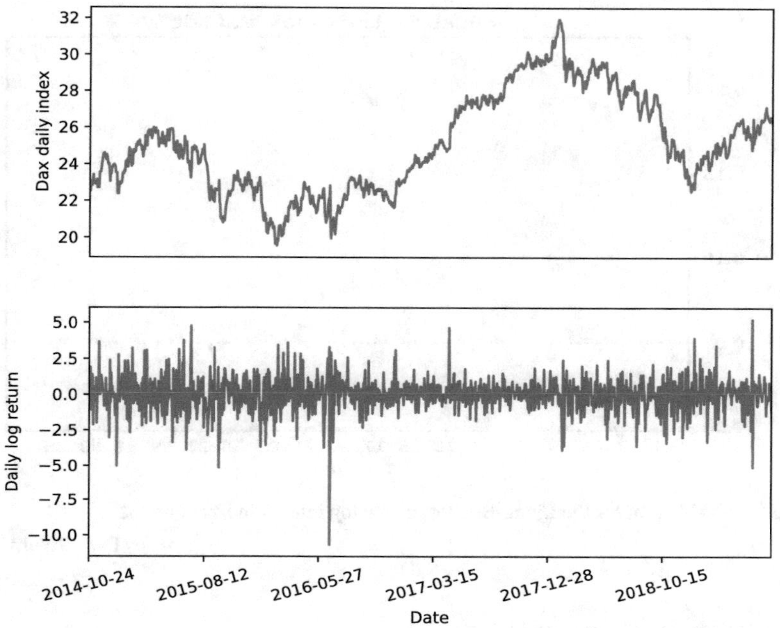


**Fig. 6.5** Time plots of the DAX daily index and daily log returns in Example 6.2

pyTSA_ReturnsDAX

*Example 6.2 (Germany DAX Daily Index from Oct 23, 2014 to July 7, 2019)* The dataset "DAX" in the folder Ptsadata is the Germany DAX daily index from October 23, 2014 to July 7, 2019. It is daily adjusted closing prices from Yahoo! Finance (shown in the top panel of Fig. 6.5). Now we transform it to the (log) returns `logret` by Eq. (6.1) in order to study the return series (shown in the bottom panel of Fig. 6.5). Using the same way as in Example 6.1, it is easy to prove that the return series is stationary and has an ARCH effect and fat tails. However it is not a white noise, which can been seen from Fig. 6.6. Therefore, we should first build an adequate ARMA model for it (see Sect. 6.2.2). More analysis of the return series is left as an exercise for the reader.

```
>>> import pandas as pd
>>> import numpy as np
>>> import matplotlib.pyplot as plt
>>> from PythonTsa.LjungBoxtest import plot_LB_pvalue
>>> from PythonTsa.datadir import getdtapath
>>> dtapath=getdtapath()
>>> dax=pd.read_csv(dtapath + 'DAX.csv', header=0)
>>> dax.rename(columns={'Adj Close' : 'index'}, inplace=True)
>>> dax['logreturns']=np.log(dax['index']/dax['index'].shift(1))
>>> dax = dax.dropna()
>>> logret = dax['logreturns']
```

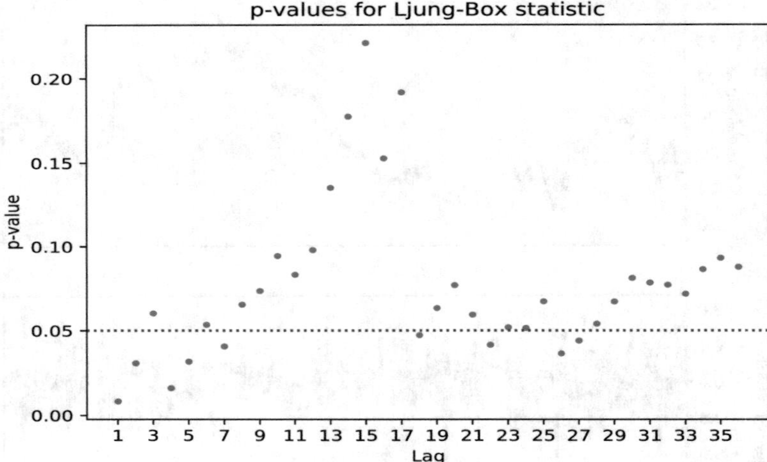


**Fig. 6.6** *p*-Value plot for the Ljung-Box test of the log returns in Example 6.2

pyTSA_ReturnsDAX

```
>>> logret.index = dax['Date']
>>> fig = plt.figure()
>>> dax['index'].plot(ax= fig.add_subplot(211))
>>> plt.ylabel('Dax daily index')
>>> plt.xticks([])
>>> logret.plot(ax= fig.add_subplot(212))
>>> plt.ylabel('Daily log return')
>>> plt.xticks(rotation=15)
>>> plt.show()
>>> plot_LB_pvalue(logret, noestimatedcoef=0, nolags=36)
>>> plt.show()
```

### 6.1.2 Stylized Facts of Financial Time Series

Now we briefly list and describe several important stylized facts (features) of financial return series:

- *Fat (heavy) tails*: The distribution density function of returns often has fatter (heavier) tails than the tails of the corresponding normal distribution density. We have observed this phenomenon (feature) from Fig. 6.4.
- *ARCH effect*: Although the return series can often be seen as a white noise, its squared (and absolute) series may usually be autocorrelated, and these autocorrelations are hardly negative. See, for example, Figures 4.19, 6.2, and 6.3.

- *Volatility clustering*: Large changes in returns tend to cluster in time, and small changes tend to be followed by small changes, just as, for example, Figures 1.17, 2.8, 4.1, and 6.5 show.
- *Asymmetry*: As we have seen in Fig. 6.4, the distribution of asset returns is slightly negatively skewed. One possible explanation could be that traders react more strongly to unfavorable information than favorable information.

Note that (1) a financial time series does not necessarily have all the stylized facts; (2) nonfinancial time series may possess one or more of the stylized facts (see, e.g., Problem 6.5; and (3) in finance, there are various definitions of volatility. We only consider the usually defined volatility, that is, the (conditional) volatility is the (conditional) standard deviation (variance) of time series.

## 6.2 GARCH Models

We summarized several stylized facts (features) of financial time series in the previous section. The question is what models can characterize those features. The pity is that the Box-Jenkins models are not able to grasp any of them. Fortunately, the ARCH models proposed firstly by Engle (1982) can capture some of the stylized facts and so had been soon extended by a large number of scholars. ARCH is the abbreviation for autoregressive conditional heteroscedasticity, that is, the ARCH models are about the time-varying variance (volatility) of time series.

### *6.2.1 ARCH Models*

Specifically, we give the definition of the ARCH model as follows.

**Definition 6.1** An ARCH($p$) model with order $p \geq 1$ is of the form

$$\begin{cases} X_t = \sigma_t \varepsilon_t \\ \sigma_t^2 = \omega + \alpha_1 X_{t-1}^2 + \alpha_2 X_{t-2}^2 + \cdots + \alpha_p X_{t-p}^2 \end{cases} \tag{6.2}$$

where $\omega \geq 0, \alpha_i \geq 0$, and $\alpha_p > 0$ are constants, $\varepsilon_t \sim \text{iid}(0, 1)$, and $\varepsilon_t$ is independent of $\{X_k; k \leq t-1\}$. A stochastic process $X_t$ is called an ARCH($p$) process if it satisfies Eq. (6.2).

By Definition 6.1, $\sigma_t^2$ (and $\sigma_t$) is independent of $\varepsilon_t$. Besides, usually it is further assumed that $\varepsilon_t \sim \text{N}(0, 1)$. Sometimes, however, we need to further suppose that $\varepsilon_t$ follows a standardized (skew) Student's T distribution or a generalized error distribution in order to capture more features of a financial time series.

Let $\mathscr{F}_s$ denote the information set generated by $\{X_k; k \leq s\}$, namely, the sigma field $\sigma(X_k; k \leq s)$. It is easy to see that $\mathscr{F}_s$ is independent of $\varepsilon_t$ for any $s <$

$t$. According to Definition 6.1 and the properties of the conditional mathematical expectation, we have that

$$\mathsf{E}(X_t|\mathscr{F}_{t-1}) = \mathsf{E}(\sigma_t\varepsilon_t|\mathscr{F}_{t-1}) = \sigma_t\mathsf{E}(\varepsilon_t|\mathscr{F}_{t-1}) = \sigma_t\mathsf{E}(\varepsilon_t) = 0 \tag{6.3}$$

and

$$\text{Var}(X_t^2|\mathscr{F}_{t-1}) = \mathsf{E}(X_t^2|\mathscr{F}_{t-1}) = \mathsf{E}(\sigma_t^2\varepsilon_t^2|\mathscr{F}_{t-1}) = \sigma_t^2\mathsf{E}(\varepsilon_t^2|\mathscr{F}_{t-1}) = \sigma_t^2\mathsf{E}(\varepsilon_t^2) = \sigma_t^2.$$

This implies that $\sigma_t^2$ is the conditional variance of $X_t$ and it evolves according to the previous values of $\{X_k^2; t-p \le k \le t-1\}$ like an AR($p$) model. And so Model (6.2) is named an ARCH($p$) model.

As an example of ARCH($p$) models, let us consider the ARCH(1) model

$$\begin{cases} X_t = \sigma_t\varepsilon_t \\ \sigma_t^2 = \omega + \alpha_1 X_{t-1}^2. \end{cases} \tag{6.4}$$

Explicitly, the unconditional mean

$$\mathsf{E}(X_t) = \mathsf{E}(\sigma_t\varepsilon_t) = \mathsf{E}(\sigma_t)\mathsf{E}(\varepsilon_t) = 0.$$

Additionally, the ARCH(1) model can be expressed as

$$X_t^2 = \sigma_t^2 + X_t^2 - \sigma_t^2 = \omega + \alpha_1 X_{t-1}^2 + \sigma_t^2\varepsilon_t^2 - \sigma_t^2 = \omega + \alpha_1 X_{t-1}^2 + \eta_t,$$

that is,

$$X_t^2 = \omega + \alpha_1 X_{t-1}^2 + \eta_t \tag{6.5}$$

where $\eta_t = \sigma_t^2(\varepsilon_t^2 - 1)$. It can been shown that $\eta_t$ is a new white noise, which is left as an exercise for reader. Hence, if $0 < \alpha_1 < 1$, Eq. (6.5) is a stationary AR(1) model for the series $X_t^2$. Thus, the unconditional variance

$$\text{Var}(X_t) = \mathsf{E}(X_t^2) = \mathsf{E}(\omega + \alpha_1 X_{t-1}^2 + \eta_t) = \omega + \alpha_1\mathsf{E}(X_t^2),$$

that is,

$$\text{Var}(X_t) = \mathsf{E}(X_t^2) = \frac{\omega}{1-\alpha_1}.$$

Moreover, for $h > 0$, in light of the properties of the conditional mathematical expectation and by (6.3), we have that

$$\mathsf{E}(X_{t+h}X_t) = \mathsf{E}(\mathsf{E}(X_{t+h}X_t|\mathscr{F}_{t+h-1})) = \mathsf{E}(X_t\mathsf{E}(X_{t+h}|\mathscr{F}_{t+h-1})) = 0.$$

In conclusion, if $0 < \alpha_1 < 1$, we have that:

- Any ARCH(1) process $\{X_t\}$ defined by Eqs. (6.4) follows a white noise $\mathrm{WN}(0, \omega/(1-\alpha_1))$.
- Since $X_t^2$ is an AR(1) process defined by (6.5), $\mathrm{Corr}(X_t^2, X_{t+h}^2) = \alpha_1^{|h|} > 0$, which reveals the ARCH effect.
- It is clear that $\mathsf{E}(\eta_t|\mathscr{F}_s) = 0$ for any $t > s$, and with Eq. (6.5), for any $k > 1$:

$$\begin{aligned}\mathrm{Var}(X_{t+k}|\mathscr{F}_t) &= \mathsf{E}(X_{t+k}^2|\mathscr{F}_t)\\ &= \mathsf{E}(\omega + \alpha_1 X_{t+k-1}^2 + \eta_{t+k}|\mathscr{F}_t)\\ &= \omega + \alpha_1 \mathrm{Var}(X_{t+k-1}|\mathscr{F}_t),\end{aligned}$$

which reflects the volatility clustering, that is, large (small) volatility is followed by large (small) one.

In addition, we are able to prove that $X_t$ defined by Eq. (6.4) has heavier tails than the corresponding normal distribution, which is left as an exercise for the reader. At last, note that these properties of the ARCH(1) model can be generalized to ARCH($p$) models.

### 6.2.2 *GARCH Models*

A natural idea for extending the ARCH model is to include a moving average part in the model, which is similar to the extension from an AR model to an ARMA model. In fact, as early as in the year of 1986, the definition of generalized ARCH models had been given by T. Bollerslev.

**Definition 6.2** A GARCH($p, q$) model with order ($p \geq 1, q \geq 0$) is of the form

$$\begin{cases} X_t = \sigma_t \varepsilon_t \\ \sigma_t^2 = \omega + \sum_{i=1}^{p} \alpha_i X_{t-i}^2 + \sum_{j=1}^{q} \beta_j \sigma_{t-j}^2 \end{cases} \tag{6.6}$$

where $\omega \geq 0$, $\alpha_i \geq 0$, $\beta_j \geq 0$, $\alpha_p > 0$, and $\beta_q > 0$ are constants, $\varepsilon_t \sim \mathrm{iid}(0, 1)$, and $\varepsilon_t$ is independent of $\{X_k; k \leq t-1\}$. A stochastic process $X_t$ is called a GARCH($p, q$) process if it satisfies Eq. (6.6).

In practice, it has been found that for some time series, the ARCH($p$) model defined by (6.2) will provide an adequate fit only if the order $p$ is large. By allowing past volatilities to affect the present volatility in (6.6), a more parsimonious model may result. That is why we need GARCH models. Besides, note the condition that the order $p \geq 1$. The GARCH model in Definition 6.2 has the properties as follows.

**Proposition 6.1** *If $X_t$ is a GARCH$(p,q)$ process defined in (6.6) and $\sum_{i=1}^{p}\alpha_i + \sum_{j=1}^{q}\beta_j < 1$, then the following propositions hold.*

- *$X_t^2$ follows the ARMA$(m,q)$ model*

$$X_t^2 = \omega + \sum_{i=1}^{m}(\alpha_i + \beta_i)X_{t-i}^2 + \eta_t - \sum_{j=1}^{q}\beta_j\eta_{t-j}$$

  *where $\alpha_i = 0$ for $i > p$, $\beta_j = 0$ for $j > q$, $m = \max(p,q)$, and $\eta_t = \sigma_t^2(\varepsilon_t^2 - 1)$.*
- *$X_t$ is a white noise with*

$$\mathsf{E}(X_t) = 0,\ \mathsf{E}(X_{t+h}X_t) = 0 \text{ for any } h \neq 0,\ \mathrm{Var}(X_t) = \frac{\omega}{1 - \sum_{i=1}^{m}(\alpha_i + \beta_i)}.$$

- *$\sigma_t^2$ is the conditional variance of $X_t$, that is, we have*

$$\mathsf{E}(X_t|\mathscr{F}_{t-1}) = 0,\ \sigma_t^2 = \mathrm{Var}(X_t^2|\mathscr{F}_{t-1}).$$

- *Model (6.6) reflects the fat tails and volatility clustering.*

Although an asset return series can usually be seen as a white noise, there exists such a return series so that it may be autocorrelated (see Example 6.2). What is more, a given original time series is not necessarily a return series, and at the same time, its values may be negative. If a time series is autocorrelated, we must first build an adequate model (e.g., an ARMA model) for the series in order to remove any autocorrelation in it. Then check whether the residual series has an ARCH effect, and if yes then we further model the residuals. In other words, if a time series $Y_t$ is autocorrelated and has ARCH effect, then a GARCH model that can capture the features of $Y_t$ should be of the form

$$\begin{cases} Y_t = f(Z_t, Y_{t-1}, X_{t-1}, Y_{t-2}, X_{t-2}, \cdots) + X_t & (6.7) \\ X_t = \sigma_t\varepsilon_t & \\ \sigma_t^2 = \omega + \sum_{i=1}^{p}\alpha_i X_{t-i}^2 + \sum_{j=1}^{q}\beta_j\sigma_{t-j}^2 & (6.8) \end{cases}$$

where Eq. (6.7) is referred to as the *mean equation* (*model*) and Eq. (6.8) is known as the *volatility* (*variance*) *equation* (*model*), and $Z_t$ is a representative of exogenous regressors. If $Y_t$ is a return series, then typically $Y_t = r + X_t$ where $r$ is a constant that means the expected returns is fixed.

Below we simulate a GARCH(1,1) process so as to inspect what a GARCH model is able to capture. We use the Python package `arch` (version 4.14) by

Sheppard (2020) to implement the simulation.[1] Of course, we should firstly install it. And later it is also used to estimate GARCH models.

*Example 6.3 (Simulating a GARCH(1,1) Process)* Given a GARCH(1, 1) process as follows

$$\begin{cases} X_t = \sigma_t \varepsilon_t, \varepsilon_t \sim \text{iidN}(0, 1) \\ \sigma_t^2 = 0.1 + 0.2X_{t-1}^2 + 0.6\sigma_{t-1}^2, \end{cases}$$

first, simulate a sample of size 1000 from this GARCH process; then check which features of financial time series can be characterized by this model. Below is the code for the simulation and analysis of the simulated sample.

```
>>> import numpy as np
>>> import statsmodels.api as sm
>>> import matplotlib.pyplot as plt
>>> from arch.univariate import arch_model
>>> sim_mod = arch_model(None, p=1, q=1)
>>> params=[0, 0.1, 0.2, 0.6]
# in general, params=[r, omega, alpha, beta]
>>> sim_data = sim_mod.simulate(params, nobs=1000)
>>> simdata=sim_data['data']
>>> simdata.plot(); plt.show()
>>> from PythonTsa.plot_acf_pacf import acf_pacf_fig
>>> acf_pacf_fig(simdata, both=False, lag=36)
>>> plt.show()
>>> acf_pacf_fig(simdata**2, both=False, lag=36)
>>> plt.show()
>>> from PythonTsa.LjungBoxtest import plot_LB_pvalue
>>> plot_LB_pvalue(simdata, noestimatedcoef=0, nolags=36)
>>> plt.show()
>>> fig = plt.figure()
>>> ax = fig.add_subplot(111)
>>> hfig=ax.hist(simdata, bins=40, density=True, label='Histogram')
>>> kde = sm.nonparametric.KDEUnivariate(simdata)
>>> kde.fit()
>>> ax.plot(kde.support, kde.density, label='KDE')
>>> smean=np.mean(simdata)
>>> scal=np.std(simdata, ddof=1)
>>> from scipy.stats import norm
>>> normden=norm.pdf(kde.support, loc=smean, scale=scal)
>>> ax.plot(kde.support, normden, label='Normal density')
>>> ax.legend(loc='best'); plt.show()
```

The time plot of the simulated series shown in Fig. 6.7 displays the volatility clustering. Its ACF plot shown in Fig. 6.8 and the $p$-values shown in Fig. 6.9 for the Ljung-Box test of the simulated series clearly suggest that the series is a white noise.

[1] The following all Python code in this chapter is also validated with Python of V. 3.9.7, statsmodels of V. 0.13.1, and arch of V. 5.1.0.

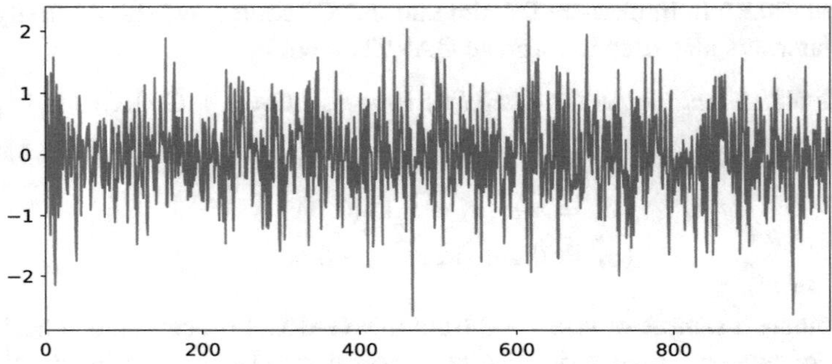


**Fig. 6.7** Time series plot of simulated data from the GARCH(1,1) in Example 6.3

pyTSA_GARCH

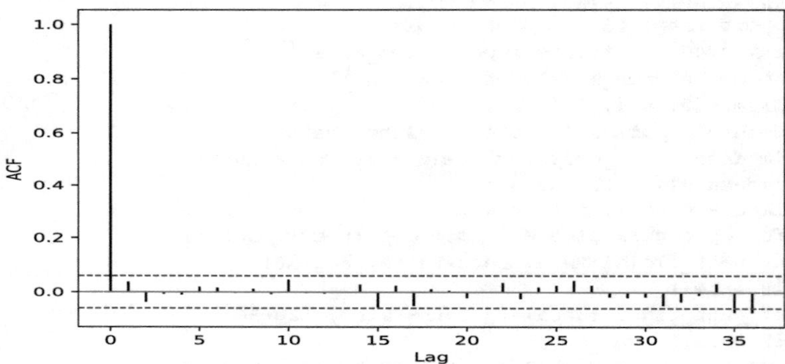


**Fig. 6.8** ACF plot of simulated data from the GARCH(1,1) in Example 6.3

pyTSA_GARCH

And the ACF plot of the squared simulated data shown in Fig. 6.10 demonstrates that the squared series is autocorrelated or the series has an ARCH effect. Furthermore, from Fig. 6.11, we see that the left tail of the distribution of the simulated series is heavier than the normal distribution. In summary, GARCH models are able to grasp several stylized features of financial time series.

### *6.2.3 Estimation and Testing*

First of all, it is always a good idea to transform a given (asset) price series into the log return series. There are four basic steps toward estimation of a GARCH model for a given time series data $Y_t$:

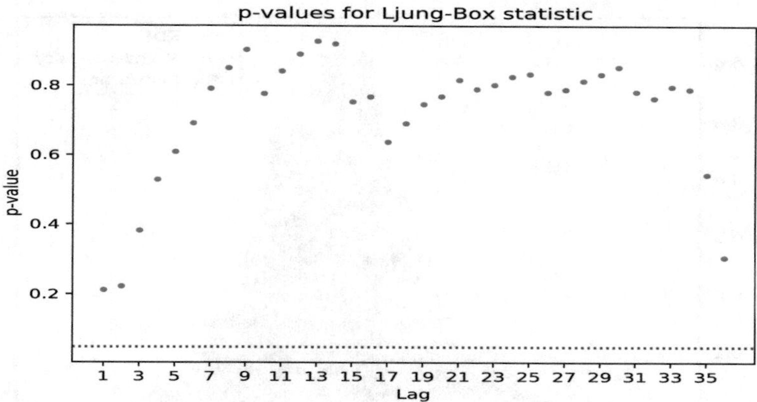


**Fig. 6.9** $p$-Value plot for the Ljung-Box test of simulated data from the GARCH(1,1) in Example 6.3

pyTSA_GARCH

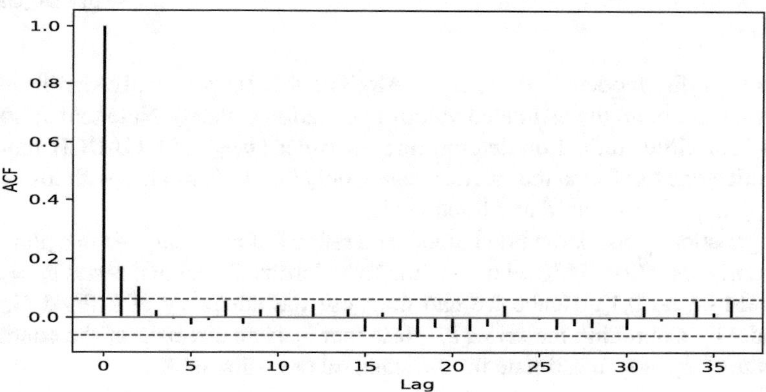


**Fig. 6.10** ACF plot of squared simulated data from the GARCH(1,1) in Example 6.3

pyTSA_GARCH

1. If $Y_t$ is autocorrelated, then in order to remove any autocorrelation in it, we can adopt the modeling procedures introduced in Chaps. 4 and 5 to build an appropriate mean model for the given data. At this point, the residual series $X_t$ of the mean equation should be a white noise.
2. Test whether $X_t$ has an ARCH effect. If yes, continue to build an adequate volatility model for it; otherwise, stop. There are several approaches to testing for the ARCH effect, for example, utilizing the squared residual series $X_t^2$: according to Proposition 6.1, $X_t^2$ should follow an ARMA model if $X_t$ is a GARCH process. Hence we can examine $X_t^2$ by its (P)ACF and the Ljung-Box testing. If $X_t^2$ is significantly autocorrelated, then $X_t$ should have an ARCH effect.

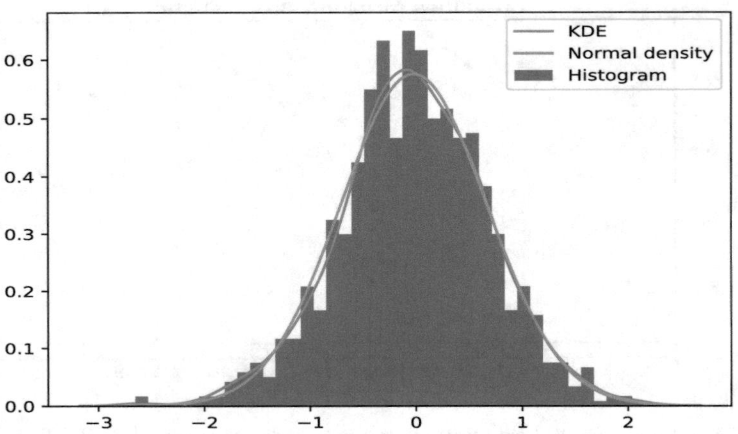


**Fig. 6.11** Histogram, KDE, and normal density for simulated data from the GARCH(1,1) in Example 6.3

pyTSA_GARCH

3. Use volatility models to fit $X_t$ if the ARCH effect is statistically significant, and so we can obtain the estimated volatility equation (model). Note that although it has been little studied on determining the order $(p, q)$ of a GARCH model, in practice one has found that in most cases, only GARCH models with lower order $(p \leq 2, q \leq 2)$ are used and fitted well.
4. Diagnostically check the fitted model and refine it if necessary. At this point, for a properly specified GARCH model, the "standardized" residual series $\tilde{\varepsilon}_t = X_t/\tilde{\sigma}_t$ should be iid(0,1). Hence we can diagnose the adequacy of a fitted GARCH model by examining the series $\tilde{\varepsilon}_t$. Note that $\tilde{\sigma}_t$ is an estimate of the conditional volatility $\sigma_t$, not an estimate for the standard deviation of $X_t$.

### 6.2.4 Examples

In this subsection, we consider how to implement estimation of GARCH models with Python and its extension packages. Let us look at the following examples.

*Example 6.4 (Example 6.1 Continued)* From Example 6.1, we see that the return series `pgret` is a white noise and has the ARCH effect. Now we are about to build an appropriate GARCH model for the return series. Due to the stationarity of the return series, the mean model is of the form $r_t = \mu + X_t$. Since the constant mean estimate `mu=1.0918` is statistically significant (see the table `Constant Mean - GARCH Model Results` below), the estimated GARCH model consists of the mean and volatility equations as follows:

$$\begin{cases} R_t = 1.0918 + X_t \\ X_t = \sigma_t \varepsilon_t,\ \varepsilon_t \sim \text{iidN}(0,\ 1) \\ \sigma_t^2 = 2.8447 + 0.1252X_{t-1}^2 + 0.7843\sigma_{t-1}^2. \end{cases} \tag{6.9}$$

Observe the $p$-value plot for the Ljung-Box test of the model residuals shown in Fig. 6.12 and the $p$-value plot for the Ljung-Box test of the residual squares shown in Fig. 6.13. It turns out that the residual is uncorrelated and neither is the residual squares. In other words, the ARCH effect is totally reflected, and the estimated GARCH model (6.9) is adequate. The only flaw is that the residuals are not normally distributed, which can be seen from the QQ plot in Fig. 6.14. It is because the return series is fat tailed as Fig. 6.4 displays.

It is well known that the Student's T distribution has heavier tails than the normal one. Therefore, now we replace the normal distribution with a standard T distribution in Model (6.9) and re-estimate this model. We obtain a new estimated model as follows:

$$\begin{cases} R_t = 1.0157 + X_t \\ X_t = \sigma_t \varepsilon_t,\ \varepsilon_t \sim \text{iidT}(0,\ 1) \\ \sigma_t^2 = 2.7251 + 0.0952X_{t-1}^2 + 0.8139\sigma_{t-1}^2 \end{cases} \tag{6.10}$$

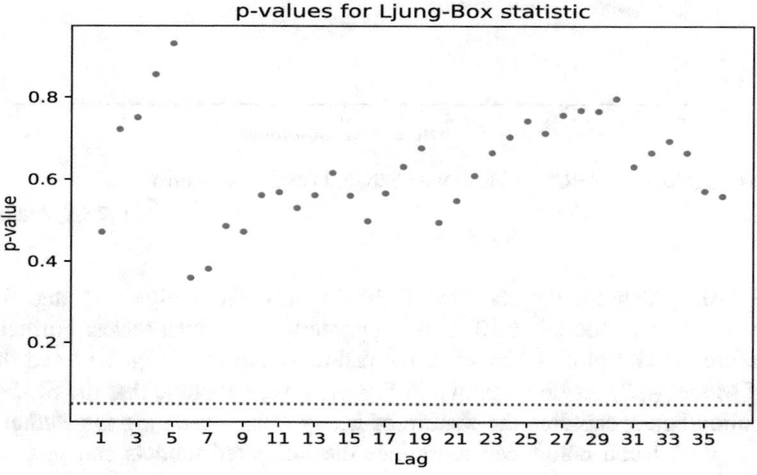


**Fig. 6.12** $p$-Values for the Ljung-Box test of GARCH residuals of the PG returns

pyTSA_ReturnsPG2

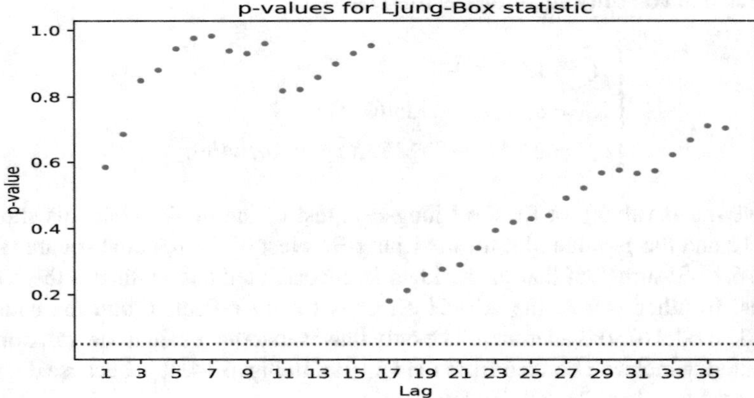


**Fig. 6.13** *p*-Values for the L-B test of squared GARCH residuals of the PG returns

pyTSA_ReturnsPG2

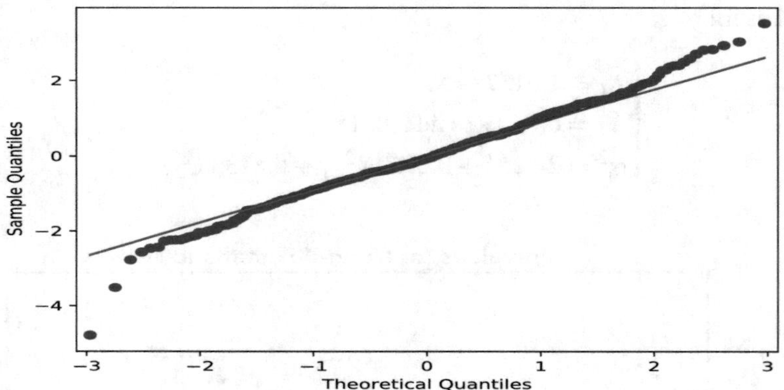


**Fig. 6.14** QQ plot of GARCH residuals with iidN(0, 1) of the PG returns

pyTSA_ReturnsPG2

where T(0, 1) denotes the standard T distribution. From Figs. 6.15 and 6.16, it is easy to see that Model (6.10) is appropriate to the return series. Furthermore, comparing the QQ plot of Model (6.10) residuals shown in Fig. 6.17 and the QQ plot of Model (6.9) residuals shown in Fig. 6.14, we conclude that the Student's T distribution better captures the feature of heavy tails. Executing the Python code in Example 6.1 and below can reproduce the estimated models and plots in this example.

```
>>> from arch import arch_model
>>> archmod = arch_model(pgret).fit()
Optimization terminated successfully.    (Exit mode 0)
            Current function value: 2078.9678389780875
```

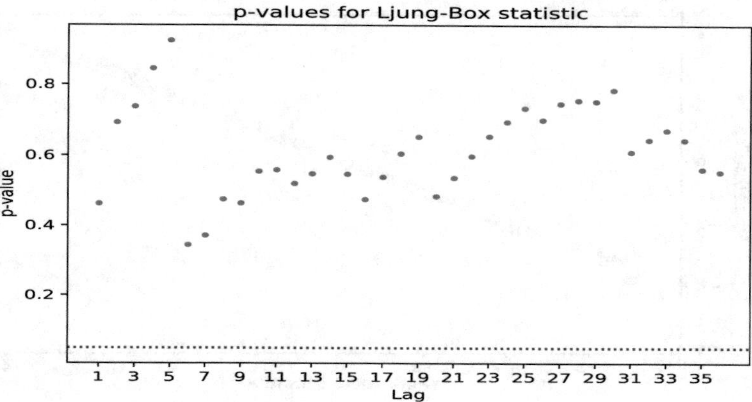


**Fig. 6.15** *p*-Values for the L-B test of GARCH residuals with iidT(0, 1) of the PG returns

pyTSA_ReturnsPG2

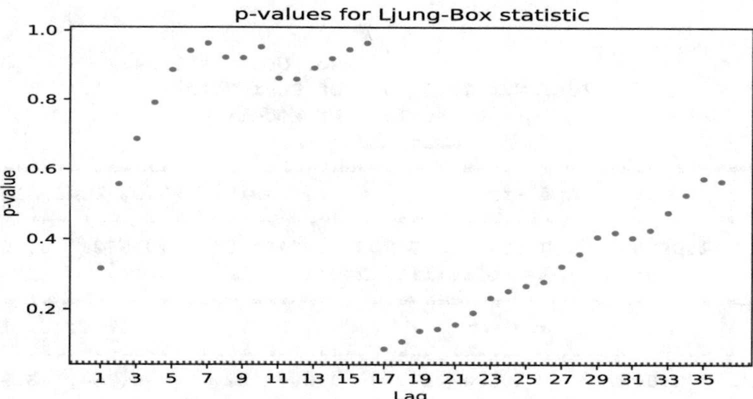


**Fig. 6.16** *p*-Values for the L-B test of squared GARCH residuals with iidT(0, 1) of the PG returns in Example 6.4

pyTSA_ReturnsPG2

```
            Iterations: 7
            Function evaluations: 49
            Gradient evaluations: 7
>>> print(archmod.summary())
                Constant Mean - GARCH Model Results
===================================================================
Dep. Variable:                    RET   R-squared:              -0.000
Mean Model:             Constant Mean   Adj. R-squared:         -0.000
Vol Model:                      GARCH   Log-Likelihood:       -2078.97
Distribution:                  Normal   AIC:                   4165.94
Method:            Maximum Likelihood   BIC:                   4183.98
```

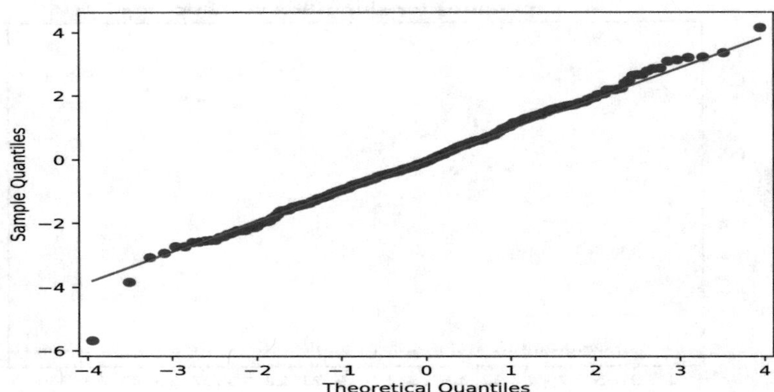


**Fig. 6.17** QQ plot of GARCH residuals with iidT(0, 1) of the PG returns in Example 6.4
pyTSA_ReturnsPG2

```
                                            No. Observations:              672
Date:                 Tue, Mar 10 2020      Df Residuals:                  668
Time:                         07:50:12      Df Model:                        4
                                Mean Model
================================================================
            coef    std err          t      P>|t|   95.0% Conf. Int.
----------------------------------------------------------------
mu        1.0918      0.209      5.225  1.746e-07   [0.682,  1.501]
                              Volatility Model
=================================================================
              coef    std err         t       P>|t|     95.0% Conf. Int.
-----------------------------------------------------------------
omega       2.8447      1.586     1.794   7.287e-02    [ -0.264,   5.953]
alpha[1]    0.1252  6.013e-02     2.083   3.726e-02 [7.390e-03,   0.243]
beta[1]     0.7843  9.104e-02     8.615   6.982e-18    [  0.606,   0.963]
=================================================================
Covariance estimator: robust
>>> archresid = archmod.std_resid
>>> plot_LB_pvalue(archresid, noestimatedcoef=0, nolags=36)
>>> plt.show()
>>> plot_LB_pvalue(archresid**2, noestimatedcoef=0, nolags=36)
>>> plt.show()
>>> from statsmodels.graphics.api import qqplot
>>> qqplot(archresid, line='q', fit=True)
>>> plt.show()
>>> garchT = arch_model(pgret, p=1,  q=1, dist='StudentsT')
>>> res = garchT.fit()
Optimization terminated successfully.    (Exit mode 0)
            Current function value: 2072.196841697201
            Iterations: 10
            Function evaluations: 78
```

```
            Gradient evaluations: 10
>>> print(res.summary())
             Constant Mean - GARCH Model Results
==================================================================
Dep. Variable:                       RET  R-squared:          -0.000
Mean Model:                Constant Mean  Adj. R-squared:     -0.000
Vol Model:                         GARCH  Log-Likelihood:   -2072.20
Distribution: Standardized Student's t  AIC:               4154.39
Method:               Maximum Likelihood  BIC:               4176.94
                                          No. Observations:      672
Date:                   Wed, Mar 11 2020  Df Residuals:          667
Time:                           06:40:42  Df Model:                5
                           Mean Model
==================================================================
              coef    std err          t      P>|t|  95.0% Conf. Int.
------------------------------------------------------------------
mu          1.0157      0.199      5.103  3.336e-07  [ 0.626,  1.406]
                        Volatility Model
==================================================================
               coef    std err         t      P>|t|     95.0% Conf. Int.
------------------------------------------------------------------
omega        2.7251      1.071     2.545  1.094e-02    [  0.626,  4.824]
alpha[1]     0.0952  3.619e-02     2.630  8.530e-03 [2.426e-02,  0.166]
beta[1]      0.8139  5.421e-02    15.014  5.913e-51    [  0.708,  0.920]
                          Distribution
==================================================================
             coef    std err              t       P>|t|   95.0% Conf. Int.
------------------------------------------------------------------
nu         9.6180      3.339          2.880   3.976e-03  [  3.073, 16.163]
==================================================================
Covariance estimator: robust
# nu denotes degree of freedom for the T distribution
>>> archresidT = res.std_resid
>>> plot_LB_pvalue(archresidT, noestimatedcoef=0, nolags=36)
>>> plt.show()
>>> plot_LB_pvalue(archresidT**2, noestimatedcoef=0, nolags=36)
>>> plt.show()
>>> from scipy import stats
>>> qqplot(archresidT, stats.t, distargs=(9.62,),line='q',fit=True)
# arguments "stats.t" and "distargs" means T distribution and its
# degree of freedom.
>>> plt.show()
```

*Example 6.5 (Standard & Poor's 500 Index Daily Returns)* The time series dataset "SP500dailyreturns" in the folder Ptsadata is the SP500 index daily returns from Jan 5, 1999 to Dec 12, 2018. Its time plot is shown in Fig. 6.18. It is easy to obtain that the return series is stationary, seeing its ACF and PACF plots in Fig. 6.19 and the KPSS test result. However it is not a white noise since it fails the Ljung-Box test, seeing the $p$-value plot shown in Fig. 6.20. Therefore, we should firstly build an adequate ARMA model for it. At this point, we find that it is tough to determine order $(p, q)$ of the ARMA model fitted to the return series. After dozens of trials,

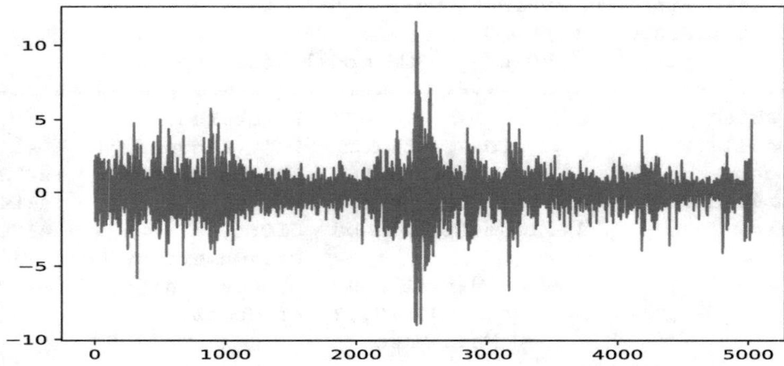

Fig. 6.18 Time series plot of the SP500 returns in Example 6.5

pyTSA_ReturnsSP500

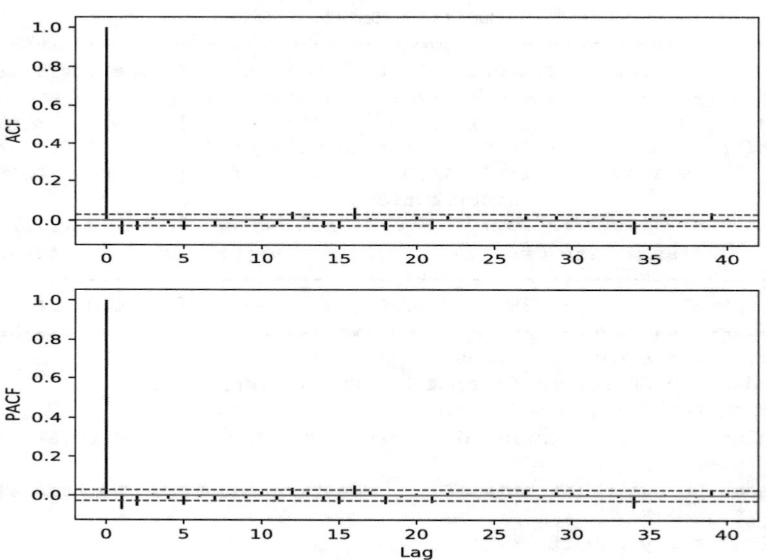


Fig. 6.19 ACF and PACF of the SP500 returns in Example 6.5

pyTSA_ReturnsSP500

we select order $(p, q) = (15, 2)$ and obtain the estimated ARMA(15, 2) model by the function `SARIMAX()`. The estimated ARMA(15, 2) model is shown in the first `SARIMAX Results` table below. Observing it carefully, we find that estimates of several parameters are not statistically significant. Thus we refine this model by fixing the parameters `intercept = ar.L8 = ar.L10 = ar.L14 = 0`. We finally have the resulting ARMA model (viz., mean model) shown in the second `SARIMAX Results` table below. Comparing the AIC, BIC, and HQIC for the second estimated model and the AIC, BIC, and HQIC for the first estimated

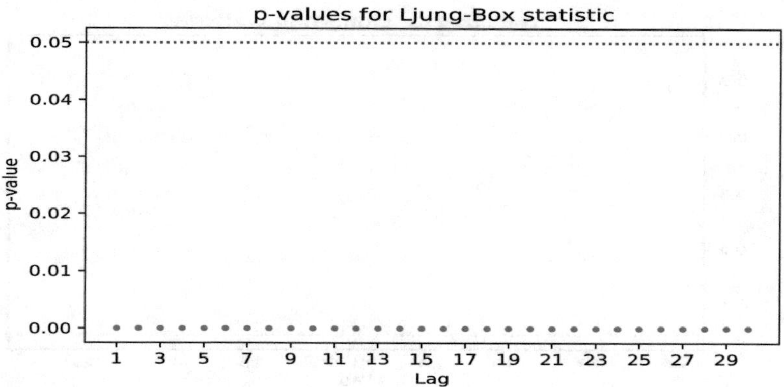


**Fig. 6.20** $p$-Value plot for the Ljung-Box test of the SP500 returns in Example 6.5

pyTSA_ReturnsSP500

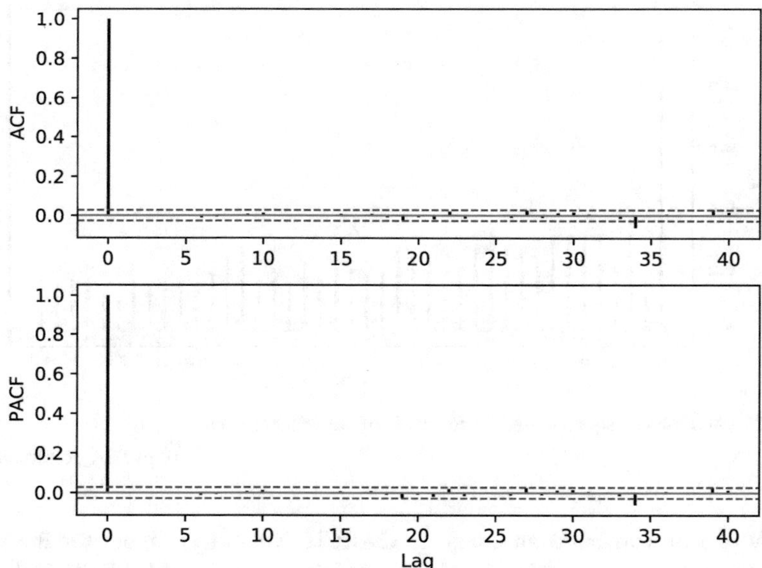


**Fig. 6.21** ACF and PACF of ARMA residuals of the SP500 returns in Example 6.5

pyTSA_ReturnsSP500

model, respectively, it follows that the second model is better than the first one. Furthermore, the resulting ARMA model is appropriate since the residuals of it pass the Ljung-Box test, seeing Figs. 6.21 and 6.22. At the same time, we find that the residual series has an ARCH effect from Fig. 6.23.

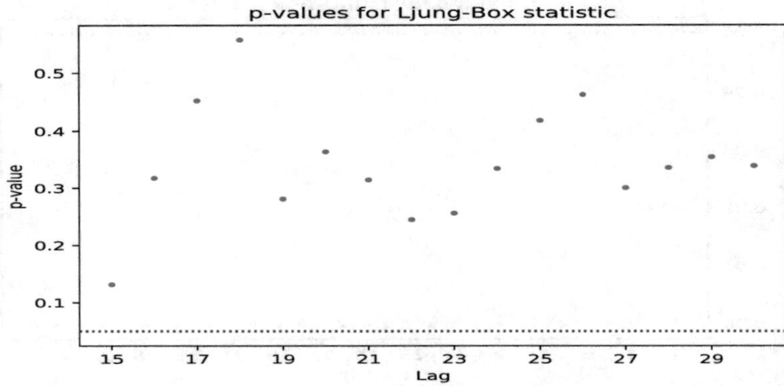


**Fig. 6.22** $p$-Values for the L-B test of ARMA residuals of the SP500 returns

pyTSA_ReturnsSP500

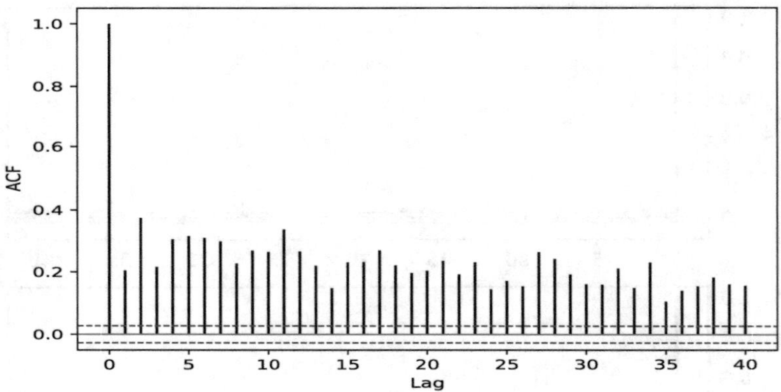


**Fig. 6.23** ACF plot of squared ARMA residuals of the SP500 returns

pyTSA_ReturnsSP500

Now we turn to build an adequate GARCH (volatility) model for the return series using the mean model residuals. We find that the order of GARCH (volatility) models should be elected as $(p, q) = (2, 2)$. The resulting volatility model is shown in the `Zero Mean - GARCH Model Results` table below. We see that the volatility model is appropriate from Figs. 6.24, 6.25, 6.26, and 6.27, that is, the GARCH model residuals are uncorrelated and have no ARCH effect anymore. A blemish in an otherwise perfect thing is that the QQ plot shown in Fig. 6.28 indicates that the left tail of the empirical distribution of the residuals is heavier than one of a normal distributions. Executing the following Python commands can reproduce all the results and plots in this example.

```
>>> import pandas as pd
>>> import numpy as np
```

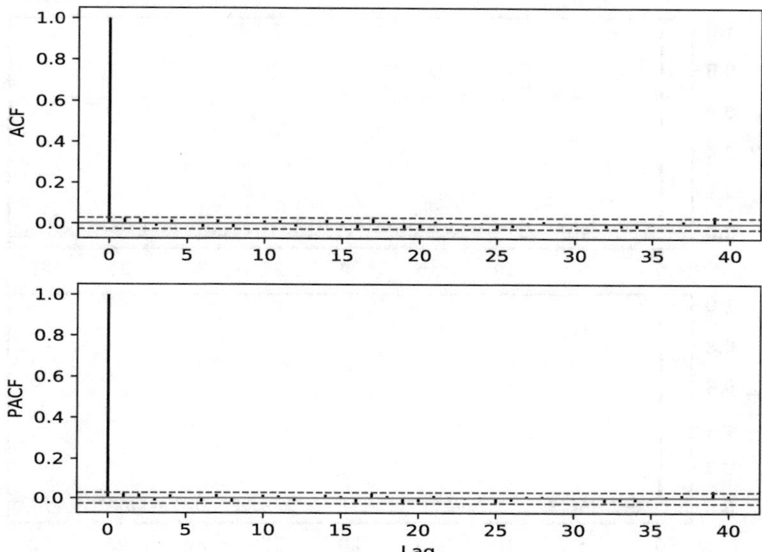


**Fig. 6.24** ACF and PACF of GARCH residuals of the SP500 returns in Example 6.5

pyTSA_ReturnsSP500

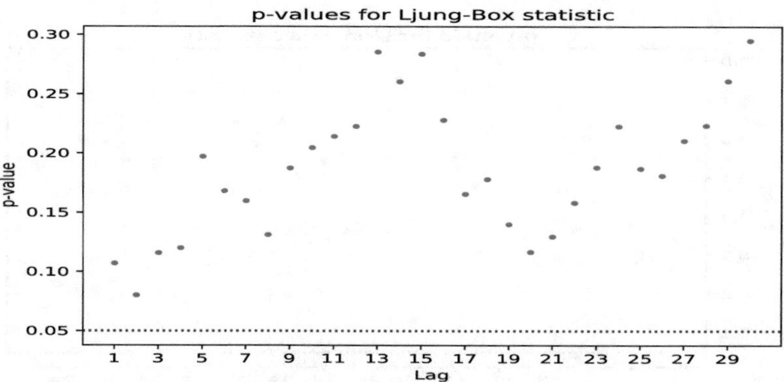


**Fig. 6.25** *p*-Values for the L-B test of GARCH residuals of the SP500 returns

pyTSA_ReturnsSP500

```
>>> import matplotlib.pyplot as plt
>>> import statsmodels.api as sm
>>> from PythonTsa.plot_acf_pacf import acf_pacf_fig
>>> from PythonTsa.LjungBoxtest import plot_LB_pvalue
>>> from PythonTsa.datadir import getdtapath
>>> dtapath=getdtapath()
>>> ret=pd.read_csv(dtapath + 'SP500dailyreturns.csv',
                    header=None)
```

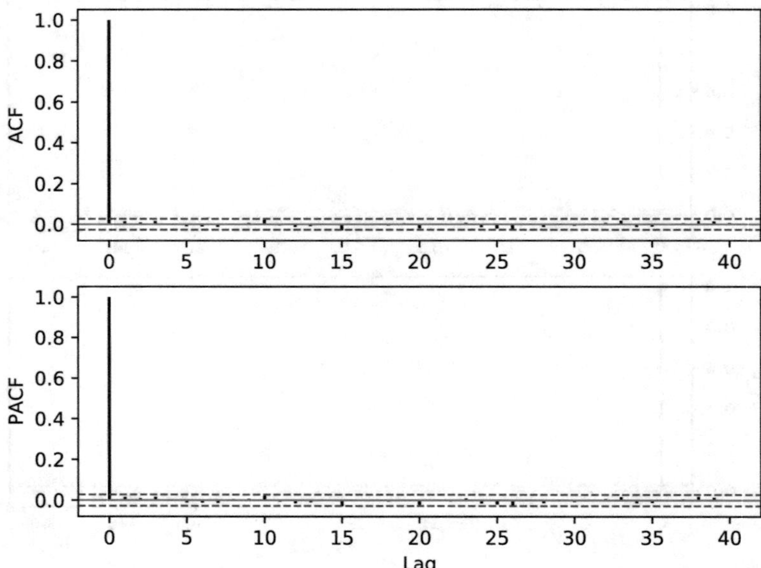


**Fig. 6.26** ACF and PACF of squared GARCH residuals of the SP500 returns

pyTSA_ReturnsSP500

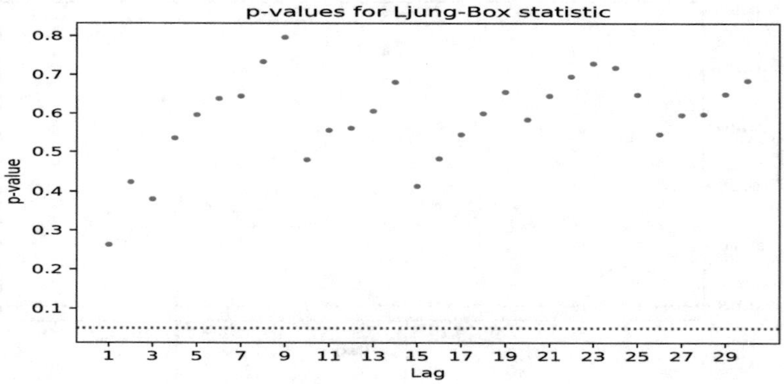


**Fig. 6.27** *p*-Values for the L-B test of squared GARCH resid. of the SP500 returns

pyTSA_ReturnsSP500

```
>>> ret.columns=['returns']
>>> ret=pd.Series(ret['returns'])
>>> ret.plot(); plt.show()
>>> acf_pacf_fig(ret, both=True, lag=40)
>>> plt.show()
>>> sm.tsa.kpss(ret, regression='c', nlags='auto')
InterpolationWarning: p-value is greater than the indicated p-value
```

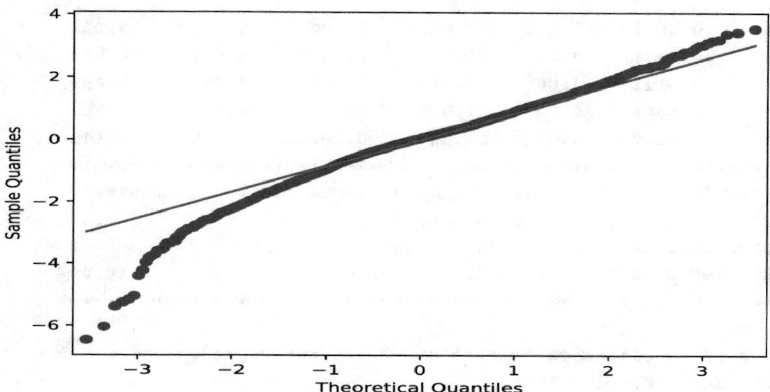


**Fig. 6.28** QQ plot of GARCH residuals with iidN(0, 1) of the SP500 returns
pyTSA_ReturnsSP500

```
(0.1501534339907281, 0.1, 22,
{'10%': 0.347, '5%': 0.463, '2.5%': 0.574, '1%': 0.739})
>>> plot_LB_pvalue(ret, noestimatedcoef=0, nolags=30)
>>> plt.show()
>>> arma152=sm.tsa.SARIMAX(ret, order=(15,0,2), trend='c').fit(disp=False)
>>> print(arma152.summary())
                              SARIMAX Results
==================================================================
Dep. Variable:              returns   No. Observations:          5030
Model:              SARIMAX(15, 0, 2)   Log Likelihood        -8009.710
Date:              Thu, 13 Feb 2020   AIC                   16057.420
Time:                      22:14:48   BIC                   16181.360
Sample:                           0   HQIC                  16100.846
                             - 5030
Covariance Type:                opg
==================================================================
                coef    std err          z      P>|z|      [0.025      0.975]
------------------------------------------------------------------
intercept     0.0575      0.041      1.414      0.157     -0.022       0.137
ar.L1        -0.5845      0.061     -9.594      0.000     -0.704      -0.465
ar.L2        -0.7568      0.059    -12.890      0.000     -0.872      -0.642
ar.L3        -0.0750      0.013     -5.799      0.000     -0.100      -0.050
ar.L4        -0.0599      0.011     -5.241      0.000     -0.082      -0.038
ar.L5        -0.0590      0.011     -5.519      0.000     -0.080      -0.038
ar.L6        -0.0494      0.012     -4.227      0.000     -0.072      -0.026
ar.L7        -0.0688      0.012     -5.510      0.000     -0.093      -0.044
ar.L8        -0.0169      0.012     -1.436      0.151     -0.040       0.006
ar.L9        -0.0339      0.012     -2.749      0.006     -0.058      -0.010
ar.L10        0.0125      0.012      1.040      0.298     -0.011       0.036
ar.L11       -0.0191      0.012     -1.556      0.120     -0.043       0.005
ar.L12        0.0419      0.012      3.590      0.000      0.019       0.065
ar.L13        0.0201      0.013      1.548      0.122     -0.005       0.046
```

```
ar.L14     0.0001    0.013     0.010     0.992    -0.025     0.026
ar.L15    -0.0538    0.011    -5.114     0.000    -0.074    -0.033
ma.L1      0.5111    0.060     8.483     0.000     0.393     0.629
ma.L2      0.6688    0.057    11.704     0.000     0.557     0.781
sigma2     1.4146    0.015    95.337     0.000     1.385     1.444
===================================================================
Ljung-Box (Q):                 40.56   Jarque-Bera (JB):   10575.16
Prob(Q):                        0.45   Prob(JB):               0.00
Heteroskedasticity (H):         0.46   Skew:                  -0.22
Prob(H) (two-sided):            0.00   Kurtosis:              10.09
===================================================================
Warnings:
[1] Covariance matrix calculated using the outer product of
    gradients (complex-step).
>>> arma122=sm.tsa.SARIMAX(ret,
    order=([1,1,1,1,1,1,1,0,1,0,1,1,1,0,1],0,2), trend='n').fit(disp=False)
>>> print(arma122.summary())
                     SARIMAX Results
=======================================================================
Dep. Variable:                          returns No. Observations:   5030
Model:
SARIMAX([1,2,3,4,5,6,7,9,11,12,13,15],0,2) Log Likelihood -8011.944
Date:                          Thu, 13 Feb 2020 AIC             16053.888
Time:                                  22:26:16 BIC             16151.736
Sample:                                       0 HQIC            16088.172
                                         - 5030
Covariance Type:                            opg
====================================================================
             coef  std err          z    P>|z|      [0.025     0.975]
--------------------------------------------------------------------
ar.L1    -0.5797    0.050    -11.513    0.000     -0.678     -0.481
ar.L2    -0.7495    0.053    -14.270    0.000     -0.852     -0.647
ar.L3    -0.0728    0.012     -6.058    0.000     -0.096     -0.049
ar.L4    -0.0580    0.011     -5.305    0.000     -0.079     -0.037
ar.L5    -0.0574    0.010     -5.589    0.000     -0.078     -0.037
ar.L6    -0.0378    0.010     -3.846    0.000     -0.057     -0.019
ar.L7    -0.0542    0.010     -5.590    0.000     -0.073     -0.035
ar.L9    -0.0319    0.008     -3.752    0.000     -0.049     -0.015
ar.L11   -0.0302    0.010     -3.086    0.002     -0.049     -0.011
ar.L12    0.0335    0.010      3.454    0.001      0.014      0.052
ar.L13    0.0209    0.010      2.151    0.031      0.002      0.040
ar.L15   -0.0537    0.008     -6.504    0.000     -0.070     -0.038
ma.L1     0.5073    0.050     10.231    0.000      0.410      0.604
ma.L2     0.6631    0.052     12.770    0.000      0.561      0.765
sigma2    1.4159    0.014     97.777    0.000      1.387      1.444
====================================================================
Ljung-Box (Q):                  43.38   Jarque-Bera (JB):   10524.97
Prob(Q):                         0.33   Prob(JB):               0.00
Heteroskedasticity (H):          0.46   Skew:                  -0.22
Prob(H) (two-sided):             0.00   Kurtosis:              10.07
====================================================================
Warnings:
```

```
[1] Covariance matrix calculated using the outer product of
    gradients (complex-step).
>>> xresid = arma122.resid
>>> acf_pacf_fig(xresid, both=True, lag=40)
>>> plt.show()
>>> plot_LB_pvalue(xresid, noestimatedcoef=14, nolags=30)
>>> plt.show()
>>> acf_pacf_fig(xresid**2, lag=40)
>>> plt.show()
>>> from arch import arch_model
>>> garch = arch_model(xresid, p=2, q=2, mean='Zero')
>>> garchmod = garch.fit(disp='off')
>>> print(garchmod.summary())
                   Zero Mean - GARCH Model Results
==================================================================
Dep. Variable:                  None   R-squared:            0.000
Mean Model:                Zero Mean   Adj. R-squared:       0.000
Vol Model:                     GARCH   Log-Likelihood:    -6941.21
Distribution:                 Normal   AIC:                13892.4
Method:           Maximum Likelihood   BIC:                13925.0
                                       No. Observations:      5030
Date:               Thu, Mar 19 2020   Df Residuals:          5025
Time:                       07:47:34   Df Model:                 5
                           Volatility Model
==================================================================
             coef    std err      t       P>|t|     95.0% Conf. Int.
------------------------------------------------------------------
omega      0.0301  8.450e-03  3.558  3.738e-04 [1.350e-02,4.662e-02]
alpha[1]   0.0607  2.217e-02  2.740  6.141e-03  [1.730e-02,  0.104]
alpha[2]   0.1090  2.058e-02  5.298  1.170e-07  [6.871e-02,  0.149]
beta[1]    0.2236      0.300  0.745      0.456    [ -0.365,  0.812]
beta[2]    0.5841      0.279  2.095  3.614e-02  [3.773e-02,  1.131]
==================================================================
Covariance estimator: robust
>>> garchresid = garchmod.std_resid
>>> acf_pacf_fig(garchresid, both=True, lag=40)
>>> plt.show()
>>> plot_LB_pvalue(garchresid, noestimatedcoef=0, nolags=30)
>>> plt.show()
>>> acf_pacf_fig(garchresid**2, both=True, lag=40)
>>> plt.show()
>>> plot_LB_pvalue(garchresid**2, noestimatedcoef=0, nolags=30)
>>> plt.show()
>>> from statsmodels.graphics.api import qqplot
>>> qqplot(garchresid, line='q', fit=True); plt.show()
```

## 6.3 Other Extensions

As we have seen above, the GARCH model successfully captures a few important features of financial time series. However, it has no ability to characterize all the stylized facts on financial time series. For instance, the financial market reacts differently to bad news than to good news. There is a certain amount of asymmetry that cannot be caught by the GARCH model. Hence, there are many extensions of the GARCH model to capture some of these features. Here we discuss two extensions.

### *6.3.1 EGARCH Models*

The exponential GARCH (EGARCH) model proposed by Nelson (1991) is used to allow for asymmetric effects between positive and negative returns. It is of the form

$$\begin{cases} X_t = \sigma_t \varepsilon_t, \varepsilon_t \sim \text{iid}(0, 1) \\ \ln \sigma_t^2 = \omega + \sum_{i=1}^{p} \alpha_i \left(|\varepsilon_{t-i}| - \mathsf{E}|\varepsilon_{t-i}|\right) + \sum_{j=1}^{o} \gamma_j \varepsilon_{t-j} + \sum_{k=1}^{q} \beta_k \ln \sigma_{t-k}^2. \end{cases} \tag{6.11}$$

Note that if $\varepsilon_t \sim \text{iidN}(0, 1)$, then $\mathsf{E}|\varepsilon_h| = \sqrt{2/\pi}$ and the coefficients of the EGARCH model may positive or negative, which is different from the GARCH model. To better understand the EGARCH model, let us consider the simple case with order $(p = 1, o = 1, q = 1)$:

$$\ln \sigma_t^2 = \omega + \alpha_1 \left(|\varepsilon_{t-1}| - \mathsf{E}|\varepsilon_{t-1}|\right) + \gamma_1 \varepsilon_{t-1} + \beta_1 \ln \sigma_{t-1}^2 = \omega + g(\varepsilon_{t-1}) + \beta_1 \ln \sigma_{t-1}^2$$

where

$$g(\varepsilon_h) = \alpha_1 \left(|\varepsilon_h| - \mathsf{E}|\varepsilon_h|\right) + \gamma_1 \varepsilon_h = \begin{cases} (\gamma_1 + \alpha_1)\varepsilon_h - \alpha_1 \mathsf{E}|\varepsilon_h| \text{ if } \varepsilon_h \geq 0, \\ (\gamma_1 - \alpha_1)\varepsilon_h - \alpha_1 \mathsf{E}|\varepsilon_h| \text{ if } \varepsilon_h < 0. \end{cases} \tag{6.12}$$

Note that $\varepsilon_h = X_h/\sigma_h$. Equation (6.12) demonstrates the asymmetry in response to positive and negative return $X_h$. In addition, as negative returns tend to have larger impacts in financial markets, we expect $\gamma_1$ to be negative.

### 6.3.2 TGARCH Models

Glosten et al. (1993) and Zakoian (1994) introduce the threshold GARCH (viz., TGARCH) model to handle leverage effects. The TGARCH model has the following form

$$\begin{cases} X_t = \sigma_t \varepsilon_t, \varepsilon_t \sim \text{iid}(0, 1) \\ \sigma_t^s = \omega + \sum_{i=1}^{p} \alpha_i |X_{t-i}|^s + \sum_{j=1}^{o} \gamma_j |X_{t-j}|^s I_{[X_{t-j}<0]} + \sum_{k=1}^{q} \beta_k \sigma_{t-k}^s \end{cases} \tag{6.13}$$

where $I_{[X_{t-j}<0]}$ is an indicator for negative $X_{t-j}$, that is,

$$I_{[X_{t-j}<0]} = \begin{cases} 1 \text{ if } X_{t-j} < 0, \\ 0 \text{ if } X_{t-j} \geq 0. \end{cases}$$

If the power $s = 2$, Model (6.13) is also called the GJRGARCH model; if $s = 1$, Model (6.13) is also known as the ZGARCH model. For illustration, consider the simple TGARCH/ZGARCH model with order $(p = 1, o = 1, q = 1)$:

$$\sigma_t = \omega + \alpha_1 |X_{t-1}| + \gamma_1 |X_{t-1}| I_{[X_{t-1}<0]} + \beta_1 \sigma_{t-1}$$

where the coefficients $\alpha_1$, $\gamma_1$, and $\beta_1$ should satisfy conditions similar to those of the GARCH model and $\gamma_1$ is sometimes known as the *leverage effect coefficient*. Since we have

$$\alpha_1 |X_{t-1}| + \gamma_1 |X_{t-1}| I_{[X_{t-1}<0]} = \begin{cases} (\alpha_1 + \gamma_1)|X_{t-1}| & \text{if } X_{t-1} < 0, \\ \alpha_1 |X_{t-1}| & \text{if } X_{t-1} \geq 0, \end{cases}$$

it is easily seen that a positive $X_{t-1}$ contributes $\alpha_1 |X_{t-1}|$ to $\sigma_t$, whereas a negative $X_{t-1}$ has a larger impact $(\alpha_1 + \gamma_1)|X_{t-1}|$. The TGARCH/ZGARCH model uses zero as its threshold to separate the impacts of past returns.

### 6.3.3 An Example

*Example 6.6 (IBM Stock Log Returns)* The dataset "ibmlogret" in the folder Ptsadata is the monthly log return series of IBM stock from January 2000 to December 2019 and shown in Fig. 6.29. It passes the KPSS stationarity test and the ACF and PACF plots shown in Fig. 6.30 also illustrate stationarity of the series. On the other hand, it can clearly be seen that it is autocorrelated from the $p$-values for the Ljung-Box test of the log returns shown in Fig. 6.31. Therefore, we should firstly build an adequate ARMA model for the return series. Using the function `choose_arma`,

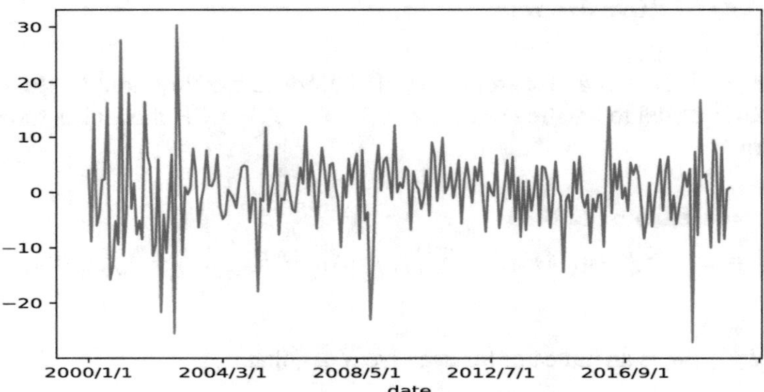


**Fig. 6.29** Time series plot of the IBM stock log returns in Example 6.6

pyTSA_ReturnsIBM

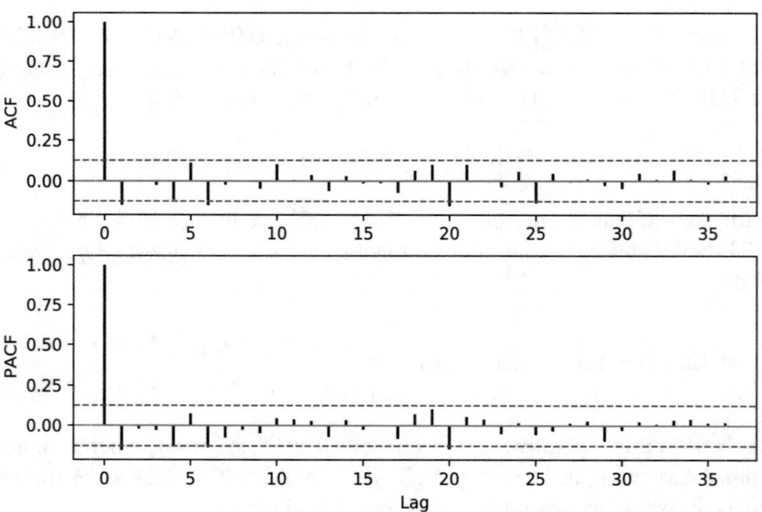


**Fig. 6.30** ACF and PACF plots of the IBM stock log returns in Example 6.6

pyTSA_ReturnsIBM

we select the order $(p, q) = (0, 1)$. From the table `ARMA Model Results`, we see that the estimate $(-0.1555)$ of the coefficient is significantly different from zero since the $p$-value is 0.017. At the same time, the residual series $\{X_t\}$ = `armaresid` of the estimated ARMA(0,1) appears a white noise by checking the $p$-value plot for the Ljung-Box test of the residuals shown in Fig. 6.32, and the residuals have the ARCH effect in it; see the $p$-value plot for the Ljung-Box test of the squared residuals shown in Fig. 6.33. Now we try to build an appropriate GARCH model for the series `armaresid`. It turns out that there is no adequate GARCH

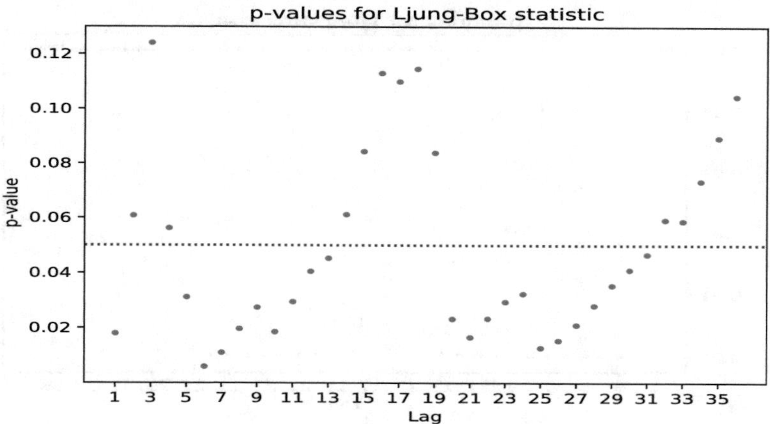


**Fig. 6.31** $p$-Value plot of the IBM stock log returns in Example 6.6

pyTSA_ReturnsIBM

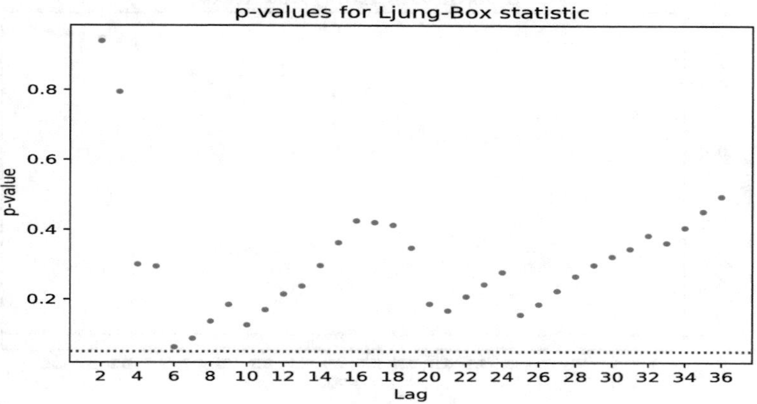


**Fig. 6.32** $p$-Value plot of residuals of ARMA(0,1) model of the IBM log returns

pyTSA_ReturnsIBM

model for it; see that in the table `Zero Mean - GARCH Model Results`, the estimate of $\alpha_1$ is zero. Thus, we decide to give the EGARCH model a whirl. From the table `Zero Mean - EGARCH Model Results`, we observe that all the estimates of the coefficients are statistically significant. Furthermore, the $p$-value plot for the Ljung-Box test of the residuals of the estimated EGARCH model shown in Fig. 6.34 indicates that the residual series is uncorrelated; the $p$-value plot for the Ljung-Box test of the squared residuals shown in Fig. 6.35 demonstrates that the residuals have no ARCH effect anymore. Besides, the QQ plot of the residuals shown in Fig. 6.36 is satisfactory except for the left tail. In summary, the estimated EGARCH model fits well to the series $\{X_t\}$ = `armaresid`. Finally, the complete

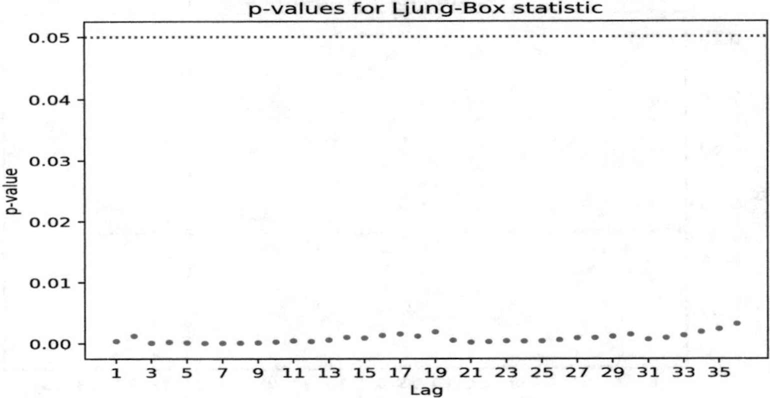


**Fig. 6.33** $p$-Values of squared residuals of ARMA model for the IBM log returns

pyTSA_ReturnsIBM

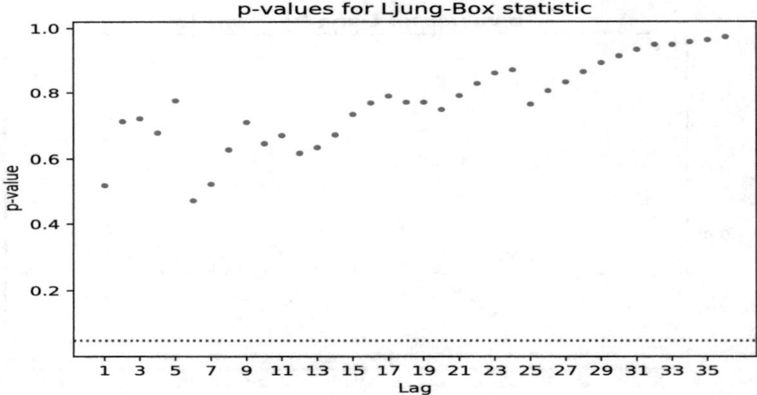


**Fig. 6.34** $p$-Value plot for LB test for the residuals of EGARCH model for the series `armaresid` in Example 6.6

pyTSA_ReturnsIBM

resulting model for the log return series of IBM stock is as follows:

$$\begin{cases} R_t = X_t - 0.1555X_{t-1} \\ X_t = \sigma_t \varepsilon_t, \ \varepsilon_t \sim \text{iidN}(0, 1) \\ \ln \sigma_t^2 = 0.1072 - 0.0901\varepsilon_{t-1} + 0.9716 \ln \sigma_{t-1}^2. \end{cases}$$

where $R_t$ denotes the log return series. For reproducing the results and plots in this example, run the Python code below.

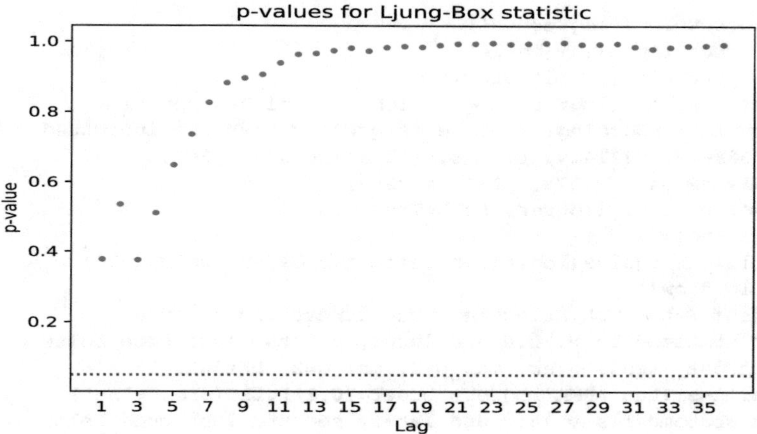


**Fig. 6.35** *p*-Value plot for LB test for the squared residuals of EGARCH model for the series `armaresid` in Example 6.6

pyTSA_ReturnsIBM

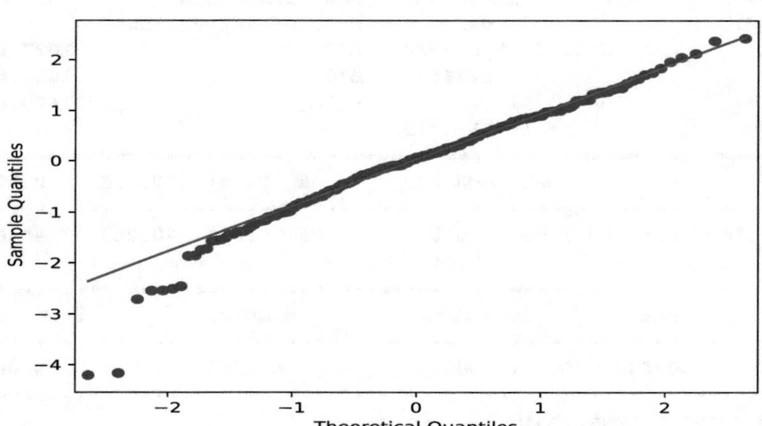


**Fig. 6.36** QQ plot of the residuals of EGARCH model for the series `armaresid` in Example 6.6

pyTSA_ReturnsIBM

```
>>> import pandas as pd
>>> import numpy as np
>>> import matplotlib.pyplot as plt
>>> import statsmodels.api as sm
>>> from PythonTsa.plot_acf_pacf import acf_pacf_fig
>>> from PythonTsa.LjungBoxtest import plot_LB_pvalue
>>> from PythonTsa.datadir import getdtapath
>>> dtapath=getdtapath()
>>> ibm=pd.read_csv(dtapath + 'ibmlogret.csv',header=0)
```

```
>>> logret=ibm['logreturn']
>>> logret.index=ibm['date']
>>> logret.plot(); plt.show()
>>> sm.tsa.kpss(logret, regression='c', nlags='auto')
InterpolationWarning: p-value is greater than the indicated p-value
(0.07583649910331416, 0.1, 5, {'10%': 0.347, '5%':
 0.463, '2.5%': 0.574, '1%': 0.739})
>>> acf_pacf_fig(logret, both=True, lag=36)
>>> plt.show()
>>> plot_LB_pvalue(logret, noestimatedcoef=0, nolags=36)
>>> plt.show()
>>> from PythonTsa.Selecting_arma import choose_arma
# for statsmodels 0.13.0 and later, see the last code below
>>> choose_arma(logret, max_p=2, max_q=2, ctrl=1.01)
>>> arma=sm.tsa.ARMA(logret, order=(0,1)).fit(trend='nc')
# for statsmodels 0.13.0 and later, see the last code below
>>> print(arma.summary())
                              ARMA Model Results
==============================================================================
Dep. Variable:              logreturn   No. Observations:                  240
Model:                     ARMA(0, 1)   Log Likelihood                -811.524
Method:                       css-mle   S.D. of innovations              7.116
Date:                Sat, 04 Apr 2020   AIC                           1627.047
Time:                        08:36:03   BIC                           1634.009
Sample:                    01-01-2000   HQIC                          1629.852
                         - 12-01-2019
==============================================================================
                   coef   std err        z  P>|z|   [0.025     0.975]
------------------------------------------------------------------------------
ma.L1.logreturn -0.1555     0.065   -2.389  0.017   -0.283     -0.028
                                    Roots
==============================================================================
            Real          Imaginary           Modulus          Frequency
------------------------------------------------------------------------------
MA.1      6.4303          +0.0000j            6.4303             0.0000
------------------------------------------------------------------------------
>>> armaresid=arma.resid
>>> plot_LB_pvalue(armaresid, noestimatedcoef=1, nolags=36)
>>> plt.show()
>>> plot_LB_pvalue(armaresid**2, noestimatedcoef=0, nolags=36)
>>> plt.show()
>>> from arch import arch_model
>>> garch = arch_model(armaresid, p=1, q=1,
    mean='Zero').fit(disp='off')
>>> print(garch.summary())
                        Zero Mean - GARCH Model Results
==============================================================================
Dep. Variable:                   None   R-squared:                       0.000
Mean Model:                 Zero Mean   Adj. R-squared:                  0.004
Vol Model:                      GARCH   Log-Likelihood:               -798.690
Distribution:                  Normal   AIC:                           1603.38
Method:            Maximum Likelihood   BIC:                           1613.82
```

```
                                       No. Observations:              240
Date:              Sun, Apr 05 2020    Df Residuals:                  237
Time:                      07:23:29    Df Model:                        3
                          Volatility Model
==========================================================================
              coef    std err          t      P>|t|      95.0% Conf. Int.
--------------------------------------------------------------------------
omega       0.8071      0.841      0.959      0.337     [ -0.842,  2.456]
alpha[1] 0.0000 3.142e-02   0.000      1.000 [-6.159e-02,6.159e-02]
beta[1]  0.9756 4.809e-02  20.286 1.721e-91     [  0.881,   1.070]
==========================================================================
Covariance estimator: robust
>>> egarch = arch_model(armaresid, p=0, o=1, q=1, mean='Zero',
    vol='EGARCH').fit(disp='off')
>>> print(egarch.summary())
                 Zero Mean - EGARCH Model Results
==========================================================================
Dep. Variable:                 None    R-squared:                   0.000
Mean Model:               Zero Mean    Adj. R-squared:              0.004
Vol Model:                   EGARCH    Log-Likelihood:           -792.663
Distribution:                Normal    AIC:                       1591.33
Method:          Maximum Likelihood    BIC:                       1601.77
                                       No. Observations:              240
Date:              Sat, Apr 04 2020    Df Residuals:                  237
Time:                      17:52:19    Df Model:                        3
                          Volatility Model
==========================================================================
              coef    std err          t      P>|t|      95.0% Conf. Int.
--------------------------------------------------------------------------
omega       0.1072  6.157e-02   1.741 8.165e-02 [-1.347e-02,  0.228]
gamma[1]   -0.0901  3.596e-02  -2.506 1.222e-02 [ -0.161,-1.963e-02]
beta[1]     0.9716  1.588e-02  61.182     0.000 [  0.940,   1.003]
==========================================================================
Covariance estimator: robust
>>> egarchresid=egarch.std_resid
>>> plot_LB_pvalue(egarchresid, noestimatedcoef=0, nolags=36)
>>> plt.show()
>>> plot_LB_pvalue(egarchresid**2, noestimatedcoef=0, nolags=36)
>>> plt.show()
>>> from statsmodels.graphics.api import qqplot
>>> qqplot(egarchresid, line='q', fit=True); plt.show()
# the following is for statsmodels of v. 0.13.0 and later
>>> from PythonTsa.Selecting_arma2 import choose_arma2
>>> choose_arma2(logret, max_p=2, max_q=2, ctrl=1.01)
>>> from statsmodels.tsa.arima.model import ARIMA
>>> arma=ARIMA(logret, order=(0,0,1),trend='n').fit()
```

In finance, an extremely important measure is the risk associated with an investment portfolio, and asset return volatility (variance) is a key factor used in risk measure. Therefore, the pursuit for understanding changes in the volatility is critical for financial markets. Investors require higher expected returns as a compensation for holding riskier assets. In this chapter we have discussed the ARCH and GARCH models and their some extensions. There are other alternatives that can be used to capture the heteroscedastic effects and other stylized features, for example, the GARCH-M model, stochastic volatility model, and so on.

More theoretical and technological details of financial time series analysis can be found in, for example, Campbell et al. (2012), Tsay (2010), and Fan and Yao (2003). As to applications of financial time series analysis in financial industry, Franke et al. (2019) is an excellent textbook for both university students and practitioners. Hilpisch (2015) is a recently published book where analytical technology is implemented with Python.

## Problems

**6.1** First descriptively analyze the return series `logret` in Example 6.2 like Example 6.1. Is `logret` a white noise? Then build a complete and appropriate model for it.

**6.2** Prove that if $\varepsilon_t \sim \text{iidN}(0, 1)$ and $0 < \alpha_1 < 1$, (1) $\eta_t = \sigma_t^2(\varepsilon_t^2 - 1)$ in Eq. (6.5) is a white noise; (2) the kurtosis of $X_t$, namely, $\mathsf{E}(X_t^4)/[\mathsf{E}(X_t^2)]^2$ is greater than 3. (Hint: $\mathsf{E}(\varepsilon_t^4) = 3$ and that the kurtosis is greater than 3 means that the tails are heavier than the corresponding normal distribution.)

**6.3** Given a GARCH(2,1) process as follows

$$\begin{cases} X_t = \sigma_t \varepsilon_t, \varepsilon_t \sim \text{iidN}(0, 1) \\ \sigma_t^2 = 0.1 + 0.2X_{t-1}^2 + 0.1X_{t-2}^2 + 0.6\sigma_{t-1}^2, \end{cases}$$

simulate a sample of size 1000, and then analyze the simulated data like Example 6.3 as well as build an adequate GARCH model for the simulated series. (Suggestion: do the exercise several times!)

**6.4** Using $\eta_t = \sigma_t^2(\varepsilon_t^2 - 1)$, (1) rewrite the GARCH(2,1) model defined by Eq. (6.6) with $p = 2, q = 1$ as an ARMA(2,1) model; (2) rewrite the GARCH(1,2) model defined by Eq. (6.6) with $p = 1, q = 2$ as an ARMA(2,2) model.

**6.5** For the Southern Hemisphere Temperature Volatility Data Series in Example 3.1, does it have an ARCH effect? Build a complete and adequate model for this time series.

**6.6** Under the conditions in Proposition 6.1, show that if $X_t$ is a stationary process, then we have

$$\mathsf{E}(X_t^2) = \frac{\omega}{1 - \sum_{i=1}^{m}(\alpha_i + \beta_i)}.$$

**6.7** Verify that the absolute series of the returns `pgret` in Example 6.1 is autocorrelated, and so is the absolute series of the returns `logret` in Example 6.2. Then answer the question: do return series tend to be independent?

**6.8** In Example 6.5, if we replace $\varepsilon_t \sim \text{iidN}(0, 1)$ with $\varepsilon_t \sim \text{iidT}(0, 1)$ and refit the GARCH(2,2) model to the series `xresid`, can we obtain a better model?

**6.9** In Example 6.6, (1) for the series `armaresid`, give the ZGARCH model a try, and can we build an adequate ZGARCH model for it? (2) If $\varepsilon_t \sim \text{iidN}(0, 1)$ is replaced with $\varepsilon_t \sim \text{iidT}(0, 1)$, can the EGARCH or ZGARCH models fit better to the series `armaresid`?

# Chapter 7
# Multivariate Time Series Analysis

In this chapter, we consider multivariate (vector) time series analysis and forecasting problems. Unlike the univariate case, we now have two difficulties with multivariate time series: identifiability and curse of dimensionality. Thus, this chapter focuses on a special and useful VAR models. First, basic concepts on multivariate time series and general VARMA models are introduced. Then, we elaborate on VAR model building, forecasting, Granger causality test, and impulse response analysis.

## 7.1 Basic Concepts

In previous chapters, we have elaborated on methodology and technology for univariate time series analysis and forecasting. However, in practice, we also often encounter multivariate time series cases. For example, a simple macroeconomic system has at least three time series variables, consume, invest, and export, which tend to be dependent on each other and need to be studied together as a system. Multivariate time series analysis simultaneously considers multiple time series, investigates the dynamic relationships over time among the variables, and forecasts future values for the multivariate time series. Generally speaking, most of the basic concepts from univariate time series analysis can extend to the multivariate case. We should particularly pay attention to the same concept differences between the multivariate case and univariate case.

### *7.1.1 Covariance and Correlation Matrix Functions*

First of all, we have a look at an example: a portfolio. It consists of $K$ equities. And the returns of the portfolio can be expressed as a $K$-dimensional vector time series

 
C. Huang, A. Petukhina, *Applied Time Series Analysis and Forecasting with Python*, Statistics and Computing,
https://doi.org/10.1007/978-3-031-13584-2_7

$\mathbf{X}_t = (X_{t1}, \cdots, X_{tK})'$ where $X_{tk}$ represents the return at time $t$ of equity $X_k$ in the portfolio and $1 \le k \le K$. Such time series not only have similar properties as univariate time series but also possess distinguishing features.

**Definition 7.1** A $K$-dimensional vector process $\mathbf{X}_t = (X_{t1}, \cdots, X_{tK})'$ is said to be a $K$-dimensional vector (multivariate) time series if each component process $X_{tk}$ $(1 \le k \le K)$ is a univariate time series.

For multivariate time series, it is important to discover relationship between two different component time series although each component time series can independently be studied. This leads to the following definition.

**Definition 7.2**

(1) The mean vector (function) of a $K$-dimensional vector time series $\mathbf{X}_t = (X_{t1}, \cdots, X_{tK})'$ is

$$\boldsymbol{\mu}_t = \mathsf{E}(\mathbf{X}_t) = (\mu_{t1}, \cdots, \mu_{tK})' = (\mathsf{E}X_{t1}, \cdots, \mathsf{E}X_{tK})'.$$

(2) The covariance matrix (function) is

$$\boldsymbol{\Gamma}(t+h, t) = \mathrm{Cov}(\mathbf{X}_{t+h}, \mathbf{X}_t) = \mathsf{E}[(\mathbf{X}_{t+h} - \boldsymbol{\mu}_{t+h})(\mathbf{X}_t - \boldsymbol{\mu}_t)'] = \left[\gamma_{ij}(t+h, t)\right]_{K \times K}$$

where $\gamma_{ij}(t+h, t) = \mathrm{Cov}(X_{t+h,i}, X_{t,j})$ is the autocovariance of the component series $X_{ti}$ when $i = j$; the cross-covariance of the two component series are $X_{ti}$ and $X_{tj}$ when $i \neq j$.

(3) The correlation matrix (function) is

$$\boldsymbol{\rho}(s, t) = \mathrm{Corr}(\mathbf{X}_s, \mathbf{X}_t) = \left[\rho_{ij}(s, t)\right]_{K \times K} = \left[\frac{\gamma_{ij}(s, t)}{\sqrt{\gamma_{ii}(s, s)\gamma_{jj}(t, t)}}\right]_{K \times K}$$

where (i) $\rho_{ij}(s, t) = \gamma_{ij}(s, t)/\sqrt{\gamma_{ii}(s, s)\gamma_{jj}(t, t)}$ is the autocorrelation of the component series $X_{ti}$ when $i = j$; the cross-correlation of the two component series are $X_{ti}$ and $X_{tj}$ when $i \neq j$. (ii) $\gamma_{ii}(s, s)$ is the variance at time $s$ of the component series $X_{ti}$, and $\gamma_{jj}(t, t)$ is the variance at time $t$ of the component series $X_{tj}$.

Note that in some literature, the covariance (correlation) matrix function is called the cross-covariance (cross-correlation) matrix function, and in some other literature, it is called the autocovariance (autocorrelation) matrix function. In fact, the diagonal elements of the covariance (correlation) matrix are of auto, and the off-diagonal elements of the covariance (correlation) matrix are of cross.

### 7.1.2 Stationarity and Vector White Noise

We know that the stationarity of a univariate time series is important in building an appropriate model for it. This is also true in the multivariate case. So we extend the concept of stationarity into the multivariate situation.

**Definition 7.3** A multivariate (vector) time series $\mathbf{X}_t = (X_{t1}, \cdots, X_{tK})'$ is said to be (weakly) stationary if its mean vector $\mathsf{E}(\mathbf{X}_t)$ and covariance matrix $\mathrm{Cov}(\mathbf{X}_{t+h}, \mathbf{X}_t)$ are both independent of $t$. That is, they can be expressed as

$$\mathsf{E}(\mathbf{X}_t) = \boldsymbol{\mu} \text{ and } \mathrm{Cov}(\mathbf{X}_{t+h}, \mathbf{X}_t) = \boldsymbol{\Gamma}(h) = \left[\gamma_{ij}(h)\right]_{K\times K} \text{ foralltimepoint} t,$$

where $\boldsymbol{\Gamma}(h)$ is known as the lag $h$ covariance matrix of the vector time series.

This definition looks similar to one in the univariate case. However, it requires that $\gamma_{ij}(t+h, t) = \gamma_{ij}(h)$ for any two component series $X_{ti}$ and $X_{tj}$. It is not difficult to prove the following properties of a stationary vector time series.

**Proposition 7.1** *If the vector time series $\mathbf{X}_t = (X_{t1}, \cdots, X_{tK})'$ is stationary, then we have that:*

*(1) Each of its component series $X_{tk}$ $(k = 1, \cdots, K)$ is a univariate stationary time series.*

*(2) The correlation matrix can be written as*

$$\boldsymbol{\rho}(h) = \left[\rho_{ij}(h)\right]_{K\times K} = \left[\frac{\gamma_{ij}(h)}{\sqrt{\gamma_{ii}(0)\gamma_{jj}(0)}}\right]_{K\times K} = \mathbf{D}^{-1/2}\boldsymbol{\Gamma}(h)\mathbf{D}^{-1/2}$$

*where $\mathbf{D} = \mathrm{Diag}[\gamma_{11}(0), \gamma_{22}(0), \cdots, \gamma_{KK}(0)]$ is a diagonal matrix and $\gamma_{ii}(0) = \mathrm{Var}(X_{ti})$.*

*(3) $\boldsymbol{\Gamma}(h) = \boldsymbol{\Gamma}(-h)'$ and $\boldsymbol{\rho}(h) = \boldsymbol{\rho}(-h)'$ for any integer $h$.*

*(4) $|\gamma_{ij}(h)| \le \sqrt{\gamma_{ii}(0)\gamma_{jj}(0)}$ and $|\rho_{ij}(h)| \le 1$ for all $1 \le i, j \le K$ and any integer $h$. In particular, $\rho_{ii}(0) = 1$ for $1 \le i \le K$.*

*(5) The covariance and correlation matrices are both positive semidefinite so that for any positive integer $n$ and any set of K-dimensional real vectors $\mathbf{a}_1, \mathbf{a}_2, \cdots, \mathbf{a}_n$,*

$$\sum_{i,j=1}^{n} \mathbf{a}_i'\boldsymbol{\Gamma}(i-j)\mathbf{a}_j \ge 0 \text{ and } \sum_{i,j=1}^{n} \mathbf{a}_i'\boldsymbol{\rho}(i-j)\mathbf{a}_j \ge 0.$$

As we know, the white noise series is used as a building block in modeling univariate time series. There is a counterpart that has the same function in the multivariate case. Now we give the definition for it.

**Definition 7.4** The vector time series $\boldsymbol{\varepsilon}_t = (\varepsilon_{t1}, \cdots, \varepsilon_{tK})'$ is called a vector white noise with mean vector $\mathbf{0}$ and covariance matrix $\boldsymbol{\Sigma}$ if (1) $\mathsf{E}(\boldsymbol{\varepsilon}_t) = \mathbf{0}$, (2) $\mathsf{E}(\boldsymbol{\varepsilon}_t \boldsymbol{\varepsilon}_t') = \boldsymbol{\Sigma}$, and (3) $\mathsf{E}(\boldsymbol{\varepsilon}_s \boldsymbol{\varepsilon}_t') = \mathbf{0}(s \neq t)$ where $\boldsymbol{\Sigma}$ is a symmetric positive definite matrix.

Although this definition looks like one in the univariate case, the following remarks on it are worthy of note:

- The vector white noise series is clearly stationary, and each of its components $\varepsilon_{tk}$ $(k = 1, \cdots, K)$ is a univariate white noise series.
- Although the components of the white noise series are uncorrelated at different time points, they may be contemporaneously correlated except that the covariance matrix $\boldsymbol{\Sigma}$ is a diagonal matrix.
- We denote the white noise series as $\boldsymbol{\varepsilon}_t \sim \mathrm{WN}(\mathbf{0}, \boldsymbol{\Sigma})$. Sometimes we further require the white noise series to be independent and identically distributed (iid) and then distinguish this by writing $\boldsymbol{\varepsilon}_t \sim \mathrm{iid}(\mathbf{0}, \boldsymbol{\Sigma})$. Moreover, if the white noise $\boldsymbol{\varepsilon}_t \sim \mathrm{N}(\mathbf{0}, \boldsymbol{\Sigma})$ ($K$-dimensional normal distribution ), then we denote $\boldsymbol{\varepsilon}_t \sim \mathrm{iidN}(\mathbf{0}, \boldsymbol{\Sigma})$.
- The covariance matrix function

$$\boldsymbol{\Gamma}(h) = \begin{cases} \boldsymbol{\Sigma}, & \text{if } h = 0, \\ \mathbf{0}, & \text{otherwise.} \end{cases}$$

- For any $h \neq 0$, the correlation matrix function $\boldsymbol{\rho}(h) = \mathbf{0}$. As for $\boldsymbol{\rho}(0)$, its all diagonal elements are ones, and off-diagonal elements may be nonzero.

*Example 7.1 (A Bivariate Stationary Time Series)* Suppose that $\varepsilon_t \sim \mathrm{WN}(0, 1)$ and the vector time series

$$\mathbf{X}_t = \begin{bmatrix} X_{t1} \\ X_{t2} \end{bmatrix} = \begin{bmatrix} \varepsilon_t \\ \varepsilon_t + 0.75\varepsilon_{t-10} \end{bmatrix}.$$

Then it is easy to see that the mean vector $\boldsymbol{\mu} = \mathbf{0}$. The covariance matrix function is

$$\boldsymbol{\Gamma}(0) = \mathsf{E}(\mathbf{X}_t \mathbf{X}_t') = \begin{bmatrix} \mathsf{E}\varepsilon_t^2 & \mathsf{E}(\varepsilon_t^2 + 0.75\varepsilon_t\varepsilon_{t-10}) \\ \mathsf{E}(\varepsilon_t^2 + 0.75\varepsilon_t\varepsilon_{t-10}) & \mathsf{E}(\varepsilon_t + 0.75\varepsilon_{t-10})^2 \end{bmatrix} = \begin{bmatrix} 1 & 1 \\ 1 & 1.5625 \end{bmatrix},$$

similarly,

$$\boldsymbol{\Gamma}(-10) = \boldsymbol{\Gamma}(10)' = \begin{bmatrix} 0 & 0.75 \\ 0 & 0.75 \end{bmatrix},$$

and $\boldsymbol{\Gamma}(h) = \mathbf{0}$ for $h \neq 0, 10, -10$. The variance vector is

$$\boldsymbol{\sigma}^2 = \mathrm{Var}(\mathbf{X}_t) = \begin{bmatrix} \mathrm{Var}(X_{t1}) \\ \mathrm{Var}(X_{t2}) \end{bmatrix} = \begin{bmatrix} 1 \\ 1.5625 \end{bmatrix}.$$

The correlation matrix function is

$$\boldsymbol{\rho}(0) = \begin{bmatrix} 1 & 0.8 \\ 0.8 & 1 \end{bmatrix}, \ \boldsymbol{\rho}(-10) = \boldsymbol{\rho}(10)' = \begin{bmatrix} 0 & 0.6 \\ 0 & 0.48 \end{bmatrix},$$

and $\boldsymbol{\rho}(h) = \mathbf{0}$ for $h \neq 0, 10, -10$.

### 7.1.3 Sample Covariance and Correlation Matrices

If we have sample $\mathbf{X}_{1:T} = \{\mathbf{X}_1, \mathbf{X}_2, \cdots, \mathbf{X}_T\}$ of size $T$ from a stationary $K$-dimensional time series, then we can compute:

(1) Sample mean vector

$$\overline{\mathbf{X}} = \frac{1}{T}\sum_{t=1}^{T}\mathbf{X}_t = \frac{1}{T}\left(\sum_{t=1}^{T}X_{t1}, \sum_{t=1}^{T}X_{t2}, \cdots, \sum_{t=1}^{T}X_{tK}\right)' = (\overline{X}_1, \overline{X}_2, \cdots, \overline{X}_K)'. \tag{7.1}$$

(2) Sample covariance matrix function

$$\hat{\boldsymbol{\Gamma}}(h) = \begin{cases} \frac{1}{T}\sum_{t=1}^{T-h}(\mathbf{X}_{t+h} - \overline{\mathbf{X}})(\mathbf{X}_t - \overline{\mathbf{X}})' = \left[\hat{\gamma}_{ij}(h)\right]_{K\times K}, & 0 \le h \le T-1, \\ \hat{\boldsymbol{\Gamma}}(-h)', \ 1-T \le h < 0 \end{cases} \tag{7.2}$$

where $\hat{\gamma}_{ij}(h) = \frac{1}{T}\sum_{t=1}^{T-h}(X_{t+h,i} - \overline{X}_i)(X_{t,j} - \overline{X}_j)$.

(3) Sample correlation matrix function

$$\hat{\boldsymbol{\rho}}(h) = \left[\hat{\rho}_{ij}(h)\right]_{K\times K} = \left[\frac{\hat{\gamma}_{ij}(h)}{\sqrt{\hat{\gamma}_{ii}(0)\hat{\gamma}_{jj}(0)}}\right]_{K\times K} = \hat{\mathbf{D}}^{-1/2}\hat{\boldsymbol{\Gamma}}(h)\hat{\mathbf{D}}^{-1/2} \tag{7.3}$$

where $\hat{\mathbf{D}} = \text{Diag}[\hat{\gamma}_{11}(0), \hat{\gamma}_{22}(0), \cdots, \hat{\gamma}_{KK}(0)]$.

Naturally, $\overline{\mathbf{X}}$, $\hat{\boldsymbol{\Gamma}}(h)$ and $\hat{\boldsymbol{\rho}}(h)$ are, respectively, estimators for the mean vector $\boldsymbol{\mu}$, covariance matrix function $\boldsymbol{\Gamma}(h)$, and correlation matrix function $\boldsymbol{\rho}(h)$. And obviously $\overline{\mathbf{X}}$ is an unbiased estimator of $\boldsymbol{\mu}$. What is more, under certain conditions, $\overline{\mathbf{X}}$, $\hat{\boldsymbol{\Gamma}}(h)$, and $\hat{\boldsymbol{\rho}}(h)$ are, respectively, consistent estimators of $\boldsymbol{\mu}$, $\boldsymbol{\Gamma}(h)$, and $\boldsymbol{\rho}(h)$. More details on their large-sample properties can be found in, for example, Wei (2019), Brockwell and Davis (2016, Chapter 8), and Tsay (2014). Besides, as in the univariate case, for a nonstationary vector time series sample, we may also compute so-called sample mean, sample covariance matrix, and sample correlation matrix by

(7.1), (7.2), and (7.3), respectively. They surely do not have the same properties of stationary vector time series samples.

Like ACF plots in the univariate case, the sample correlation function $\hat{\rho}_{ij}(h)$ plots are helpful for us to judge stationarity and whiteness for the vector time series or resulting residual series. The function `multi_ACFfig(x, nlags)` in the package `PythonTsa` is a tool for plotting such plots. The argument `nlags` is actually the lag $h$ in (7.3). If the dimension is $K$, then there are $K^2$ such plots. Thus, when $K$ is large, it becomes cumbersome to plot the sample correlation functions. At this point, an alternative is to summarize the information of $\hat{\boldsymbol{\Gamma}}(h)$ ($\hat{\boldsymbol{\rho}}(h)$). For any $h > 0$, under the null hypothesis $H_0$: $\boldsymbol{\Gamma}(h) = \mathbf{0}$ (viz., $\boldsymbol{\rho}(h) = \mathbf{0}$), the test statistic

$$\eta(h) = \frac{T^2}{T-h}\text{tr}[\hat{\boldsymbol{\Gamma}}(h)'\hat{\boldsymbol{\Gamma}}(0)^{-1}\hat{\boldsymbol{\Gamma}}(h)\hat{\boldsymbol{\Gamma}}(0)^{-1}]$$

is asymptotically chi-squared distributed with $K^2$ degrees of freedom where $\text{tr}[\mathbf{A}] = \sum_{i=1}^{K} a_{ii}$ is called the trace of the matrix $\mathbf{A}$ and $a_{11}, \cdots, a_{KK}$ are the diagonal elements of $\mathbf{A}$. The statistic $\eta(h)$ can be called the *extended trace of correlation matrix at the lag h*. Thus, we can compute the $p$-value of the $\chi^2(K^2)$ statistic for this test and plot the $p$-value against the lag $h$. See, for example, Tsay (2014, Chapter 1). The function `MultiTrCorrPvalue` in the package `PythonTsa` is used to implement this alternative solution. Examples will be given later.

### 7.1.4 Multivariate Portmanteau Test

As in univariate case, it is critical both to check if a $K$-dimensional stationary vector time series $\mathbf{X}_t$ is correlated and test whether a resulting residual series is a white noise series. The portmanteau test for univariate time series has been extended to the vector case by several authors, for example, Hosking (1980), Li and McLeod (1981). And up to now, the topic has still been under investigation, seeing, for instance, Mahdi and Mcleod (2012) and Thu (2017).

Suppose that the sample $\mathbf{X}_{1:T} = \{\mathbf{X}_1, \mathbf{X}_2, \cdots, \mathbf{X}_T\}$ of size $T$ comes from a stationary $K$-dimensional time series $\mathbf{X}_t$ with mean vector $\mathbf{0}$ and covariance matrix $\boldsymbol{\Sigma} = \boldsymbol{\Gamma}(0)$. Evidently, $\mathbf{X}_t$ is a white noise series if and only if $\boldsymbol{\rho}(h) = \mathbf{0}$ for any integer $h > 0$. The *multivariate portmanteau test* is designed for testing $H_0 : \boldsymbol{\rho}(1) = \boldsymbol{\rho}(2) = \cdots = \boldsymbol{\rho}(m) = \mathbf{0}$ versus $H_1 : \boldsymbol{\rho}(s) \neq \mathbf{0}$ for some $1 \leq s \leq m$ where $m$ is a positive integer. The original test statistic is

$$Q(m) = T\sum_{h=1}^{m}\text{tr}[\hat{\boldsymbol{\Gamma}}(h)'\hat{\boldsymbol{\Gamma}}(0)^{-1}\hat{\boldsymbol{\Gamma}}(h)\hat{\boldsymbol{\Gamma}}(0)^{-1}]. \tag{7.4}$$

There are two widely used modifications of (7.4). One is

$$Q_{LB}(m) = T^2 \sum_{h=1}^{m} \frac{1}{T-h} \mathrm{tr}[\hat{\boldsymbol{\Gamma}}(h)' \hat{\boldsymbol{\Gamma}}(0)^{-1} \hat{\boldsymbol{\Gamma}}(h) \hat{\boldsymbol{\Gamma}}(0)^{-1}] \tag{7.5}$$

which is similar to the Ljung-Box test statistic in the univariate case and often known as the multivariate Ljung-Box test statistic and introduced by Hosking (1980). The other is

$$Q_{LM}(m) = T \sum_{h=1}^{m} \mathrm{tr}[\hat{\boldsymbol{\Gamma}}(h)' \hat{\boldsymbol{\Gamma}}(0)^{-1} \hat{\boldsymbol{\Gamma}}(h) \hat{\boldsymbol{\Gamma}}(0)^{-1}] + \frac{K^2 m(m+1)}{2T} \tag{7.6}$$

which is given by Li and McLeod (1981). In general, both (7.5) and (7.6) provide considerable improvements over (7.4). See, for example, Lütkepohl (2005) and Li (2004). On the other hand, in the large-sample case, both $Q_{LB}(m)$ and $Q_{LM}(m)$ are asymptotically equivalent to $Q(m)$. And under the null hypothesis and certain other mild conditions, the three portmanteau statistics all are asymptotically chi-squared distributed with degrees of freedom $mK^2$.

There is a function in the package `PythonTsa` to implement the portmanteau test. The function is

```
MultiQpvalue_plot(x,p,q,noestimatedcoef,nolags,modified)
```

where the arguments `p` and `q` are the orders of a VARMA model (see Sect. 6.2) and zeros by default; `noestimatedcoef` is the number of estimated coefficients in the model and zero by default; `nolags` is a positive integer and means that the portmanteau test is conducted for $m = 1, 2, \cdots$ , `nolags`; `modified` is boolean and by default, `modified=True`, that means that $Q_{LB}(m)$ is computed; and else if `modified=False`, then $Q_{LM}(m)$ is calculated. Once the function is executed, the $P$-value plot is automatically produced. If you wish, the test statistic values and $P$-values can be obtained as well.

*Example 7.2 (Four-Dimensional Gauss White Noise)* If a white noise $\boldsymbol{\varepsilon}_t \sim \mathrm{N}(\mathbf{0}, \boldsymbol{\Sigma})$, then it is a Gauss (normal) white noise. Now suppose that the four-dimensional white noise $\boldsymbol{\varepsilon}_t \sim \mathrm{N}(\mathbf{0}, \boldsymbol{\Sigma})$ where

$$\boldsymbol{\Sigma} = \begin{bmatrix} 1.0 & 0.6 & 0.2 & 0.1 \\ 0.6 & 1.0 & 0.1 & 0.4 \\ 0.2 & 0.1 & 1.0 & 0.5 \\ 0.1 & 0.4 & 0.5 & 1.0 \end{bmatrix}$$

is a symmetric positive definite matrix. Thus, we have that $\boldsymbol{\rho}(0) = \boldsymbol{\Sigma}$ and for any integer $h > 0$, $\boldsymbol{\Gamma}(h) = \boldsymbol{\rho}(h) = \mathbf{0}$. We firstly simulate a sample of size 10,000 from the white noise using the function `multivariate_normal` in the

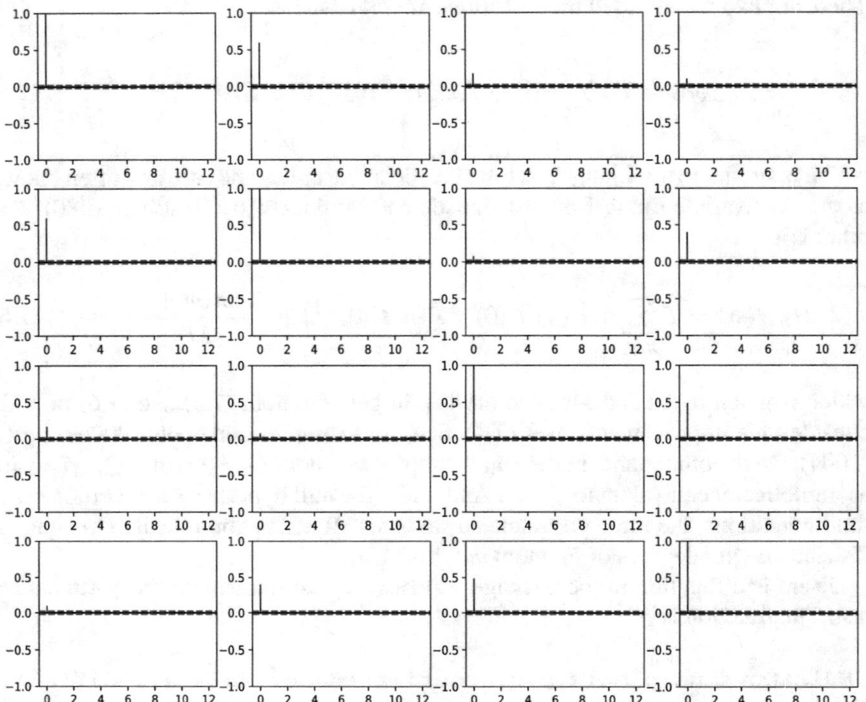

Fig. 7.1 Sample correlation plots of four-dimensional Gauss white noise

pyTSA_MultiDimGN

package `numpy`. And then apply the function `multi_ACFfig` to plot the sample correlation functions. Figure 7.1 shows the sample correlation function plots and exhibits as well the features of the correlation functions of a white noise. However, the values of the sample correlation functions $\hat{\rho}_{ij}(h)$ could not be clearly seen since space is limited. Now we plot the $P$-value plot of the extended trace of the sample correlation matrix. From Fig. 7.2, we observe that all the $P$-values are much greater than 0.05, which means that the sample is from a white noise process. Finally, the $P$-value plot for the portmanteau test of this sample is plotted by the function `MultiQpvalue_plot` and shown in Fig. 7.3. The label "Lag" on the horizontal axis is actually $m$ in (7.5). We also see that all the $P$-values are much greater than 0.05, that is, we should again accept that the sample is from a white noise process. Below is the Python code for this example.

```
>>> import numpy as np
>>> import matplotlib.pyplot as plt
>>> mean = [0,0,0,0]
>>> cov = [[1.0,0.6,0.2,0.1],
           [0.6,1.0,0.1,0.4],
           [0.2,0.1,1.0,0.5],
```

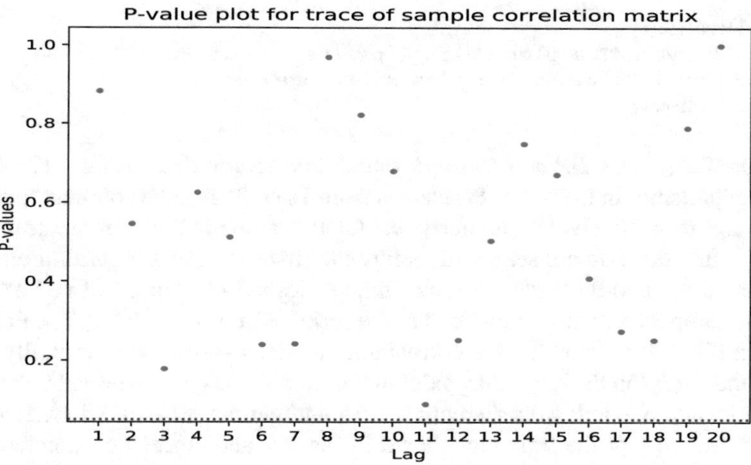


**Fig. 7.2** $P$-values of extended trace of correlation matrix of four-dimensional Gauss white noise pyTSA_MultiDimGN

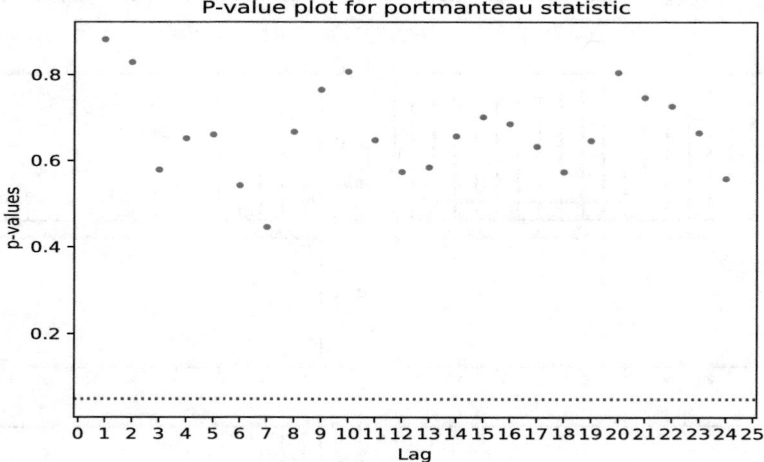


**Fig. 7.3** $P$-value plot for portmanteau test of sample of four-dimensional Gauss white noise pyTSA_MultiDimGN

```
          [0.1,0.4,0.5,1.0]]
>>> np.random.seed(1517)
>>> x= np.random.multivariate_normal(mean, cov, size=10000)
>>> from PythonTsa.plot_multi_ACF import multi_ACFfig
>>> multi_ACFfig(x, nlags=12)
>>> plt.show()
>>> from PythonTsa.MultiCorrPvalue import MultiTrCorrPvalue
>>> tr_st, pv = MultiTrCorrPvalue(x, lags=20)
```

```
>>> plt.show()
>>> from PythonTsa.plot_multi_Q_pvalue import MultiQpvalue_plot
>>> qs, pv = MultiQpvalue_plot(x, nolags=24)
>>> plt.show()
```

*Example 7.3 (USA GDP and Unemployment Rate Vector Time Series)* The dataset "USQgdpunemp" in the folder Ptsadata is from Tsay (2014). the columns "gdp" and "rate" are, respectively, US quarterly real GDP and unemployment rate from 1948 to 2011, and the data are seasonally adjusted. First, we take a logarithm on "gdp" and obtain the two-dimensional vector time series `mda['lgdp','rate']`. We plot the sample correlation functions of the series `mda`, shown in Fig. 7.4. From the plots in Fig. 7.4, we see that the correlation functions very slowly or hardly tapers off as the lag $h$ (on the horizontal axis) increases and thus the series `mda` should be nonstationary. Second, we differentiate `mda` and get the differenced series `dmda`. For `dmda`, we plot the time plot shown in Fig. 7.5 and correlation function plots shown in Fig. 7.6. We observe that the correlation function plots in Fig. 7.6 are very different from ones in Fig. 7.4 and the former correlation function values quickly enter into the 95% confidence (interval) band. This plus the pattern of the time plot in Fig. 7.5 suggests that the differenced series `dmda` should be stationary. Third, we

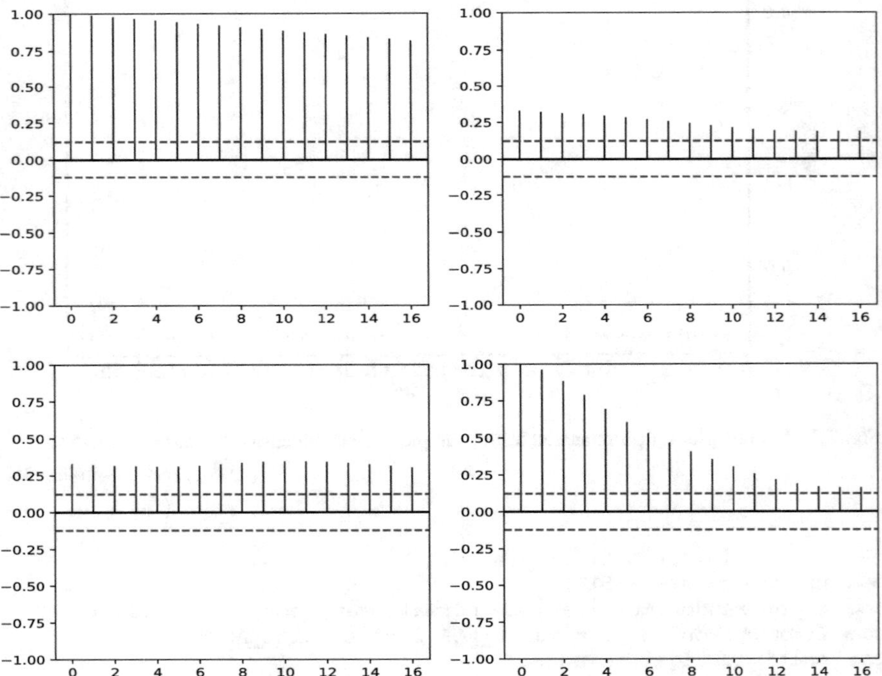


**Fig. 7.4** Sample correlation plots of US GDP (in log) and unemployment rates

pyTSA_USUnempGDP

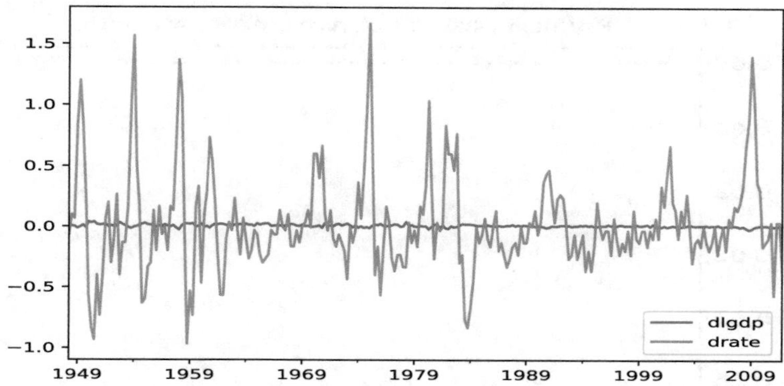


**Fig. 7.5** Time series plot of differences of US GDP (in log) and unemployment rate

pyTSA_USUnempGDP

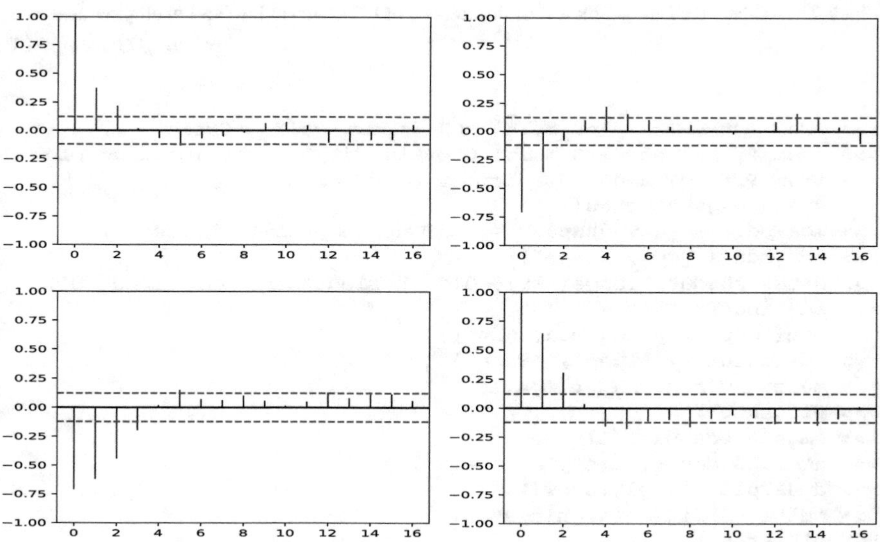

**Fig. 7.6** Sample correlation plots of differences of US GDP (in log) and unemployment rate

pyTSA_USUnempGDP

conduct the portmanteau (Ljung-Box) test of the series `dmda` and have the $p$-value plot shown in Fig. 7.7. We see that all the $p$-values are almost zero. This illustrates that the series `dmda` is not a vector white noise but correlated and so modeling it is worthy of consideration.

The Python code for this example is as follows.

```
>>> import numpy as np
>>> import pandas as pd
>>> import matplotlib.pyplot as plt
```

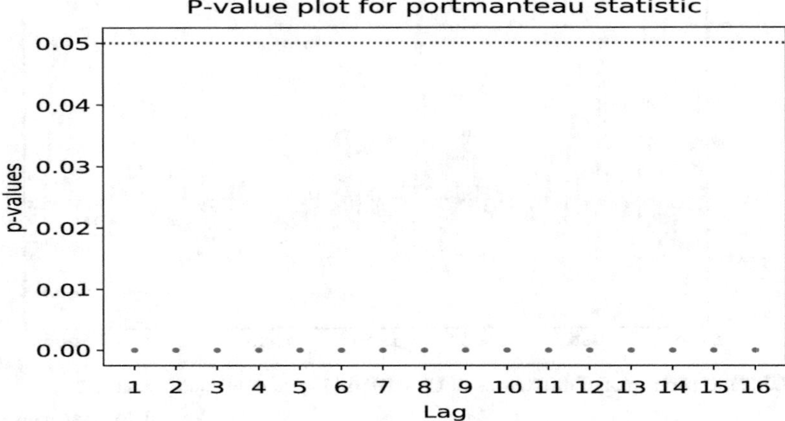


**Fig. 7.7** $p$-Values for Ljung-Box test of differences of US GDP (in log) and unemployment rate pyTSA_USUnempGDP

```
>>> from PythonTsa.plot_multi_ACF import multi_ACFfig
>>> from PythonTsa.plot_multi_Q_pvalue import MultiQpvalue_plot
>>> from PythonTsa.datadir import getdtapath
>>> dtapath=getdtapath()
>>> mda=pd.read_csv(dtapath + 'USQgdpunemp.csv', header=0)
>>> mda=mda[['gdp', 'rate']]
>>> dates=pd.date_range('1948-01', periods=len(mda), freq='Q')
>>> mda.index=dates
>>> mda['gdp']=np.log(mda['gdp'])
>>> mda.columns=['lgdp', 'rate']
>>> multi_ACFfig(mda, nlags=16)
>>> plt.show()
>>> dmda = mda.diff(1).dropna()
>>> dmda.columns=['dlgdp', 'drate']
>>> dmda.plot(); plt.show()
>>> multi_ACFfig(dmda, nlags=16)
>>> plt.show()
>>> qs, pv = MultiQpvalue_plot(dmda, nolags=16)
>>> plt.show()
```

## 7.2 VARMA Models

As in the univariate situation, we can define autoregressive moving average models for stationary vector (multivariate) time series. They are a widely used class of multivariate time series models. They formally look like ARMA models in the univariate case. However, there are some new issues with them.

### 7.2.1 Definitions

**Definition 7.5** (1) Suppose that $p$, $q$ are both nonnegative integers, and then the following equation is called the $K$-dimensional vector (multivariate) autoregressive moving average model of order $(p, q)$ and denoted by VARMA$(p, q)$:

$$\mathbf{X}_t - \boldsymbol{\Phi}_1 \mathbf{X}_{t-1} - \cdots - \boldsymbol{\Phi}_p \mathbf{X}_{t-p} = \boldsymbol{\nu} + \boldsymbol{\varepsilon}_t + \boldsymbol{\Theta}_1 \boldsymbol{\varepsilon}_{t-1} + \cdots + \boldsymbol{\Theta}_q \boldsymbol{\varepsilon}_{t-q} \tag{7.7}$$

where $\boldsymbol{\nu}$ is a $K$-dimensional constant vector; $\boldsymbol{\varepsilon}_t \sim \text{WN}(\mathbf{0}, \boldsymbol{\Sigma})$; $\boldsymbol{\Phi}_1, \cdots, \boldsymbol{\Phi}_p$ are AR part $K \times K$ matrix parameters (coefficients) with $\boldsymbol{\Phi}_p \neq \mathbf{0}$; $\boldsymbol{\Theta}_1, \cdots, \boldsymbol{\Theta}_q$ are MA part $K \times K$ matrix parameters (coefficients) with $\boldsymbol{\Theta}_q \neq \mathbf{0}$. (2) If a vector time series $\{\mathbf{X}_t\}$ is stationary and satisfies such an equation as (7.7), then we call it a VARMA$(p, q)$ process.

We give some remarks on Definition 7.5:

- If the vector time series $\{\mathbf{X}_t\}$ is a VARMA$(p, q)$ process or VARMA$(p, q)$ model (7.7) is stationary, then its mean vector

$$\boldsymbol{\mu} = \mathsf{E}(\mathbf{X}_t) = (I - \boldsymbol{\Phi}_1 - \cdots - \boldsymbol{\Phi}_p)^{-1} \boldsymbol{\nu}$$

  where $I$ is an identity matrix.
- Using the backshift (lag) operator $B$, the VARMA$(p, q)$ model can be rewritten as

$$\boldsymbol{\Phi}(B)\mathbf{X}_t = \boldsymbol{\nu} + \boldsymbol{\Theta}(B)\boldsymbol{\varepsilon}_t$$

  where $\boldsymbol{\Phi}(z) = I_K - \boldsymbol{\Phi}_1 z - \cdots - \boldsymbol{\Phi}_p z^p$ and $\boldsymbol{\Theta}(z) = I_K + \boldsymbol{\Theta}_1 z + \cdots + \boldsymbol{\Theta}_q z^q$ are, respectively, VAR and VMA matrix polynomials. Note that each element of the matrices $\boldsymbol{\Phi}(z)$, $\boldsymbol{\Theta}(z)$ is a polynomial with real coefficients and degree less than or equal to $p$, $q$, respectively.
- If $p > 0, q = 0$, then the VARMA$(p, q)$ model reduces to the VAR$(p)$ model

$$\mathbf{X}_t = \boldsymbol{\nu} + \boldsymbol{\Phi}_1 \mathbf{X}_{t-1} + \cdots + \boldsymbol{\Phi}_p \mathbf{X}_{t-p} + \boldsymbol{\varepsilon}_t, \quad \text{namely } \boldsymbol{\Phi}(B)\mathbf{X}_t = \boldsymbol{\nu} + \boldsymbol{\varepsilon}_t.$$

  This is an extremely useful subclass of VARMA models and is also called the VAR part of VARMA model (7.7).
- If $p = 0, q > 0$, then the VARMA$(p, q)$ model reduces to the VMA$(q)$ model

$$\mathbf{X}_t = \boldsymbol{\nu} + \boldsymbol{\varepsilon}_t + \boldsymbol{\Theta}_1 \boldsymbol{\varepsilon}_{t-1} + \cdots + \boldsymbol{\Theta}_q \boldsymbol{\varepsilon}_{t-q}, \quad \text{namely } \mathbf{X}_t = \boldsymbol{\nu} + \boldsymbol{\Theta}(B)\boldsymbol{\varepsilon}_t.$$

  It is also known as the VMA part of VARMA model (7.7).

Now we give an example of VAR models (processes) and then plot its time series plot and theoretical correlation function plots with Python.

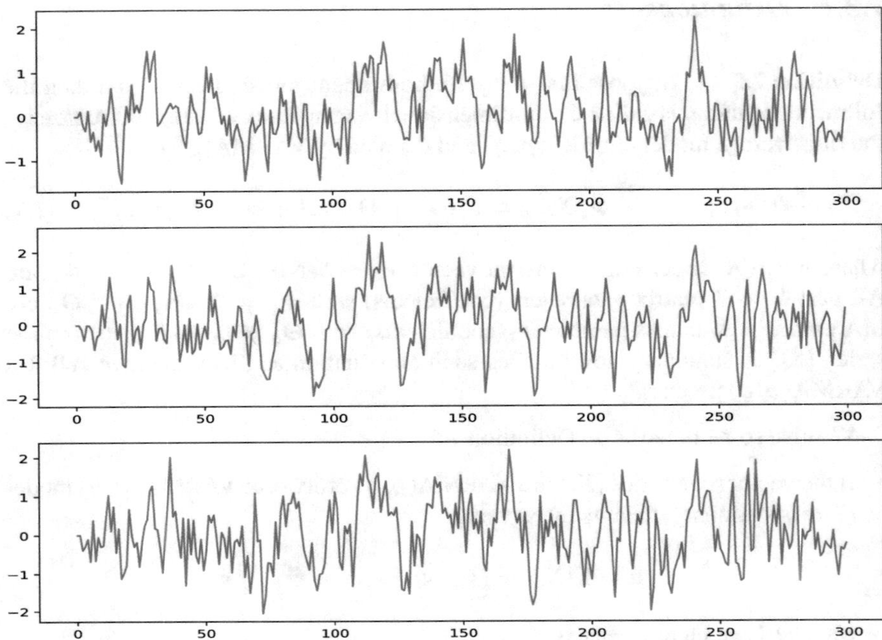


**Fig. 7.8** Time series plot of the VAR(2) process in Example 7.4

pyTSA_SimVAR

*Example 7.4 (Simulating a VAR(2) Process)* Suppose that we have a three-dimensional VAR(2) process like this

$$\mathbf{X}_t = \boldsymbol{\Phi}_1 \mathbf{X}_{t-1} + \boldsymbol{\Phi}_2 \mathbf{X}_{t-2} + \boldsymbol{\varepsilon}_t$$

where $\boldsymbol{\varepsilon}_t \sim \text{iidN}(\mathbf{0}, \boldsymbol{\Sigma})$ and

$$\boldsymbol{\Phi}_1 = \begin{bmatrix} 0.5 & 0.2 & 0.0 \\ 0.4 & 0.3 & 0.5 \\ 0.5 & 0.2 & 0.3 \end{bmatrix}, \quad \boldsymbol{\Phi}_2 = \begin{bmatrix} 0.0 & 0.01 & 0.0 \\ -0.19 & -0.2 & 0.0 \\ -0.31 & 0.01 & -0.1 \end{bmatrix}, \quad \boldsymbol{\Sigma} = \begin{bmatrix} 0.28 & 0.03 & 0.07 \\ 0.03 & 0.3 & 0.14 \\ 0.07 & 0.14 & 0.4 \end{bmatrix}.$$

Running the function (actually class) `VARProcess` in the package `statsmodels` and its methods, we can simulate a sample of size 300 from the VAR(2) process and plot its time series plot shown in Fig. 7.8. In addition, we can plot the theoretical correlation functions for the VAR(2) process shown Fig. 7.9. Both Figs. 7.8 and 7.9 suggest that the VAR(2) process is stationary as it is.

Execute the following Python code in order to reproduce the plots.

```
>>> from statsmodels.tsa.vector_ar.var_model import VARProcess
>>> import numpy as np
>>> import matplotlib.pyplot as plt
```

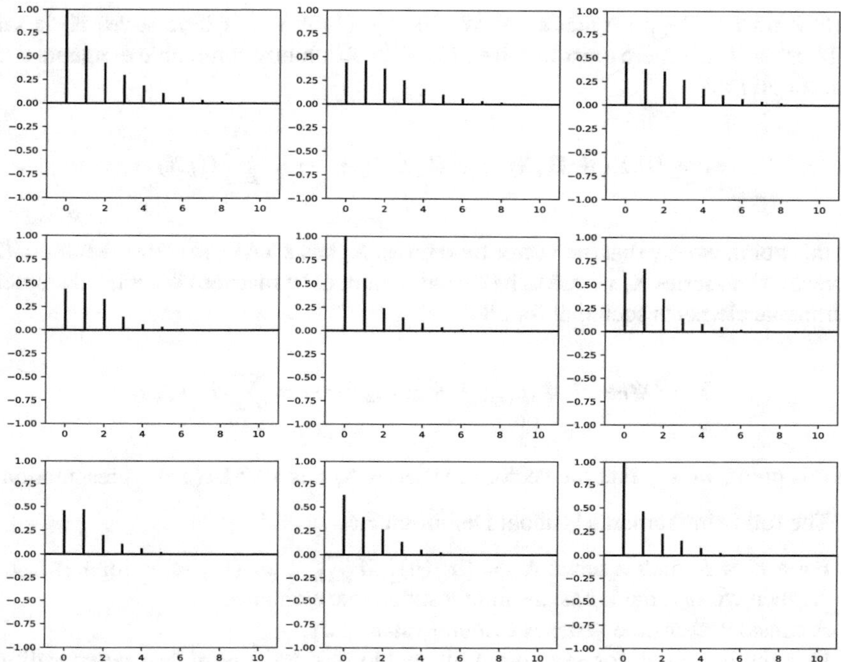

**Fig. 7.9** Plots of theoretical correlation functions for the VAR(2) process

pyTSA_SimVAR

```
>>> coefs=np.array([[[0.5,0.2,0.0],[0.4,0.3,0.5],[0.5,0.2,0.3]],
       [[0.0,0.01,0.0],[-0.19,-0.2,0.0],[-0.31,0.01,-0.1]]])
>>> sigma_u=np.array([[0.28,0.03,0.07],[0.03,0.3,0.14],
             [0.07,0.14,0.4]])
>>> coefs_exog=np.array([0,0,0])
# the constant vector is zero
>>> varProcess=VARProcess(coefs, coefs_exog, sigma_u)
>>> varSimul=varProcess.simulate_var(steps=300, seed=1237)
# Draw a sample of size 300 from the VAR(2) process
>>> varProcess.plotsim(steps=300, seed=1237); plt.show()
# Plot the sample
>>> varProcess.plot_acorr(linewidth=2, nlags=10); plt.show()
# Plot theoretical correlation functions
```

### 7.2.2 *Properties*

First of all, we generalize the two concepts of invertibility and causality to the multivariate time series.

**Definition 7.6** Suppose that $\boldsymbol{\varepsilon}_t \sim \text{WN}(\mathbf{0}, \boldsymbol{\Sigma})$. (1) A vector time series $\mathbf{X}_t$ is said to be invertible if there exist matrices $\boldsymbol{\Pi}_h$ with absolutely summable elements such that for all $t$

$$\boldsymbol{\varepsilon}_t = \boldsymbol{\Pi}_0 \mathbf{X}_1 + \boldsymbol{\Pi}_1 \mathbf{X}_{t-1} + \boldsymbol{\Pi}_2 \mathbf{X}_{t-2} + \cdots = \sum_{h=0}^{\infty} \boldsymbol{\Pi}_h \mathbf{X}_{t-h}.$$

At this point, we say that the vector time series $\mathbf{X}_t$ has a VAR($\infty$) representation. (2) A vector time series $\mathbf{X}_t$ is said to be causal if there exist matrices $\boldsymbol{\Psi}_h$ with absolutely summable elements such that for all $t$

$$\mathbf{X}_t = \boldsymbol{\Psi}_0 \boldsymbol{\varepsilon}_t + \boldsymbol{\Psi}_1 \boldsymbol{\varepsilon}_{t-1} + \boldsymbol{\Psi}_2 \boldsymbol{\varepsilon}_{t-2} + \cdots = \sum_{h=0}^{\infty} \boldsymbol{\Psi}_h \boldsymbol{\varepsilon}_{t-h}.$$

At this point, we say that the vector time series $\mathbf{X}_t$ has a VMA($\infty$) representation.

The following remarks is about Definition 7.6:

- For a $K \times K$ matrix series $\mathbf{A}_h = \left[a_{ij}(h)\right]$, if $\sum_{h=1}^{\infty} |a_{ij}(h)| < \infty$ for $1 \leq i, j \leq K$, then we say that it has absolutely summable elements.
- A causal vector time series is evidently stationary.
- In some literature, for example, Lütkepohl (2005), the term "causal(causality)" is replaced with "stable(stability)," which is perhaps for avoiding confusing "causality" with "Granger causality."
- As in the univariate case, resulting models of multivariate series tend to be required both causal and invertible.

Without loss of generality, in the following two theorems, we assume that the VARMA process $\{\mathbf{X}_t\}$ has zero mean vector. Otherwise, we can consider $\{\mathbf{X}_t - \boldsymbol{\mu}\}$. The proofs of the two theorems can be found in, for example, Brockwell and Brockwell and Davis (1991), among others.

**Theorem 7.1 (Multivariate Invertibility Theorem)** *(1) The VARMA($p, q$) process (model) (7.7) is invertible if and only if the determinant of its VMA matrix polynomial* $\det \boldsymbol{\Theta}(z) = |I_K + \boldsymbol{\Theta}_1 z + \cdots + \boldsymbol{\Theta}_q z^q| \neq 0$ *for all $z \in \mathscr{C}$ (the complex field) such that $|z| \leq 1$. (2) The VARMA($p, q$) process (7.7) is invertible if and only if its VMA part is invertible. (3) The VAR($p$) process is always invertible. (4) If the VARMA($p, q$) process (7.7) is invertible, then the coefficient matrices of its VAR($\infty$) representation can be computed recursively from the equations*

$$\boldsymbol{\Pi}_h = -\boldsymbol{\Phi}_h - \sum_{k=1}^{\infty} \boldsymbol{\Theta}_k \boldsymbol{\Pi}_{h-k}, \; h = 0, 1, \cdots$$

*where* $\boldsymbol{\Phi}_0 = -\mathbf{I}$, $\boldsymbol{\Phi}_h = \mathbf{0}$ for $h > p$, $\boldsymbol{\Theta}_h = \mathbf{0}$ for $h > q$, *and* $\boldsymbol{\Pi}_h = \mathbf{0}$ for $h < 0$.

**Theorem 7.2 (Multivariate Causality Theorem)** *(1) The VARMA(p, q) process (model) (7.7) is causal if and only if the determinant of its VAR matrix polynomial* $\det \boldsymbol{\Phi}(z) = |I_K - \boldsymbol{\Phi}_1 z - \cdots - \boldsymbol{\Phi}_p z^p| \neq 0$ *for all* $z \in \mathscr{C}$ *(the complex field) such that* $|z| \leq 1$. *(2) The VARMA(p, q) process (7.7) is causal if and only if its VAR part is causal. (3) The VMA(q) process is always causal. (4) If the VARMA(p, q) process (7.7) is causal, then the coefficient matrices of its VMA(∞) representation can be computed recursively from the equations*

$$\boldsymbol{\Psi}_h = \boldsymbol{\Theta}_h + \sum_{k=1}^{\infty} \boldsymbol{\Phi}_k \boldsymbol{\Psi}_{h-k},\ h = 0, 1, \cdots$$

*where* $\boldsymbol{\Theta}_0 = \mathbf{I}$, $\boldsymbol{\Theta}_h = \mathbf{0}$ for $h > q$, $\boldsymbol{\Phi}_h = \mathbf{0}$ for $h > p$, *and* $\boldsymbol{\Psi}_h = \mathbf{0}$ for $h < 0$.

Note that (1) in the two theorems, we do not use the statements "all the roots (zeros) of $\det \boldsymbol{\Theta}(z)$ are greater than 1 in modulus or outside the unit circle" and "all the roots (zeros) of $\det \boldsymbol{\Phi}(z)$ are greater than 1 in modulus or outside the unit circle" since $\det \boldsymbol{\Theta}(z)$ may be a nonzero constant for all $z \in \mathscr{C}$ and so may $\det \boldsymbol{\Phi}(z)$. See Example 7.6. (2) For a causal VR($p$) process, the coefficient matrices of its VMA(∞) representation reduce to

$$\boldsymbol{\Psi}_0 = \mathbf{I}_K,\ \boldsymbol{\Psi}_h = \sum_{k=1}^{\min(h,p)} \boldsymbol{\Phi}_k \boldsymbol{\Psi}_{h-k},\ \text{for}\ h = 1, 2, \cdots \tag{7.8}$$

*Example 7.5 (Example 7.4 Continued)* Executing the following Python code, we can determine whether a VAR model is causal (stable). Now the result "True" illustrates that the VAR(2) model in Example 7.4 is causal. Furthermore, the coefficient matrices of its VMA(∞) representation are estimated as follows:

$$\boldsymbol{\Psi}_1 = \begin{bmatrix} 0.5\ 0.2\ 0.0 \\ 0.4\ 0.3\ 0.5 \\ 0.5\ 0.2\ 0.3 \end{bmatrix},\ \boldsymbol{\Psi}_2 = \begin{bmatrix} 0.33\ 0.17\ 0.10 \\ 0.38\ 0.07\ 0.30 \\ 0.17\ 0.23\ 0.09 \end{bmatrix},\ \boldsymbol{\Psi}_3 = \begin{bmatrix} 0.245\ 0.102\ 0.115 \\ 0.156\ 0.106\ 0.075 \\ 0.091\ 0.089\ 0.112 \end{bmatrix},\ \cdots.$$

```
>>> varProcess.is_stable()
True
>>> varProcess.ma_rep(maxn=3)
array([[[1.    , 0.    , 0.    ],
        [0.    , 1.    , 0.    ],
        [0.    , 0.    , 1.    ]],

       [[0.5   , 0.2   , 0.    ],
        [0.4   , 0.3   , 0.5   ],
        [0.5   , 0.2   , 0.3   ]],

       [[0.33  , 0.17  , 0.1   ],
        [0.38  , 0.07  , 0.3   ],
```

```
       [0.17 , 0.23 , 0.09 ]],

      [[0.245, 0.102, 0.115],
       [0.156, 0.106, 0.075],
       [0.091, 0.089, 0.112]]])
```

*Example 7.6 (Identifiability Problem)* Consider the following VARMA(1, 1) model

$$\mathbf{X}_t = \boldsymbol{\Phi}_1 \mathbf{X}_{t-1} + \boldsymbol{\varepsilon}_t + \boldsymbol{\Theta}_1 \boldsymbol{\varepsilon}_{t-1} \tag{7.9}$$

where

$$\boldsymbol{\Phi}_1 = \begin{bmatrix} 0 & a+b \\ 0 & 0 \end{bmatrix}, \ \boldsymbol{\Theta}_1 = \begin{bmatrix} 0 & -b \\ 0 & 0 \end{bmatrix}$$

and $a, b$ are any given real numbers.

First, a little interesting finding is $\det(\mathbf{I}-\boldsymbol{\Phi}_1 z) \equiv 1 \equiv \det(\mathbf{I}+\boldsymbol{\Theta}_1 z)$ for all $z \in \mathscr{C}$. Second, it is easy to prove that the VARMA(1,1) model is both causal and invertible for any real numbers $a, b$, which is left as an exercise to the reader. Third, for any given $b$, since $(\mathbf{I} - \boldsymbol{\Phi}_1 B)^{-1} = (\mathbf{I} + \boldsymbol{\Phi}_1 B)$, then the VMA representation of (7.9) is

$$\begin{aligned} \mathbf{X}_t =& (\mathbf{I} - \boldsymbol{\Phi}_1 B)^{-1} (\mathbf{I} + \boldsymbol{\Theta}_1 B) \boldsymbol{\varepsilon}_t \\ =& (\mathbf{I} + \boldsymbol{\Phi}_1 B)(\mathbf{I} + \boldsymbol{\Theta}_1 B) \boldsymbol{\varepsilon}_t \\ =& \boldsymbol{\varepsilon}_t + \boldsymbol{\Psi}_1 \boldsymbol{\varepsilon}_{t-1} \end{aligned}$$

where

$$\boldsymbol{\Psi}_1 = \begin{bmatrix} 0 & a \\ 0 & 0 \end{bmatrix}.$$

That is, this VMA representation corresponds to an infinite number of causal and invertible VARMA(1,1) models. Thus, it is not always true that we can identify a VARMA model uniquely by a sample dataset from a given VMA representation. Further restrictions need to be imposed. More details about these issues can be found in Tsay (2014), Lütkepohl (2005), and so on.

In multivariate time series analysis, another big difficulty is the high-dimensional problem (called the curse of dimensionality). For example, consider a ten-dimensional VARMA(5, 5) model with the zero constant vector. The dimensions are just ten and not so large. The number of its coefficients, however, is $10^2(5 + 5) = 1000$. Sometimes, the length of the time series sample may be much shorter than the length (number) of the parameters of the model. Traditional time series methods are not designed to handle these kinds of high-dimensional variables. Even with today's computer power and speed, there are many tough problems that remain unsolved. See, for example, Wei (2019).

## 7.3 VAR Model Building and Analysis

In the previous section, we have seen that without more restrictions it is not always possible to identify a VARMA model uniquely by a sample dataset. Fortunately, in the class of VAR models, identifying a VAR model uniquely is no problem. Moreover, because of its easier interpretation, a VAR model is often used in practice. Hence we focus on VAR model building.

### 7.3.1 *VAR(1) Representation of VARMA Processes*

Any VARMA process can be represented by a VAR(1) model. Here we give an example, and for the general case, see, for example, Lütkepohl (2005, p 426). Suppose that $\mathbf{X}_t$ is a $K$-dimensional VARMA(1,1) process

$$\mathbf{X}_t = \boldsymbol{\Phi}_1 \mathbf{X}_{t-1} + \boldsymbol{\varepsilon}_t + \boldsymbol{\Theta}_1 \boldsymbol{\varepsilon}_{t-1}. \tag{7.10}$$

Let

$$\mathbf{Y}_t = \begin{bmatrix} \mathbf{X}_t \\ \boldsymbol{\varepsilon}_t \end{bmatrix}, \ \mathbf{A} = \begin{bmatrix} \boldsymbol{\Phi}_1 & \boldsymbol{\Theta}_1 \\ \mathbf{0} & \mathbf{0} \end{bmatrix}, \ \mathbf{U}_t = \begin{bmatrix} \boldsymbol{\varepsilon}_t \\ \boldsymbol{\varepsilon}_t \end{bmatrix}.$$

Then it is easy to show

$$\mathbf{Y}_t = \mathbf{A}\mathbf{Y}_{t-1} + \mathbf{U}_t. \tag{7.11}$$

What is more,

$$\det(\mathbf{I}_{2K} - \mathbf{A}z) = \left| \begin{bmatrix} \mathbf{I}_K & \mathbf{0} \\ \mathbf{0} & \mathbf{I}_K \end{bmatrix} - \begin{bmatrix} \boldsymbol{\Phi}_1 & \boldsymbol{\Theta}_1 \\ \mathbf{0} & \mathbf{0} \end{bmatrix} z \right| = \begin{vmatrix} \mathbf{I}_K - \boldsymbol{\Phi}_1 z & -\boldsymbol{\Theta}_1 z \\ \mathbf{0} & \mathbf{I}_K \end{vmatrix} = \det(\mathbf{I}_K - \boldsymbol{\Phi}_1 z).$$

This illustrates that VAR(1) model (7.11) is causal if and only if VARMA(1, 1) model (7.10) is causal. (7.11) is a VAR(1) representation of (7.10).

### 7.3.2 *VAR Model Building Steps*

In principle, the VAR model building procedure resembles one in the univariate case. Now we are interested in fitting the following $K$-dimensional VAR($p$) model

$$\mathbf{X}_t = \boldsymbol{\nu} + \boldsymbol{\Phi}_1 \mathbf{X}_{t-1} + \cdots + \boldsymbol{\Phi}_p \mathbf{X}_{t-p} + \boldsymbol{\varepsilon}_t, \ \boldsymbol{\varepsilon}_t \sim \mathrm{WN}(\mathbf{0}, \boldsymbol{\Sigma}) \tag{7.12}$$

to a given $K$-dimensional multivariate time series data with the length (size) $T$. Then we should take the following steps.

First, check if the series data is stationary by plotting time plots of all the component series, correlation functions, and so forth. For nonstationary data or nonstationary component data (sometimes some components are stationary and the others are not), transforming (e.g., taking a log) and/or differencing the data or component data make them stationary.

Second, identify models. That is, select the order $p$ for the VAR model. We continue using the information criteria AIC, BIC, and HQIC to help determine the order $p$. In the case of VAR modeling, they have the forms, respectively,

$$\text{AIC}(p) = \ln[\det \tilde{\Sigma}(p)] + \frac{2}{T} pK^2, \ \text{BIC}(p) = \ln[\det \tilde{\Sigma}(p)] + \frac{\ln(T)}{T} pK^2$$

and

$$\text{HQIC}(p) = \ln[\det \tilde{\Sigma}(p)] + \frac{2\ln[\ln(T)]}{T} pK^2$$

where $\tilde{\Sigma}(p)$ is an estimator of $\boldsymbol{\Sigma}$. In addition, we also introduce the FPE criterion in selecting the order $p$. Akaike (1969) proposes a criterion called the final prediction error (FPE) for selecting the order $p$. FPE finds the order $p$ for the VAR mode that minimizes the approximate mean squared one-step-ahead prediction error. It can be derived as

$$\text{FPE}(p) = \left(\frac{T + Kp + 1}{T - Kp - 1}\right)^K \det \tilde{\Sigma}(p).$$

If $\hat{p}$ minimizes FPE($p$), then we choose $\hat{p}$ as the order of the VAR model. Lütkepohl (2005) also compares these order selection criteria. For more details, see Akaike (1969) and Lütkepohl (2005), among others. Note that different criteria may choose different orders for the same vector time series, and we sometimes have to try several models for the same time series.

Third, estimate the parameters of the VAR($p$) model. They can be estimated by the LS or ML or Bayesian methods. For the properties and large-sample properties of the LS or ML or Bayesian estimators of the parameters, see, for example, Lütkepohl (2005) and Tsay (2014).

Fourth, diagnose the estimated model to determine whether it is adequate. To complete the step, we need to analyze the residual series and sometimes test the significance of the estimators. At this point, note that for the portmanteau test in Sect. 7.1.4, the degrees of freedom of the asymptotical chi-squared distribution are adjusted to $K^2(m - p)$. Note $K^2 p$ is actually the number of estimated parameters in $\boldsymbol{\Phi}_1, \cdots, \boldsymbol{\Phi}_p$ in Model (7.12). Naturally, if the resulting model is appropriate, then we can use it to forecast.

Fifth, refine a fitted model. Vector time series models may contain many parameters if the dimension $K$ is moderate or large. In practice, we often find that some of the estimates for parameters are not statistically significant and yet they have an impact on adequacy of the fitted model. Hence it is advantageous to refine the model by removing some insignificant parameters. Unfortunately, there exists no optimal approach to simplifying an estimated model. In principle, we firstly delete the most insignificant parameters (e.g., the $p$-values are greater than 0.5), then re-estimate the model, and check appropriateness of the re-estimated model. See the examples in Sect. 7.4.

### *7.3.3 Granger Causality*

To be easy to understand, we consider the two-dimensional VAR(1) model

$$\mathbf{X}_t = \boldsymbol{\nu} + \boldsymbol{\Phi}_1 \mathbf{X}_{t-1} + \boldsymbol{\varepsilon}_t, \ \boldsymbol{\varepsilon}_t \sim \mathrm{WN}(\mathbf{0}, \boldsymbol{\Sigma})$$

where

$$\mathbf{X}_t = \begin{bmatrix} X_{t1} \\ X_{t2} \end{bmatrix}, \ \boldsymbol{\nu} = \begin{bmatrix} \nu_1 \\ \nu_2 \end{bmatrix}, \ \boldsymbol{\Phi}_1 = \begin{bmatrix} \varphi_{11}(1) & \varphi_{12}(1) \\ \varphi_{21}(1) & \varphi_{22}(1) \end{bmatrix}, \ \boldsymbol{\varepsilon}_t = \begin{bmatrix} \varepsilon_{t1} \\ \varepsilon_{t2} \end{bmatrix}.$$

This model can be explicitly rewritten as

$$\begin{cases} X_{t1} = \nu_1 + \varphi_{11}(1) X_{t-1,1} + \varphi_{12}(1) X_{t-1,2} + \varepsilon_{t1} \\ X_{t2} = \nu_2 + \varphi_{21}(1) X_{t-1,1} + \varphi_{22}(1) X_{t-1,2} + \varepsilon_{t2}. \end{cases}$$

If all the off-diagonal elements of $\boldsymbol{\Phi}_1$ are 0, namely, $\varphi_{12}(1) = \varphi_{21}(1) = 0$, then $X_{t1}$ and $X_{t2}$ are not correlated. In this case, we say that the two component series are *uncoupled*. If $\varphi_{12}(1) = 0$ but $\varphi_{21}(1) \neq 0$, then we have a unidirectional relationship with $X_{t1}$ acting as the input variable and $X_{t2}$ as the output variable. In the econometric literature, the model implies the existence of *Granger causality* between the two series with $X_{t1}$ causing $X_{t2}$, but not being caused by $X_{t2}$. The formal definition of Granger causality and more details can be found in Granger (1969) and Lütkepohl (2005), among others. Note that if $\boldsymbol{\Sigma}$ is not a diagonal matrix, then $X_{t1}$ and $X_{t2}$ are contemporaneously (or instantaneously) correlated. At this point, we say that $X_{t1}$ and $X_{t2}$ have contemporaneous (or instantaneous) Granger causality. The `test_causality` function in the `statsmodels.tsa.vector_ar` module of the Python package `statsmodels` is used to test whether the null hypothesis that there is no Granger causality for indicated variables may be rejected.

### 7.3.4 Impulse Response Analysis

In the previous subsection, we have briefly discussed the Granger causality among variables in a system of multivariate time series. We can also consider the relationship between variables from another aspect. In practice, especially in econometrics, we are often interested in knowing the effect of changes in one variable on another variable. In other words, we will investigate the *response* of one variable to an *impulse* in another variable in the system. This kind of study is called the *impulse response analysis* and sometimes also the *multiplier analysis*. Without loss of generality, assume that a $K$-dimensional causal VR($p$) process is as follows:

$$\mathbf{X}_t = \boldsymbol{\Phi}_1\mathbf{X}_{t-1} + \boldsymbol{\Phi}_2\mathbf{X}_{t-2} + \cdots + \boldsymbol{\Phi}_p\mathbf{X}_{t-p} + \boldsymbol{\varepsilon}_t, \ \boldsymbol{\varepsilon}_t \sim \text{WN}(\mathbf{0}, \boldsymbol{\Sigma}).$$

To investigate the effects of impulse (shock or change) in $X_{t1}$ on $\mathbf{X}_{t+h}$, $h > 0$, we can suppose that $\mathbf{X}_t = \mathbf{0}$ for $t < 0$ and $\boldsymbol{\varepsilon}_0 = (1, 0, \cdots, 0)'$, which implies that $\mathbf{X}_0 = \boldsymbol{\varepsilon}_0$ and $X_{01}$ increases by 1. Now we can trace out what happens to the system during time $t = 1, 2, \cdots$ if no further impulse occur, that is, $\boldsymbol{\varepsilon}_t = \mathbf{0}$ for $t > 0$. Using the MA representation of the causal VAR($p$) model with coefficient matrix $\boldsymbol{\Psi}_h$ given in (7.8), we obtain

$$\mathbf{X}_1 = \boldsymbol{\Psi}_1\boldsymbol{\varepsilon}_0 = \begin{bmatrix} \Psi_{11}(1) \\ \Psi_{21}(1) \\ \vdots \\ \Psi_{K1}(1) \end{bmatrix} = \boldsymbol{\Psi}_{\cdot 1}(1), \ \mathbf{X}_2 = \boldsymbol{\Psi}_2\boldsymbol{\varepsilon}_0 = \begin{bmatrix} \Psi_{11}(2) \\ \Psi_{21}(2) \\ \vdots \\ \Psi_{K1}(2) \end{bmatrix} = \boldsymbol{\Psi}_{\cdot 1}(2), \ \cdots$$

where $\boldsymbol{\Psi}_{\cdot 1}(h)$ is the first column of $\boldsymbol{\Psi}_h$. Similarly, to study the effects of unit impulse in $X_{ti}$ on $\mathbf{X}_{t+h}$, $h > 0$, let $\boldsymbol{\varepsilon}_0 = \mathbf{e}_i = (0, \cdots, 0, 1, 0 \cdots, 0)'$ where the $i$th element of $\mathbf{e}_i$ is 1 and all the others are zero. We obtain

$$\mathbf{X}_0 = \mathbf{e}_i, \ \mathbf{X}_1 = \boldsymbol{\Psi}_{\cdot i}(1), \ \mathbf{X}_2 = \boldsymbol{\Psi}_{\cdot i}(2), \ \cdots$$

where $\boldsymbol{\Psi}_{\cdot i}(h)$ is the $i$th column of the coefficient matrix $\boldsymbol{\Psi}_h$ of the MA representation of the causal VAR($p$) model. Thus, $\boldsymbol{\Psi}_h$ is known as the *coefficients of impulse response*. Sometimes interest focuses on the accumulated effect over several or more periods of an impulse in one variable. For this reason, the summation

$$\widetilde{\boldsymbol{\Psi}}(n) = \sum_{h=0}^{n} \boldsymbol{\Psi}_h$$

is referred to as the *accumulated (cumulative) responses* over $n$ periods to a unit impulse to $\mathbf{X}_t$. The elements of $\widetilde{\boldsymbol{\Psi}}(n)$ are called the $n$th *interim multipliers*. The total accumulated responses for all future periods are defined as

$$\widetilde{\boldsymbol{\Psi}}(\infty) = \sum_{h=0}^{\infty} \boldsymbol{\Psi}_h$$

and are called the *total multipliers* or *long-run effects*. It is not hard to prove

$$\widetilde{\boldsymbol{\Psi}}(\infty) = (\mathbf{I}_K - \boldsymbol{\Phi}_1 - \cdots - \boldsymbol{\Phi}_p)^{-1}.$$

For more details, see, for example, Lütkepohl (2005) and Tsay (2014).

We can implement an impulse response analysis by calling the `irf` function (method) in the `statsmodels.tsa.vector_ar` module. These can be visualized using the plot function. Asymptotic standard errors are plotted by default at the 95% confidence level. By the way, impulse response analysis is often used in macro econometric analysis.

## 7.4 Examples

*Example 7.7 (West German Macroeconomic Data)* The dataset "EconGermany" in the folder Ptsadata is the quarterly, seasonally adjusted West German fixed investment (inv.), disposable income ( inc.), and consumption expenditures (cons.) from the first quarter of 1960 to the fourth quarter of 1982. It comes from Lütkepohl (2005) and constitutes a tiny macroeconomic system. Its time series plot is shown in Fig. 7.10 and displays a linear increasing trend. Thus we take a logarithm on it and then first-difference the logarithms (Fig. 7.11). Observing the time series plot in Fig. 7.11 and correlation function plots in Fig. 7.12, we think that the first differences of the logarithms of the West German economic time series should be stationary and can build an adequate VAR model for the series. We take the steps below.

First, we use the function `select_order( )` through the `VAR` class to select order $p$ for the VAR($p$) model. The pity is that different criteria choose different orders. We select the order $p = 2$ by trial and error at last.

Second, using the `fit` and `summary` methods, we obtain the table

```
Summary of Regression Results,
```

which exhibits the estimates of the parameters of the VAR(2) model and other useful information.

Third, using the `is_stable` method, we see that the fitted VAR(2) model is causal (stable). In addition, the correlation function plots of the model residuals in Fig. 7.13 and $p$-value plot for portmanteau test of the model residuals in Fig. 7.14 suggest that the model residuals could be regarded as a white noise. In other words, the estimated model is appropriate although the estimates of some parameters in it

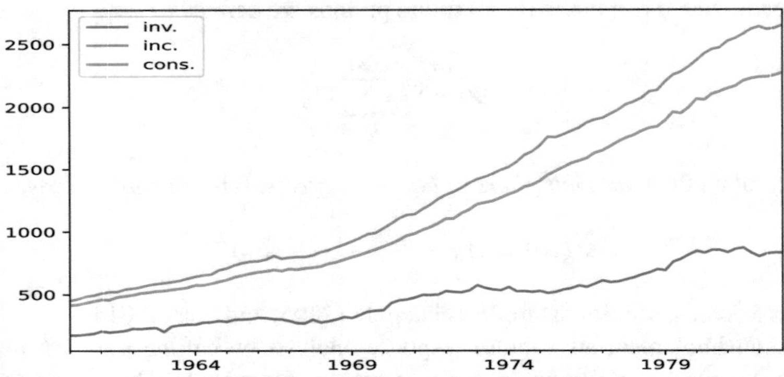


**Fig. 7.10** Time series plot of West German investment, income, and consumption

pyTSA_MacroDE

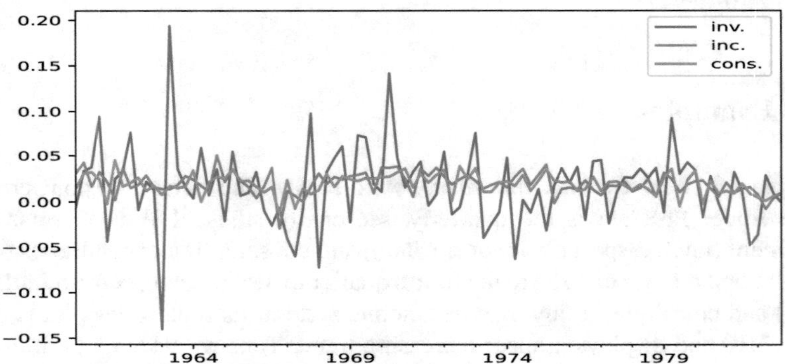


**Fig. 7.11** First differences of logarithms of West German economic time series

pyTSA_MacroDE

are statistically insignificant. We can write the VAR(2) model as follows:

$$\begin{bmatrix} X_{t1} \\ X_{t2} \\ X_{t3} \end{bmatrix} = \begin{bmatrix} -0.012 \\ 0.016 \\ 0.014 \end{bmatrix} + \begin{bmatrix} -0.299 & 0.379 & 0.666 \\ 0.039 & -0.146 & 0.256 \\ -0.005 & 0.281 & -0.321 \end{bmatrix} \begin{bmatrix} X_{t-1,1} \\ X_{t-1,2} \\ X_{t-1,3} \end{bmatrix}$$

$$+ \begin{bmatrix} -0.139 & 0.182 & 0.668 \\ 0.054 & 0.007 & 0.018 \\ 0.041 & 0.363 & -0.128 \end{bmatrix} \begin{bmatrix} X_{t-2,1} \\ X_{t-2,2} \\ X_{t-2,3} \end{bmatrix} + \begin{bmatrix} \varepsilon_{t1} \\ \varepsilon_{t2} \\ \varepsilon_{t3} \end{bmatrix}$$

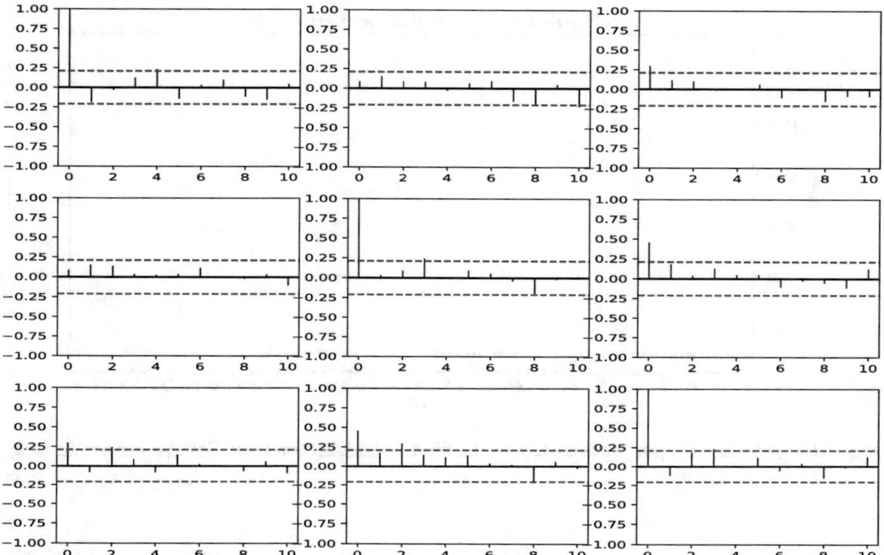

**Fig. 7.12** Correlation plots of first differences of logs of West German economic time series

pyTSA_MacroDE

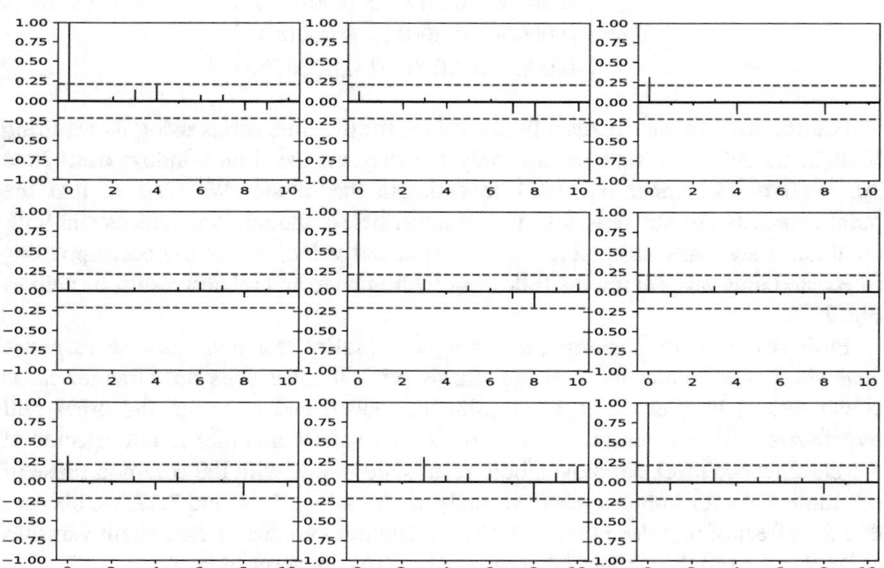

**Fig. 7.13** Correlation plots of the model residuals for West German economic time series

pyTSA_MacroDE

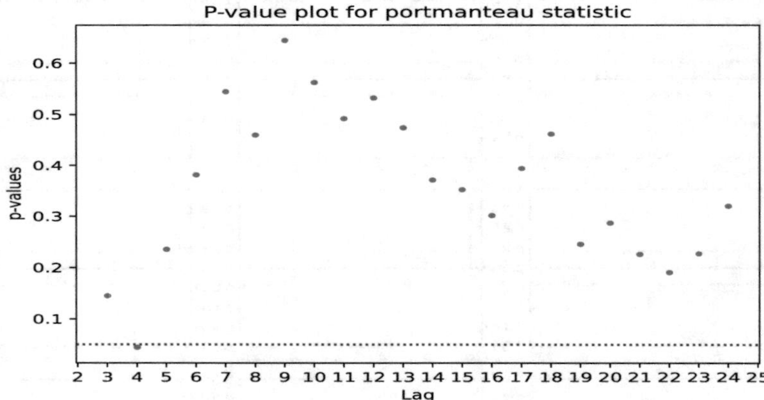


**Fig. 7.14** $p$-Values for portmanteau test of the model residuals for West German economic data pyTSA_MacroDE

where $(X_{t1}, X_{t2}, X_{t3})' = (\texttt{inv.}, \texttt{inc.}, \texttt{cons.})'$, $(\varepsilon_{t1}, \varepsilon_{t2}, \varepsilon_{t3})' = \boldsymbol{\varepsilon}_t \sim \text{WN}(\mathbf{0}, \boldsymbol{\Sigma})$, and

$$\boldsymbol{\Sigma} = \begin{bmatrix} 0.002004 & 0.000065 & 0.000140 \\ 0.000065 & 0.000125 & 0.000061 \\ 0.000140 & 0.000061 & 0.000098. \end{bmatrix}$$

Fourth, now we can forecast future values for the time series using its resulting VAR(2) model. Note that we use only the data on the time window from June 30, 1960 to December 31, 1981 to estimate the model. We observe that the point forecasts are not satisfactory, and nevertheless the interval forecasts at 95% confidence level are acceptable since most actual values fall in the corresponding forecasted intervals, seeing the following Python code and related results as well as Fig. 7.15.

Fifth, we turn to perform the Granger causality test and impulse response analysis. For example, the null hypothesis (viz., income does not Granger-cause consumption) is rejected at 5% significance level, and however, the other null hypothesis (viz., consumption does not Granger-cause income) is not rejected at 5% significance level. Moreover, both results are in line with the common sense of economics. As for impulse response analysis, from Figs. 7.16 and 7.17, we observe that the effect of impulse in one variable on another variable in the system vanishes at lag 9 or so and the accumulated responses $\widetilde{\boldsymbol{\Psi}}(n)$ converge at lag 6.

To reproduce the results in this example, run the following Python code.

```
>>> import numpy as np
>>> import pandas as pd
>>> import matplotlib.pyplot as plt
>>> from statsmodels.tsa.api import VAR
```

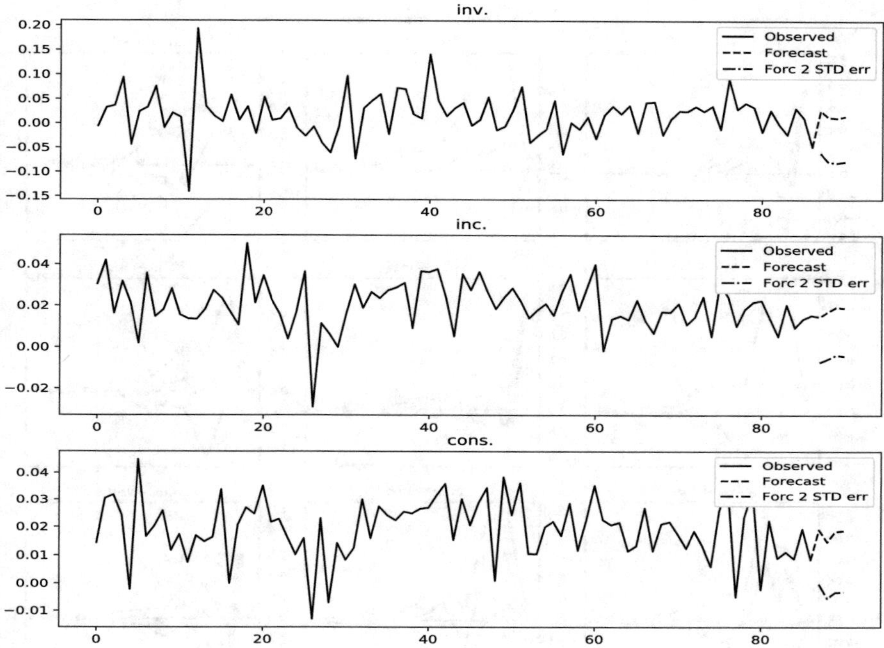


**Fig. 7.15** Forecasts of first differences of logarithms of West German economic time series

pyTSA_MacroDE

```
>>> from PythonTsa.datadir import getdtapath
>>> dtapath=getdtapath()
>>> gEco=pd.read_csv(dtapath + 'EconGermany.dat',
                     header=0,sep='\s+')
# Extension name of the data is dat, not csv or txt.
# argument sep='\s+' needed.
>>> dates=pd.date_range('1960-03', periods=len(gEco), freq='Q')
>>> gEco.index=dates
>>> gEco=gEco[['inv.', 'inc.', 'cons.']]
>>> gEco.plot(); plt.show()
>>> dlge=np.log(gEco).diff(1).dropna()
>>> dlge.plot(); plt.show()
>>> dlge.tail(4)
                inv.       inc.     cons.
1982-03-31 -0.035565  0.007226  0.004484
1982-06-30  0.028310 -0.007989  0.000894
1982-09-30  0.008459  0.003812  0.005795
1982-12-31 -0.001204  0.008714  0.009290
>>> dlgem=dlge['1960-06-30':'1981-12-31']
# leave the last four data for forecasting comparison
>>> dlgem.tail(4)
                inv.       inc.     cons.
1981-03-31 -0.022553  0.020437  0.011252
```

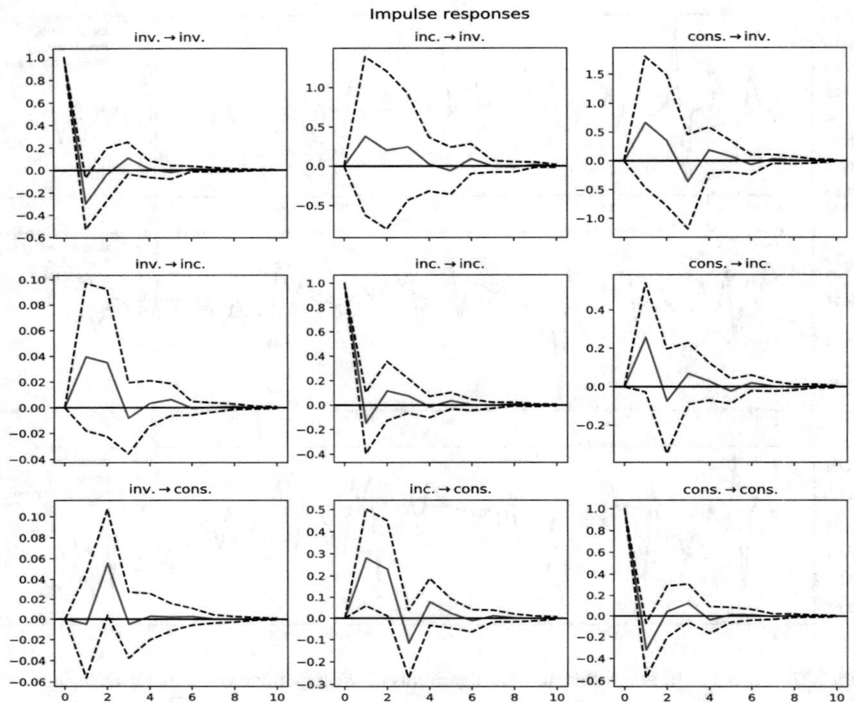


**Fig. 7.16** Impulse responses of first differences of logarithms of West German economic data

pyTSA_MacroDE

```
1981-06-30   0.031899  0.009475  0.008819
1981-09-30   0.011561  0.013659  0.019223
1981-12-31 -0.047068  0.015385  0.008576
>>> from PythonTsa.plot_multi_ACF import multi_ACFfig
>>> multi_ACFfig(dlgem, nlags=10); plt.show()
>>> dlgemMod=VAR(dlgem)
>>> print(dlgemMod.select_order(maxlags=4))
<statsmodels.tsa.vector_ar.var_model.LagOrderResults object.
Selected orders are: AIC -> 2, BIC -> 0, FPE -> 2, HQIC -> 1>
>>> print(dlgemMod.select_order(maxlags=9))
<statsmodels.tsa.vector_ar.var_model.LagOrderResults object.
Selected orders are: AIC -> 1, BIC -> 0, FPE -> 1, HQIC -> 0>
>>> dlgemRes = dlgemMod.fit(maxlags=2, ic=None, trend='c')
# ic=None since the order has been selected.
>>> print(dlgemRes.summary())
 Summary of Regression Results
==================================
Model:                         VAR
Method:                        OLS
Date:             Tue, 05, May, 2020
Time:                     11:35:35
```

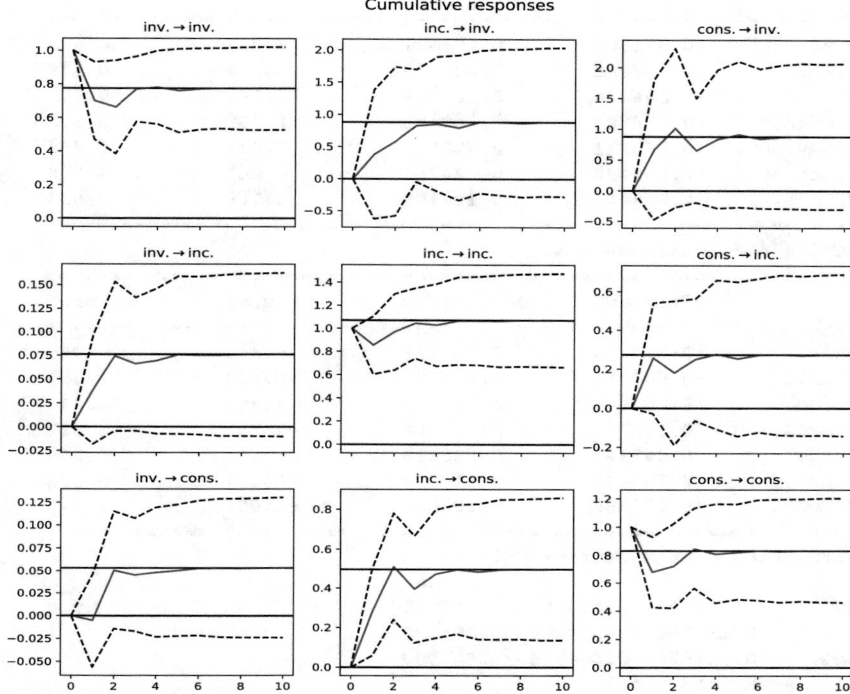


**Fig. 7.17** Cumulative responses for first differences of logs of West German economic data

pyTSA_MacroDE

```
---------------------------------------------------------------------
No. of Equations:         3.00000    BIC:                   -24.0674
Nobs:                     85.0000    HQIC:                  -24.4282
Log likelihood:           707.684    FPE:                1.93219e-11
AIC:                     -24.6709    Det(Omega_mle):     1.52386e-11
---------------------------------------------------------------------
Results for equation inv.
=====================================================================
            coefficient       std. error          t-stat        prob
---------------------------------------------------------------------
const         -0.011545         0.014813          -0.779       0.436
L1.inv.       -0.298530         0.117464          -2.541       0.011
L1.inc.        0.379067         0.512410           0.740       0.459
L1.cons.       0.665986         0.581375           1.146       0.252
L2.inv.       -0.138900         0.116888          -1.188       0.235
L2.inc.        0.182435         0.499537           0.365       0.715
L2.cons.       0.668114         0.577267           1.157       0.247
=====================================================================
Results for equation inc.
=====================================================================
            coefficient       std. error          t-stat        prob
```

```
------------------------------------------------------------------
const          0.015567         0.003699           4.208          0.000
L1.inv.        0.039380         0.029335           1.342          0.179
L1.inc.       -0.146334         0.127969          -1.144          0.253
L1.cons.       0.256283         0.145192           1.765          0.078
L2.inv.        0.053811         0.029192           1.843          0.065
L2.inc.        0.007450         0.124754           0.060          0.952
L2.cons.       0.018477         0.144166           0.128          0.898
==================================================================
Results for equation cons.
==================================================================
            coefficient      std. error          t-stat           prob
------------------------------------------------------------------
const          0.014256         0.003273           4.355          0.000
L1.inv.       -0.005134         0.025956          -0.198          0.843
L1.inc.        0.280842         0.113229           2.480          0.013
L1.cons.      -0.321276         0.128468          -2.501          0.012
L2.inv.        0.041347         0.025829           1.601          0.109
L2.inc.        0.363444         0.110384           3.293          0.001
L2.cons.      -0.128282         0.127561          -1.006          0.315
==================================================================
Correlation matrix of residuals
             inv.       inc.      cons.
inv.     1.000000  0.130754  0.316526
inc.     0.130754  1.000000  0.554385
cons.    0.316526  0.554385  1.000000
>>> dlgemRes.is_stable()
True
>>> resid=dlgemRes.resid
>>> multi_ACFfig(resid, nlags=10)
>>> plt.show()
>>> from PythonTsa.plot_multi_Q_pvalue import MultiQpvalue_plot
>>> q,p=MultiQpvalue_plot(resid,p=2,q=0,noestimatedcoef=18,nolags=24)
>>> plt.show()
>>> coefMat=dlgemRes.coefs
>>> coefMat
array([[[-0.29852954,  0.37906743,  0.66598562],
        [ 0.03938029, -0.14633432,  0.25628306],
        [-0.00513406,  0.28084152, -0.32127566]],

       [[-0.13889995,  0.18243526,  0.66811412],
        [ 0.05381098,  0.00744979,  0.01847657],
        [ 0.04134668,  0.36344357, -0.12828162]]])
>>> sigma_u=dlgemRes.sigma_u
>>> sigma_u
           inv.       inc.      cons.
inv.   0.002004  0.000065  0.000140
inc.   0.000065  0.000125  0.000061
cons.  0.000140  0.000061  0.000098
>>> dlgem=dlgem.values
# transferred to 'numpy.ndarray' class
>>> type(dlgem)
<class 'numpy.ndarray'>
# for forecast, the data needs to belong to the class 'ndarray'
```

```
>>> fore_interval=dlgemRes.forecast_interval(dlgem, steps=4)
>>> point,lower,upper=dlgemRes.forecast_interval(dlgem,steps=4)
>>> point
array([[0.02777827, 0.01473927, 0.01903946],
       [0.01350363, 0.01712411, 0.01468125],
       [0.01224347, 0.01931209, 0.01834218],
       [0.01539323, 0.01904964, 0.01862259]])
>>> lower
array([[-0.05997249, -0.0071755 , -0.00035114],
       [-0.07852948, -0.00562412, -0.0056024 ],
       [-0.08028986, -0.00372197, -0.00345532],
       [-0.0776575 , -0.0041226 , -0.00329656]])
>>> upper
array([[0.11552903, 0.03665404, 0.03843006],
       [0.10553675, 0.03987234, 0.03496491],
       [0.1047768 , 0.04234616, 0.04013969],
       [0.10844396, 0.04222187, 0.04054173]])
>>> dlgemRes.plot_forecast(steps=4); plt.show()
>>> g1=dlgemRes.test_causality(caused='cons.', causing='inc.',
      kind='f', signif=0.05)
>>> print(g1)
<statsmodels.tsa.vector_ar.hypothesis_test_results.
CausalityTestResults object.
H_0: inc. does not Granger-cause cons.:
reject at 5% significance level.
Test statistic: 7.038, critical value: 3.034>, p-value: 0.001>
>>> g2=dlgemRes.test_causality(caused='inc.', causing='cons.',
      kind='f', signif=0.05)
>>> print(g2)
<statsmodels.tsa.vector_ar.hypothesis_test_results.
CausalityTestResults object.
H_0: cons. does not Granger-cause inc.:
fail to reject at 5% significance level.
Test statistic: 1.718, critical value: 3.034>, p-value: 0.182>
>>> irf=dlgemRes.irf(periods=10)
>>> irf.plot(); plt.show()
>>> irf.plot_cum_effects(); plt.show()
```

*Example 7.8 (United States Macroeconomic Time Series)* This US quarterly macroeconomic dataset is built in the Python package `statsmodels`. It comes from the Federal Reserve Bank of St. Louis, USA. We extract the variables `realgdp, realcons, and realinv` from the dataset to constitute a three-dimensional vector time series (Fig. 7.18). Taking a logarithm on the series and then first-differencing the logarithms, we obtain the new series `dLdata` (Fig. 7.19). Both the time plot in Fig. 7.19 and the correlation function plots of `dLdata` in Fig. 7.20 suggest that the series `dLdata` (`myd`) should be stationary. We select $p = 3$ as the order of VAR($p$) model, and the alternative order $p = 1$ is left to the reader as an exercise. In addition, we adopt the `VARMAX` class instead of the `VAR` class since we want to refine estimated models, and there seems to be no such a mechanism in the `VAR` class. In the `VARMAX` class, due to estimating by state space methods, its computing results are known as the `Statespace Model Results`.

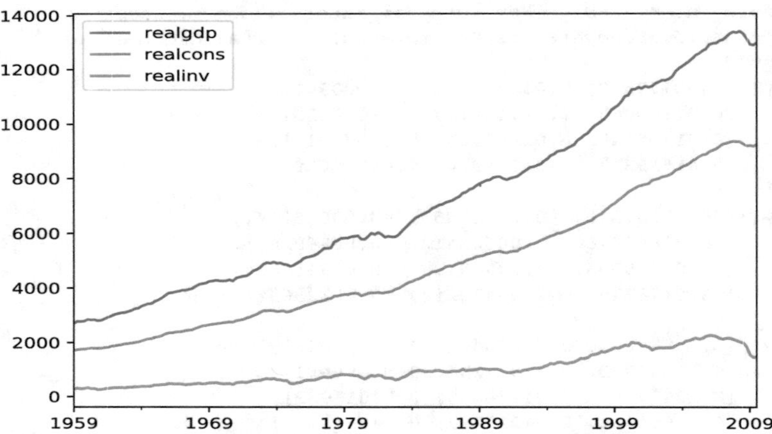


**Fig. 7.18** Time series plot of US macroeconomic data in Example 7.8

pyTSA_MacroUS

We find that there are a number of the estimates of the parameters in the firstly estimated VAR(3) model that are not statistically significant, seeing the following first table `Statespace Model Results`. What is more, the firstly fitted model residual series fails to pass the portmanteau test although its correlation plots resemble ones of a white noise, seeing Figs. 7.21 and 7.22. Therefore, we continue to refine the fitted VAR(3) model. We fix the parameter zero if its $p$-value for the estimate is greater than 0.2, that is, `P` $>$ `|z|` $>$ 0.2. But, by trial and error, we unfix the parameters (`coef`) for the variables `L1.realinv, L2.realcons` in the table `Results for equation realinv`. Note that we elect the estimating method: `method='bfgs'` since other estimating methods in the `VARMAX` class do not work for the present fitting. Finally, we obtain the secondly estimated VAR(3) model, seeing the following second table `Statespace Model Results`. Moreover, we observe that the secondly estimated VAR(3) model residual series passes the portmanteau test, and its correlation function plots look like ones of a vector white noise, seeing Figs. 7.23 and 7.24. Thus, the secondly estimated VAR(3) model is adequate and acceptable. Besides, AIC $= -3878.198$, BIC $= -3809.039$, and HQIC $= -3850.207$ for the secondly estimated VAR(3) model are, respectively, smaller than AIC $= -3856.548$, BIC $= -3737.989$, and HQIC $= -3808.564$ for the firstly estimated VAR(3) model. That is, the second is more appropriate than the first. Now it is time to predict the time series. Checking the predicted (or fitted) values in Fig. 7.25, we see that the predictions are satisfactory in general. But for some actual values at spikes, the fitted values are not so good.

Below is the Python code and some results for this example.

```
>>> import numpy as np
>>> import pandas as pd
>>> import matplotlib.pyplot as plt
```

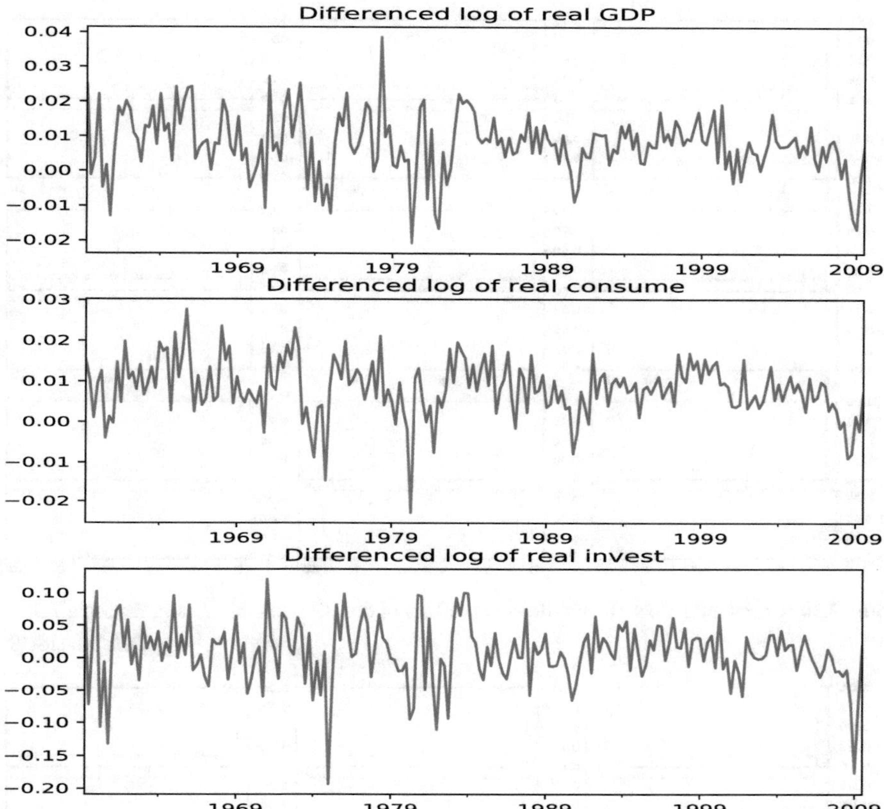


**Fig. 7.19** Time series plot of the first differences of logs of the US economic data in Example 7.8
pyTSA_MacroUS

```
>>> from statsmodels.tsa.api import VAR
>>> from statsmodels.tsa.api import VARMAX
>>> import statsmodels.api as sm
>>> mdata = sm.datasets.macrodata.load_pandas().data
>>> mdata = mdata[['realgdp','realcons','realinv']]
>>> dates=pd.date_range('1959-01', periods=len(mdata), freq='Q')
>>> mdata.index=dates
>>> mdata.plot(); plt.show()
>>> dLdata = np.log(mdata).diff(1).dropna()
# log, difference and then drop NAN
>>> fig = plt.figure()
>>> dLdata['realgdp'].plot(ax= fig.add_subplot(311))
>>> plt.title('Differenced log of real GDP')
>>> dLdata['realcons'].plot(ax= fig.add_subplot(312))
>>> plt.title('Differenced log of real consume')
>>> dLdata['realinv'].plot(ax= fig.add_subplot(313))
```

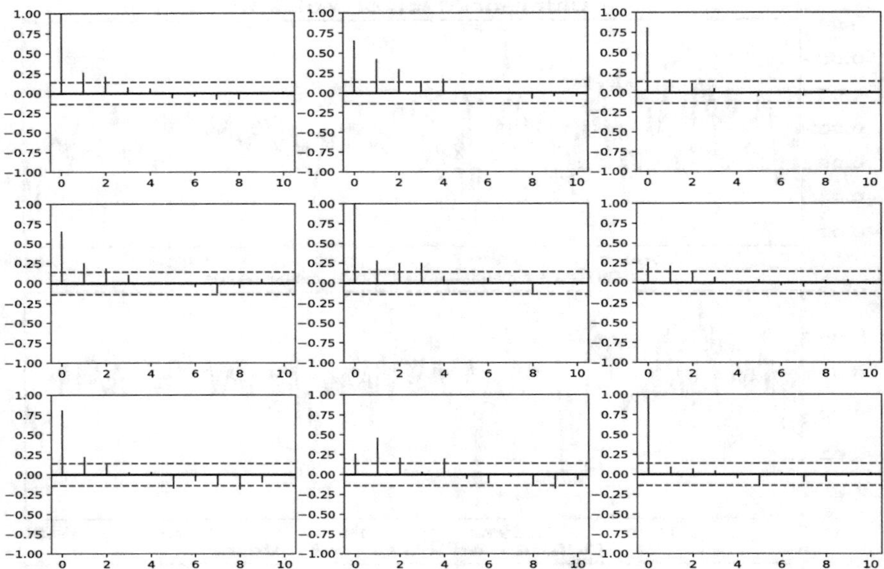

**Fig. 7.20** Correlation plots of the differences of logs of the US economic data in Example 7.8

pyTSA_MacroUS

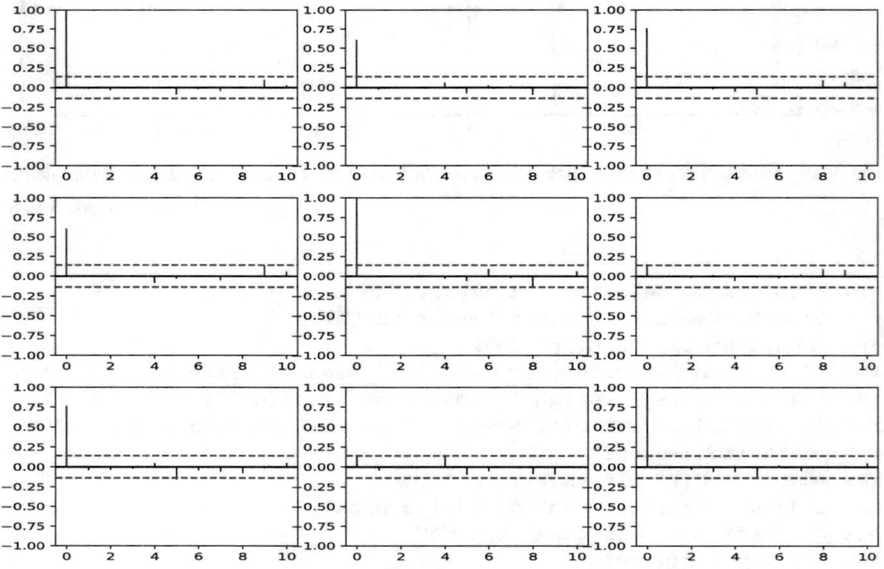

**Fig. 7.21** Residual correlation plots of the firstly estimated model in Example 7.8

pyTSA_MacroUS

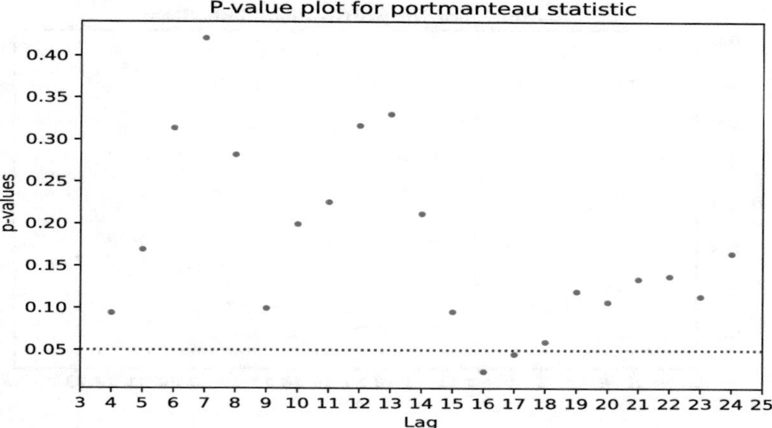


**Fig. 7.22** *p*-Values for portmanteau test of the firstly estimated model residuals in Example 7.8

pyTSA_MacroUS

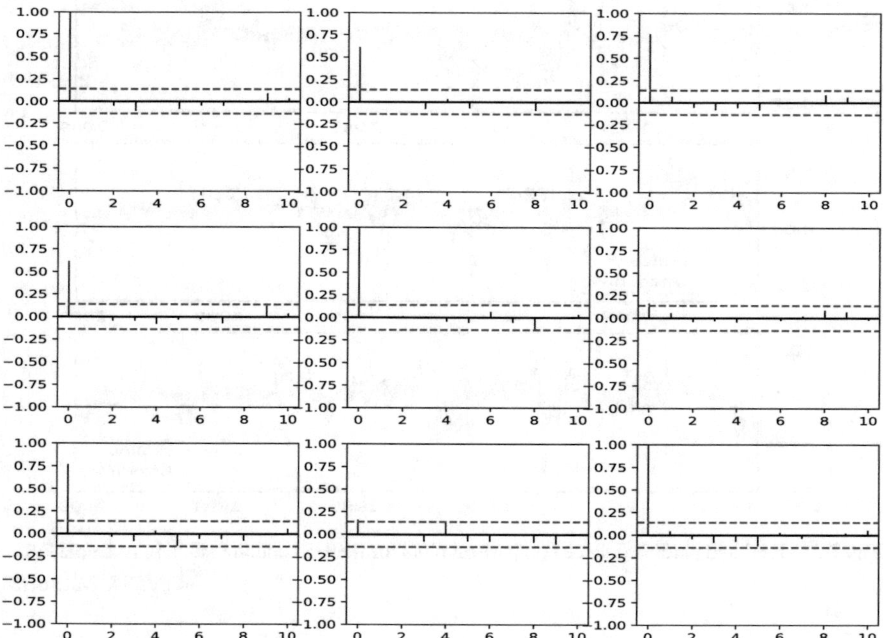

**Fig. 7.23** Residual correlation plots of the refined model in Example 7.8

pyTSA_MacroUS

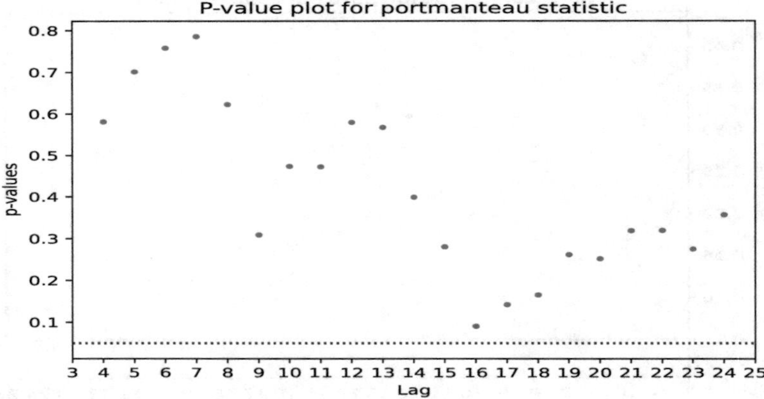


**Fig. 7.24** *p*-Value plot for portmanteau test of the refined model residuals in Example 7.8

pyTSA_MacroUS

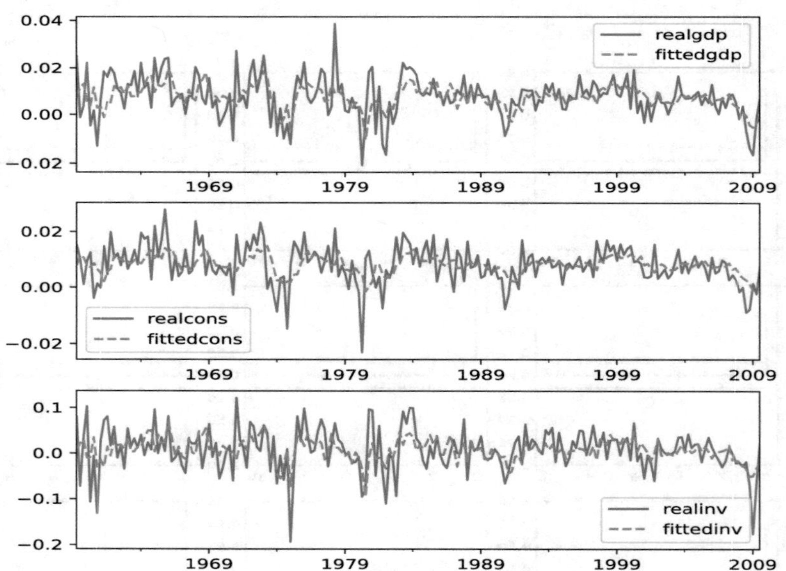


**Fig. 7.25** Real and predicted value comparison plots for the US economic data in Example 7.8

pyTSA_MacroUS

```
>>> plt.title('Differenced log of real invest')
>>> plt.show()
>>> dLdata.tail(3)
             realgdp  realcons   realinv
2009-03-31 -0.016612  0.001511 -0.175598
2009-06-30 -0.001851 -0.002196 -0.067561
```

```
2009-09-30  0.006862  0.007265  0.020197
>>> myd=dLdata['1959-06-30' : '2008-12-31']
# leave the last three data for forecasting comparison
>>> myd.tail(4)
             realgdp  realcons   realinv
2008-03-31 -0.001823 -0.001496 -0.019276
2008-06-30  0.003614  0.000150 -0.027435
2008-09-30 -0.006781 -0.008948 -0.017836
2008-12-31 -0.013805 -0.007843 -0.069165
>>> from PythonTsa.plot_multi_ACF import multi_ACFfig
>>> multi_ACFfig(myd, nlags=10)
>>> plt.show()
>>> mydmod1=VAR(myd)
>>> print(mydmod1.select_order(maxlags=10))
<statsmodels.tsa.vector_ar.var_model.LagOrderResults object.
Selected orders are: AIC -> 1, BIC -> 1, FPE -> 1, HQIC -> 1>
>>> print(mydmod1.select_order(maxlags=11))
<statsmodels.tsa.vector_ar.var_model.LagOrderResults object.
Selected orders are: AIC -> 3, BIC -> 1, FPE -> 3, HQIC -> 1>
# here select p=3 for VAR(p).
>>> mydmod=VARMAX(myd, order=(3,0), enforce_stationarity=True)
>>> modfit=mydmod.fit(disp=False)
>>> print(modfit.summary())
           Statespace Model Results
=================================================================
Dep. Variable:
       ['realgdp','realcons','realinv']  No. Observations:    199
Model:                           VAR(3)  Log Likelihood  1964.274
                            + intercept  AIC            -3856.548
Date:                  Fri, 01 May 2020  BIC            -3737.989
Time:                          21:10:22  HQIC           -3808.564
Sample:                      06-30-1959
                           - 12-31-2008
Covariance Type:                    opg
=================================================================
Ljung-Box (Q):34.74,44.73,40.31 Jarque-Bera (JB):15.72,0.90,11.03
Prob(Q):       0.71, 0.28, 0.46 Prob(JB):         0.00, 0.64, 0.00
Heteroskedasticity (H):
                0.25,0.38,0.51 Skew:             0.39, -0.03, -0.37
Prob(H)(two-sided):
                0.00,0.00,0.01 Kurtosis:          4.14, 3.32, 3.88
                Results for equation realgdp
=================================================================
                coef  std err        z    P>|z|   [0.025    0.975]
-----------------------------------------------------------------
intercept     0.0016    0.001    1.191    0.234   -0.001     0.004
L1.realgdp   -0.3249    0.228   -1.425    0.154   -0.772     0.122
L1.realcons   0.6898    0.162    4.271    0.000    0.373     1.006
L1.realinv    0.0348    0.038    0.907    0.364   -0.040     0.110
L2.realgdp    0.0033    0.234    0.014    0.989   -0.455     0.461
L2.realcons   0.2890    0.174    1.661    0.097   -0.052     0.630
L2.realinv   -0.0080    0.033   -0.247    0.805   -0.072     0.056
```

```
L3.realgdp  -0.1376     0.235  -0.586   0.558   -0.598         0.323
L3.realcons  0.1537     0.172   0.893   0.372   -0.184         0.491
L3.realinv   0.0076     0.037   0.205   0.837   -0.065         0.081
               Results for equation realcons
=================================================================
              coef  std err         z   P>|z|   [0.025        0.975]
-----------------------------------------------------------------
intercept    0.0048     0.001   3.861   0.000    0.002         0.007
L1.realgdp  -0.1393     0.164  -0.849   0.396   -0.461         0.182
L1.realcons  0.2724     0.112   2.429   0.015    0.053         0.492
L1.realinv   0.0249     0.027   0.912   0.362   -0.029         0.079
L2.realgdp  -0.0980     0.205  -0.477   0.633   -0.500         0.304
L2.realcons  0.2136     0.164   1.300   0.194   -0.108         0.536
L2.realinv   0.0069     0.028   0.252   0.801   -0.047         0.061
L3.realgdp  -0.3604     0.193  -1.869   0.062   -0.738         0.018
L3.realcons  0.4186     0.135   3.111   0.002    0.155         0.682
L3.realinv   0.0424     0.031   1.381   0.167   -0.018         0.103
                Results for equation realinv
=================================================================
              coef  std err         z   P>|z|   [0.025        0.975]
-----------------------------------------------------------------
intercept   -0.0163     0.008  -2.148   0.032   -0.031        -0.001
L1.realgdp  -2.1033     1.133  -1.856   0.063   -4.324         0.118
L1.realcons  4.4487     0.816   5.450   0.000    2.849         6.049
L1.realinv   0.2297     0.192   1.198   0.231   -0.146         0.605
L2.realgdp   0.0649     1.320   0.049   0.961   -2.522         2.652
L2.realcons  0.8353     0.961   0.869   0.385   -1.048         2.718
L2.realinv  -0.0336     0.173  -0.194   0.846   -0.373         0.306
L3.realgdp  -0.0558     1.136  -0.049   0.961   -2.283         2.171
L3.realcons -0.4920     0.938  -0.524   0.600   -2.330         1.346
L3.realinv  -0.0181     0.182  -0.099   0.921   -0.375         0.339
                   Error covariance matrix
=================================================================
                             coef std err    z  P>|z| [0.025 0.975]
-----------------------------------------------------------------
sqrt.var.realgdp           0.0074 0.000 16.070 0.000  0.007  0.008
sqrt.cov.realgdp.realcons  0.0038 0.000  9.800 0.000  0.003  0.005
sqrt.var.realcons          0.0050 0.000 16.969 0.000  0.004  0.006
sqrt.cov.realgdp.realinv   0.0291 0.003 10.928 0.000  0.024  0.034
sqrt.cov.realcons.realinv-0.0152 0.002 -7.900 0.000 -0.019 -0.011
sqrt.var.realinv           0.0195 0.001 14.312 0.000  0.017  0.022
=================================================================
Warnings:
[1] Covariance matrix calculated using the outer product
    of gradients (complex-step).
>>> resid=modfit.resid
>>> multi_ACFfig(resid, nlags=10)
>>> plt.show()
>>> from PythonTsa.plot_multi_Q_pvalue import MultiQpvalue_plot
>>> qs, pv = MultiQpvalue_plot(resid, p=3, q=0,
             noestimatedcoef=27, nolags=24)
>>> plt.show()
```

```
>>> param=mydmod.param_names
>>> mydmodf=VARMAX(myd, order=(3,0), enforce_stationarity=False)
# Cannot fix individual autoregressive parameters when
# 'enforce_stationarity=True'. In this case, must either
# fix all autoregressive parameters or none.
>>> with mydmodf.fix_params({param[0]: 0.0, param[5]: 0.0,
param[6]: 0.0, param[8]: 0.0, param[9]: 0.0, param[11]: 0.0,
param[12]: 0.0, param[14]: 0.0, param[15]: 0.0, param[17]: 0.0,
param[24]: 0.0, param[26]: 0.0,param[27]: 0.0, param[28]: 0.0,
param[29]: 0.0}):
        modff = mydmodf.fit(method='bfgs')

Optimization terminated successfully.
         Current function value: -9.849744
         Iterations: 44
         Function evaluations: 57
         Gradient evaluations: 57
>>> print(modff.summary())
                 Statespace Model Results
===============================================================
Dep.Variable:
     ['realgdp','realcons','realinv']  No.Observations:    199
Model:                          VAR(3)  Log Likelihood 1960.099
                           + intercept  AIC            -3878.198
Date:                 Sat, 02 May 2020  BIC            -3809.039
Time:                         05:57:48  HQIC           -3850.207
Sample:                     06-30-1959
                          - 12-31-2008
Covariance Type:                   opg
================================================================
Ljung-Box(Q):37.29,46.84,40.92  Jarque-Bera (JB):16.50,1.15,8.05
Prob(Q):        0.59,0.21,0.43  Prob(JB):          0.00,0.56,0.02
Heteroskedasticity(H):
                0.25,0.41,0.54  Skew:            0.43,-0.07,-0.34
Prob(H)(two-sided):
                0.00,0.00,0.01  Kurtosis:          4.12,3.35,3.72
                Results for equation realgdp
================================================================
                      coef  std err       z  P>|z|  [0.025  0.975]
----------------------------------------------------------------
intercept (fixed)        0      nan     nan    nan     nan     nan
L1.realgdp         -0.1618    0.069  -2.348  0.019  -0.297  -0.027
L1.realcons         0.5946    0.099   5.987  0.000   0.400   0.789
L1.realinv (fixed)       0      nan     nan    nan     nan     nan
L2.realgdp (fixed)       0      nan     nan    nan     nan     nan
L2.realcons         0.3274    0.071   4.635  0.000   0.189   0.466
L2.realinv (fixed)       0      nan     nan    nan     nan     nan
L3.realgdp (fixed)       0      nan     nan    nan     nan     nan
L3.realcons         0.1337    0.059   2.277  0.023   0.019   0.249
L3.realinv (fixed)       0      nan     nan    nan     nan     nan
                Results for equation realcons
================================================================
```

```
                       coef   std err        z   P>|z|    [0.025   0.975]
-------------------------------------------------------------------------
intercept            0.0038     0.001   5.191   0.000     0.002    0.005
L1.realgdp (fixed)        0       nan     nan     nan       nan      nan
L1.realcons          0.2016     0.065   3.088   0.002     0.074    0.330
L1.realinv (fixed)        0       nan     nan     nan       nan      nan
L2.realgdp (fixed)        0       nan     nan     nan       nan      nan
L2.realcons          0.1728     0.064   2.689   0.007     0.047    0.299
L2.realinv (fixed)        0       nan     nan     nan       nan      nan
L3.realgdp          -0.2285     0.098  -2.334   0.020    -0.420   -0.037
L3.realcons          0.3338     0.088   3.801   0.000     0.162    0.506
L3.realinv           0.0325     0.017   1.916   0.055    -0.001    0.066
                 Results for equation realinv
=========================================================================
                       coef   std err        z   P>|z|    [0.025   0.975]
-------------------------------------------------------------------------
intercept           -0.0231    0.004  -5.963    0.000    -0.031   -0.016
L1.realgdp          -1.6803    0.625  -2.690    0.007    -2.905   -0.456
L1.realcons          4.1178    0.553   7.448    0.000     3.034    5.201
L1.realinv           0.1101    0.079   1.393    0.164    -0.045    0.265
L2.realgdp (fixed)        0      nan     nan      nan       nan      nan
L2.realcons          1.0428    0.366   2.851    0.004     0.326    1.760
L2.realinv (fixed)        0      nan     nan      nan       nan      nan
L3.realgdp (fixed)        0      nan     nan      nan       nan      nan
L3.realcons (fixed)       0      nan     nan      nan       nan      nan
L3.realinv (fixed)        0      nan     nan      nan       nan      nan
                    Error covariance matrix
=========================================================================
                             coef std err      z  P>|z|  [0.025  0.975]
-------------------------------------------------------------------------
sqrt.var.realgdp           0.0076 0.000 17.988 0.000   0.007   0.008
sqrt.cov.realgdp.realcons 0.0039 0.000 10.476 0.000   0.003   0.005
sqrt.var.realcons          0.0050 0.000 19.189 0.000   0.004   0.006
sqrt.cov.realgdp.realinv  0.0296 0.003 11.829 0.000   0.025   0.035
sqrt.cov.realcons.realinv-0.0152 0.002 -9.258 0.000  -0.018  -0.012
sqrt.var.realinv           0.0197 0.001 15.735 0.000   0.017   0.022
=========================================================================
Warnings:
[1] Covariance matrix calculated using the outer product
    of gradients (complex-step).
>>> residf=modff.resid
>>> multi_ACFfig(residf, nlags=10)
>>> plt.show()
>>> qs, pv = MultiQpvalue_plot(residf, p=3, q=0,
               noestimatedcoef=13, nolags=24)
>>> plt.show()
>>> fore=modff.predict(end='2009-09-30')
>>> realgdpFitgdp=pd.DataFrame({'realgdp':dLdata['realgdp'],
                     'fittedgdp':fore['realgdp']})
>>> realconsFitcons=pd.DataFrame({'realcons':dLdata['realcons'],
                     'fittedcons':fore['realcons']})
>>> realinvFitinv=pd.DataFrame({'realinv':dLdata['realinv'],
```

```
                    'fittedinv':fore['realinv']})
>>> fig = plt.figure()
>>> realgdpFitgdp.plot(style=['-','--'], ax=fig.add_subplot(311))
>>> realconsFitcons.plot(style=['-','--'],ax=fig.add_subplot(312))
>>> realinvFitinv.plot(style=['-','--'], ax=fig.add_subplot(313))
>>> plt.show()
```

In this chapter, we have introduced the fundamental concepts and methods of multivariate time series analysis and how to perform an analysis of vector time series with Python. We see that the curse of dimensionality is still a tough problem to be solved. In recent years, several monographs on multivariate time series analysis have been published, for example, Tsay (2014), Gómez (2016), and Wei (2019), among others. Gómez (2016) is a book on multivariate time series analysis by state space methods, and it is the state space methods that are the main topic in the next chapter. And Lütkepohl (2005) has become a classic on multiple time series analysis. Interested readers can refer to them for more theoretical and technological details.

## Problems

**7.1** Suppose that $\boldsymbol{\varepsilon}_t \sim \mathrm{N}(\mathbf{0}, \boldsymbol{\Sigma})$ is a five-dimensional Gauss (normal) white noise where

$$\boldsymbol{\Sigma} = \begin{bmatrix} 1.0 & 0.6 & 0.2 & 0.1 & 0.3 \\ 0.6 & 1.0 & 0.1 & 0.4 & 0.2 \\ 0.2 & 0.1 & 1.0 & 0.5 & 0.1 \\ 0.1 & 0.4 & 0.5 & 1.0 & 0.2 \\ 0.3 & 0.2 & 0.1 & 0.2 & 1.0. \end{bmatrix}$$

Simulate a sample of size 10,000 from the white noise. Then plot the sample correlation functions, the $p$-value plot for the extended trace of the sample correlation matrix, and the $p$-value plot for the portmanteau test of this sample. Finally, explain the features of these plots. (Suggestion: for deeper understanding, without setting the seed, do the exercise more than one time!)

**7.2** Prove that the VARMA(1,1) model in Example 7.6 is both causal and invertible for any real numbers $a$, $b$, and then find its VAR representation.

**7.3** Given a three-dimensional VAR(2) process

$$\mathbf{X}_t = \boldsymbol{\nu} + \boldsymbol{\Phi}_1 \mathbf{X}_{t-1} + \boldsymbol{\Phi}_2 \mathbf{X}_{t-2} + \boldsymbol{\varepsilon}_t$$

where $\mathbf{X}_t = (X_{t1}, X_{t2}, X_{t3})'$, $\boldsymbol{\nu} = (2, 1, 0)'$, $\boldsymbol{\varepsilon}_t \sim \text{WN}(\mathbf{0}, \boldsymbol{\Sigma})$, and

$$\boldsymbol{\Phi}_1 = \begin{bmatrix} 0.7 & 0.1 & 0.0 \\ 0.0 & 0.4 & 0.1 \\ 0.9 & 0.0 & 0.8 \end{bmatrix}, \quad \boldsymbol{\Phi}_2 = \begin{bmatrix} -0.2 & 0.0 & 0.0 \\ 0.0 & 0.1 & 0.1 \\ 0.0 & 0.0 & 0.0 \end{bmatrix}, \quad \boldsymbol{\Sigma} = \begin{bmatrix} 0.26 & 0.03 & 0.0 \\ 0.03 & 0.09 & 0.0 \\ 0.0 & 0.0 & 0.81 \end{bmatrix}.$$

(1) Prove that the process $\mathbf{X}_t$ is causal (stable).
(2) Determine the mean vector of $\mathbf{X}_t$.
(3) Rewrite the process $\mathbf{X}_t$ in VAR(1) form.
(4) Estimate the coefficient matrices of its VMA($\infty$) representation.
(5) Carry out Granger causality tests of this VAR(2) process.
(6) Perform impulse response analysis of this VAR(2) process.

**7.4** In Example 7.8, by running the command `mydmod.select_order`, we see that the information criteria and FPE also select $p = 1$ for the VAR($p$) model to be fitted to the dataset `myd`. Surely, VAR(1) is more parsimonious than VAR(3). Now fit VAR(1) to the dataset `myd`, and explain whether your fitted model is adequate. Besides, try to fit VAR(2) to the dataset `myd` and see what happens.

**7.5** The dataset "dlGDPukcaus1q1980.csv" in the folder Ptsadata is the quarterly GDP growth rates of the United Kingdom, Canada, and United States from the first quarter of 1980 to the first quarter of 2011. Build a VAR model for the time series data, and diagnostically check the model residuals. If necessary, refine the estimated model.

**7.6** In Example 7.3, we have concluded that we should consider to build an adequate model for the differenced series `dmda`. Now model the series `dmda`, diagnose the fitted model, and refine it.

# Chapter 8
# State Space Models and Markov Switching Models

The state space methods or models provide a unified and flexible methodology and technology for handling a wide range of problems in time series analysis and are also applied in other fields including artificial intelligence. This chapter introduces the basic principle of state space methods and its application to SARIMAX modeling with Python, presents relationship between state space models and ARIMAX models using the local-level model, and lastly discusses the Markov switching model which is useful in econometrics and other disciplines.

## 8.1 State Space Models and Representations

The state space model or method is developed in Kalman (1960) and Kalman and Bucy (1961) for control engineering and actually is a very general model that subsumes a whole class of special cases of interest. State space modeling provides a unified approach to dealing with a wide range of problems in time series analysis. Since the 1970s, a large number of monographs and textbooks on state space modeling have been published. For example, Tsay and Chen (2019) gives a lot of examples to classical and novel applications of state space methods. Gómez (2016) uses the state space methods to analyze multivariate time series. Casals et al. (2016) discusses the theory and application of time series state space methods. Harvey (1989) has an online version published in 2014 and focuses on analysis and forecasting of structural time series models using state space approach. Durbin and Koopman (2012) is an excellent textbook on making time series analysis by state space methods. In addition, Many time series books include this topic. For example, Shumway and Stoffer (2017, Chapter 6), Brockwell and Davis (2016, Chapter 9), and Tsay (2010, Chapter 11) all have a chapter to address the state space methods and technology. In particular, for introduction to application of the



C. Huang, A. Petukhina, *Applied Time Series Analysis and Forecasting with Python*, Statistics and Computing,
https://doi.org/10.1007/978-3-031-13584-2_8

state space approach in the field of artificial intelligence, see, for instance, Russell and Norvig (2021, Chapter 14) or (Russell and Norvig, Chapter 15).

### 8.1.1 State Space Models

How is the state space model formulated? Suppose that there is a system, in which $\mathbf{X}_t$ is the (unobserved) state vector of the system at time $t$ and $\mathbf{Y}_t$ is an observation or measurement vector at time $t$. A general state space model for the system consists of two equations: one describes how the system output (observation) is realized as a linear combination of different dynamic components and the other characterize the system dynamics, that is, the evolution of the system over time. Specifically, we give the following definition.

**Definition 8.1** A linear state space model is of the form

$$\mathbf{X}_{t+1} = \mathbf{c}_t + \mathbf{F}_t\mathbf{X}_t + \mathbf{D}_t\mathbf{u}_t + \mathbf{R}_t\boldsymbol{\eta}_t, \ \boldsymbol{\eta}_t \sim \mathrm{N}(\mathbf{0}, \mathbf{Q}_t), \tag{8.1}$$

$$\mathbf{Y}_t = \mathbf{d}_t + \mathbf{Z}_t\mathbf{X}_t + \mathbf{B}_t\mathbf{u}_t + \boldsymbol{\varepsilon}_t, \ \boldsymbol{\varepsilon}_t \sim \mathrm{N}(\mathbf{0}, \mathbf{H}_t), \tag{8.2}$$

where $\mathbf{X}_t$ is an $m$-dimensional (unobserved) state vector and called the state variables, $\mathbf{Y}_t$ is a $k$-dimensional observation vector and known as the endogenous variables, $\mathbf{u}_t$ is an $n$-dimensional input vector and referred to as the control or exogenous variables, $\boldsymbol{\eta}_t$ is an $r$-dimensional state error (disturbance or noise), and $\boldsymbol{\varepsilon}_t$ is a $k$-dimensional observation error (disturbance or noise). Equations (8.1) and (8.2) are, respectively, known as the state and observation equations.

Remarks about Definition 8.1:

- Equations (8.1) and (8.2) are sometimes also called the *transition equation* and *measurement equation*, respectively.
- The state intercept $\mathbf{c}_t$ is an $m$-dimensional vector, and coefficients $\mathbf{F}_t$, $\mathbf{D}_t$, and $\mathbf{R}_t$ are, respectively, $m \times m$, $m \times n$, and $m \times r$ matrices.
- The observation intercept $\mathbf{d}_t$ is a $k$-dimensional vector, and coefficients $\mathbf{Z}_t$ and $\mathbf{B}_t$ are, respectively, $k \times m$ and $k \times n$ matrices.
- The state error covariance matrix $\mathbf{Q}_t$ and observation error covariance matrix $\mathbf{H}_t$ are, respectively, $r \times r$ and $k \times k$ positive semi-definite matrices.
- It is usually assumed that $\boldsymbol{\eta}_t$ and $\boldsymbol{\varepsilon}_t$ are uncorrelated, namely, $\mathrm{E}(\boldsymbol{\eta}_s\boldsymbol{\varepsilon}'_t) = \mathbf{0}$ for all $s$ and $t$, but both of them may be correlated if needed. See, for example, Gómez (2016, Chapter 4).
- The initial state vector $\mathbf{X}_1$ is assumed to be uncorrelated with all of the error terms $\boldsymbol{\eta}_t$ and $\boldsymbol{\varepsilon}_t$ and often $\mathbf{X}_1 \sim \mathrm{N}(\mathbf{a}_1, \mathbf{P}_1)$. Moreover, the mean $\mathbf{a}_1$ and covariance matrix $\mathbf{P}_1$ of the distribution of $\mathbf{X}_1$ must be first specified, which is called *initialization*.
- In many important special cases, including ARIMA, VARMA, and so forth, the parameters $\mathbf{c}_t$, $\mathbf{F}_t$, $\mathbf{D}_t$, $\mathbf{R}_t$, $\mathbf{Q}_t$, $\mathbf{d}_t$, $\mathbf{Z}_t$, $\mathbf{B}_t$, *and* $\mathbf{H}_t$ are all time-invariant, and

thus the time subscripts will be dropped. In this case, the state space model is also referred to as being time homogeneous.

*Example 8.1 (CAPM with Time-Varying Coefficients)* In finance, the capital asset pricing model (CAPM) provides a theoretical structure for the pricing of assets with risky returns and is used to determine an appropriate required rate of return of a financial asset. Now we consider a CAPM with time-varying coefficients as follows:

$$R_t = \alpha_t + \beta_t R_{M,t} + \varepsilon_t, \quad \varepsilon_t \sim \text{iidN}(0, \sigma_\varepsilon^2)$$

where $R_t$ is the excess return of an asset and $R_{M,t}$ is the excess return of the market. What is more, it allows for the time-varying intercept $\alpha_t$ and slope $\beta_t$ to evolve like a random walk. In other words, $\alpha_t$ and $\beta_t$, respectively, follow

$$\alpha_t = \alpha_{t-1} + \eta_t, \quad \eta_t \sim \text{iidN}(0, \sigma_\eta^2) \text{ and } \beta_t = \beta_{t-1} + \upsilon_t, \quad \upsilon_t \sim \text{iidN}(0, \sigma_\upsilon^2)$$

where the innovations $\varepsilon_t, \eta_t, and \upsilon_t$ are assumed to be mutually uncorrelated. If we let the state vector $\mathbf{X}_t = (\alpha_t, \beta_t)'$, then we can easily rewrite the CAPM as a state space model

$$\mathbf{X}_{t+1} = \mathbf{X}_t + \boldsymbol{\eta}_t,$$
$$R_t = [1, R_{M,t}]\mathbf{X}_t + \varepsilon_t$$

where $\boldsymbol{\eta}_t = (\eta_{t+1}, \upsilon_{t+1})'$.

## 8.1.2 State Space Representations of Time Series

What is a state space form or representation for a time series? We give the following definition.

**Definition 8.2** A vector time series $\mathbf{Y}_t$ is said to have a state space form or representation if there is a state space model for $\mathbf{Y}_t$ as specified by Eqs. (8.1) and (8.2).

There are a huge number of time series that can be represented in state space form. They include all the processes generated by SARIMA and VARMAX models in the previous chapters. Now let us have a look at a few examples.

*Example 8.2 (State Space Form of the MA(1) Model)* Given the following MA(1) model

$$Y_t = \mu + \varepsilon_t + \theta\varepsilon_{t-1},$$

and let $X_{t+1} = \theta\varepsilon_t$, then we obtain the representation of the MA(1) model as follows:

$$X_{t+1} = 0 \cdot X_t + \theta\varepsilon_t,$$
$$Y_t = \mu + X_t + \varepsilon_t.$$

*Example 8.3 (State Space Form of the ARMA(1,1) Model)* Consider the following causal and invertible ARMA(1,1) model

$$Y_t = \varphi_0 + \varphi Y_{t-1} + \varepsilon_t + \theta\varepsilon_{t-1}. \tag{8.3}$$

Let the state variable $X_t = Y_t - \varepsilon_t$. Then we have

$$X_{t+1} = \varphi_0 + \varphi X_t + (\theta + \varphi)\varepsilon_t, \tag{8.4}$$

$$Y_t = X_t + \varepsilon_t. \tag{8.5}$$

These two equations form a state space representation for the ARMA(1,1) model (8.3) and are equivalent to it.

Note that there exist other state space representations for the ARMA(1,1) model. In fact, let $X_t = \varphi X_{t-1} + \varepsilon_t$, that is, an AR(1) process and the state vector $\mathbf{X}_t = (X_{t-1}, X_t)'$. Then we arrive at the equivalent model

$$\begin{bmatrix} X_t \\ X_{t+1} \end{bmatrix} = \begin{bmatrix} 0 & 1 \\ 0 & \varphi \end{bmatrix} \begin{bmatrix} X_{t-1} \\ X_t \end{bmatrix} + \begin{bmatrix} 0 \\ \varepsilon_{t+1} \end{bmatrix}, \text{ namely, } \mathbf{X}_{t+1} = \begin{bmatrix} 0 & 1 \\ 0 & \varphi \end{bmatrix} \mathbf{X}_t + \begin{bmatrix} 0 \\ 1 \end{bmatrix} \eta_t,$$

where $\eta_t = \varepsilon_{t+1}$. We say that this equation is simply the state equation for (8.3), and the observation equation is

$$Y_t = \varphi_0(1 - \varphi)^{-1} + \theta X_{t-1} + X_t = \varphi_0(1 - \varphi)^{-1} + \begin{bmatrix} \theta & 1 \end{bmatrix} \mathbf{X}_t. \tag{8.6}$$

Actually, at this moment, due to $X_t = \varphi X_{t-1} + \varepsilon_t$, we have

$$\begin{aligned} Y_t - \varphi Y_{t-1} &= \varphi_0(1 - \varphi)^{-1} + \theta X_{t-1} + X_t - \varphi[\varphi_0(1 - \varphi)^{-1} + \theta X_{t-2} + X_{t-1}] \\ &= \varphi_0(1 - \varphi)^{-1}(1 - \varphi) + (\theta - \varphi)X_{t-1} + X_t - \varphi\theta X_{t-2} \\ &= \varphi_0 + (\theta - \varphi)X_{t-1} + X_t - \theta(X_{t-1} - \varepsilon_{t-1}) \\ &= \varphi_0 + X_t - \varphi X_{t-1} + \theta\varepsilon_{t-1} \\ &= \varphi_0 + \varepsilon_t + \theta\varepsilon_{t-1}. \end{aligned}$$

This means that Eq. (8.3) holds. Thus Eq. (8.6) is the observation equation for (8.3).

Example 8.3 illustrates that there may exist a few state space form for a single time series. In other words, state space representations are not unique. Furthermore, it is worthy of noting that any linear state space model can be conversely translated into an equivalent VARMAX representation. See, for example, Casals et al. (2016, Chapter 9) and Gómez (2016).

## 8.2 Kalman Recursions

In this section, for simplicity and highlighting key ideas, we do not consider including the control variables in the state space model (8.1)–(8.2). That is, the state space model is now as follows:

$$\mathbf{X}_{t+1} = \mathbf{c}_t + \mathbf{F}_t\mathbf{X}_t + \mathbf{R}_t\boldsymbol{\eta}_t, \quad \boldsymbol{\eta}_t \sim \mathrm{N}(\mathbf{0}, \mathbf{Q}_t), \tag{8.7}$$

$$\mathbf{Y}_t = \mathbf{d}_t + \mathbf{Z}_t\mathbf{X}_t + \boldsymbol{\varepsilon}_t, \quad \boldsymbol{\varepsilon}_t \sim \mathrm{N}(\mathbf{0}, \mathbf{H}_t), \tag{8.8}$$

$$\mathbf{X}_1 \sim \mathrm{N}(\mathbf{a}_1, \mathbf{P}_1), \tag{8.9}$$

which is adequate for most purposes.

Once a state space model is specified for the observations $\mathbf{Y}_{1:t} = \{\mathbf{Y}_1, \mathbf{Y}_2, \cdots, \mathbf{Y}_t\}$, we can then consider a number of important algorithms and their applications. These algorithms are all recursion ones and often called the *Kalman recursions* due to the seminal papers by Kalman (1960) and Kalman and Bucy (1961). There are three fundamental problems associated with the state space model (8.7)–(8.9) as follows:

- *Kalman filtering*: Given $\mathbf{Y}_{1:t}$, to recover or estimate $\mathbf{X}_t$
- *Kalman forecasting*: Given $\mathbf{Y}_{1:t}$, to forecast $\mathbf{X}_s$ where $s > t$
- *Kalman smoothing*: Given $\mathbf{Y}_{1:t}$, to estimate $\mathbf{X}_s$ where $s < t$

For simplicity, we introduce the following notations. Let $\mathbf{X}_{h|j} = \mathsf{E}(\mathbf{X}_h|\mathbf{Y}_{1:j})$ and $\mathbf{P}_{h|j} = \mathrm{Var}(\mathbf{X}_h|\mathbf{Y}_{1:j}) = \mathsf{E}[(\mathbf{X}_h - \mathbf{X}_{h|j})(\mathbf{X}_h - \mathbf{X}_{h|j})'|\mathbf{Y}_{1:j}]$ be, respectively, the conditional mean vector (viz., an estimate) and covariance matrix of $\mathbf{X}_h$ given $\mathbf{Y}_{1:j}$, where we define the starting values $\mathbf{X}_{1|0} = \mathbf{a}_1$ and $\mathbf{P}_{1|0} = \mathbf{P}_1$ for $h = 1$, $j = 0$. Note that if $\mathbf{X}_{h|j}$ are viewed as the estimates of $\mathbf{X}_h$ given $\mathbf{Y}_{1:j}$, then $\mathbf{P}_{h|j}$ is the covariance matrix of the estimation error. Furthermore, let $\boldsymbol{\nu}_h = \mathbf{Y}_h - \mathbf{Z}_h\mathbf{X}_{h|h-1} - \mathbf{d}_h$ be the estimate of $\boldsymbol{\varepsilon}_h$ and $\mathbf{V}_h = \mathbf{Z}_h\mathbf{P}_{h|h-1}\mathbf{Z}_h' + \mathbf{H}_h$ the conditional covariance matrix of $\boldsymbol{\nu}_h$ given $\mathbf{Y}_{1:(h-1)}$. The following three theorems give a set of Kalman recursions to solve the three fundamental problems above.

**Theorem 8.1 (Kalman Filtering)** *Kalman filtering has the recursion algorithm*

$$\mathbf{X}_{t|t} = \mathbf{X}_{t|t-1} + \mathbf{P}_{t|t-1}\mathbf{Z}_t'\mathbf{V}_t^{-1}\boldsymbol{\nu}_t, \tag{8.10}$$

$$\mathbf{P}_{t|t} = \mathbf{P}_{t|t-1} - \mathbf{P}_{t|t-1}\mathbf{Z}_t'\mathbf{V}_t^{-1}\mathbf{Z}_t\mathbf{P}_{t|t-1} \tag{8.11}$$

*where* $\mathbf{X}_{t|t}$ *are the filtered estimates and* $\mathbf{P}_{t|t}$ *are the corresponding error covariance matrices. Besides, the conditional distribution of* $\mathbf{X}_t$ *given* $\mathbf{Y}_{1:t}$ *is* $N(\mathbf{X}_{t|t}, \mathbf{P}_{t|t})$.

**Theorem 8.2 (Kalman Forecasting)** *The one-step-ahead predictors* $\mathbf{X}_{t+1|t}$ *of* $\mathbf{X}_{t+1}$ *and error covariance matrices* $\mathbf{P}_{t+1|t}$ *are determined by the recursions*

$$\mathbf{X}_{t+1|t} = \mathbf{c}_t + \mathbf{F}_t\mathbf{X}_{t|t}, \tag{8.12}$$

$$\mathbf{P}_{t+1|t} = \mathbf{F}_t\mathbf{P}_{t|t}\mathbf{F}_t' + \mathbf{R}_t\mathbf{Q}_t\mathbf{R}_t'. \tag{8.13}$$

*In this case, the conditional distribution of* $\mathbf{X}_{t+1}$ *given* $\mathbf{Y}_{1:t}$ *is* $N(\mathbf{X}_{t+1|t}, \mathbf{P}_{t+1|t})$.

**Theorem 8.3 (Kalman Smoothing)** *For* $s = t, \cdots, 2, 1$, *the smoothed estimates* $\mathbf{X}_{s|t}$ *and error covariance matrices* $\mathbf{P}_{s|t}$ *are determined by the following backward recursions*

$$\begin{aligned}
\mathbf{r}_{s-1} &= \mathbf{Z}_s'\mathbf{V}_s^{-1}\boldsymbol{v}_s + \mathbf{L}_s'\mathbf{r}_s \quad \textit{with } \mathbf{r}_t = \mathbf{0}, \\
\mathbf{X}_{s|t} &= \mathbf{X}_{s|s-1} + \mathbf{P}_{s|s-1}\mathbf{r}_{s-1}, \\
\mathbf{M}_{s-1} &= \mathbf{Z}_s'\mathbf{V}_s^{-1}\mathbf{Z}_s + \mathbf{L}_s'\mathbf{M}_s\mathbf{L}_s \quad \textit{with } \mathbf{M}_t = \mathbf{0}, \\
\mathbf{P}_{s|t} &= \mathbf{P}_{s|s-1} - \mathbf{P}_{s|s-1}\mathbf{M}_{s-1}\mathbf{P}_{s|s-1},
\end{aligned}$$

*where* $\mathbf{L}_s = \mathbf{F}_s - \mathbf{K}_s\mathbf{Z}_s$ *and* $\mathbf{K}_s = \mathbf{F}_s\mathbf{P}_{s|s-1}\mathbf{Z}_s'\mathbf{V}_s^{-1}$, *which is called the Kalman gain (matrix) at time s.*

Note that plugging Eq. (8.10) into Eq. (8.12), we obtain the one-step-ahead predictors of $\mathbf{X}_{t+1}$

$$\mathbf{X}_{t+1|t} = \mathbf{c}_t + \mathbf{F}_t(\mathbf{X}_{t|t-1} + \mathbf{P}_{t|t-1}\mathbf{Z}_t'\mathbf{V}_t^{-1}\boldsymbol{v}_t) = \mathbf{c}_t + \mathbf{F}_t\mathbf{X}_{t|t-1} + \mathbf{K}_t\boldsymbol{v}_t.$$

From the equation above, we see that the Kalman gain as a weight gives a kind of improvement of forecasting. In addition, considering the one-step-ahead forecast of the observation variable $\mathbf{Y}_{t+1}$ given $\mathbf{Y}_{1:t}$ and using Eq. (8.8), we arrive at

$$\mathbf{Y}_{t+1|t} = \mathsf{E}(\mathbf{Y}_{t+1}|\mathbf{Y}_{1:t}) = \mathbf{d}_{t+1} + \mathbf{Z}_{t+1}\mathbf{X}_{t+1|t},$$

and the covariance matrix of the forecast error is

$$\begin{aligned}
\mathrm{Var}(\mathbf{Y}_{t+1}|\mathbf{Y}_{1:t}) &= \mathsf{E}[(\mathbf{Y}_{t+1} - \mathbf{Y}_{t+1|t})(\mathbf{Y}_{t+1} - \mathbf{Y}_{t+1|t})'|\mathbf{Y}_{1:t}] \\
&= \mathsf{E}[(\mathbf{Z}_{t+1}(\mathbf{X}_{t+1} - \mathbf{X}_{t+1|t}) + \boldsymbol{\varepsilon}_{t+1})((\mathbf{X}_{t+1} - \mathbf{X}_{t+1|t})'\mathbf{Z}_{t+1}' + \boldsymbol{\varepsilon}_{t+1}')|\mathbf{Y}_{1:t}] \\
&= \mathbf{Z}_{t+1}\mathbf{P}_{t+1|t}\mathbf{Z}_{t+1}' + \mathbf{H}_{t+1} \\
&= \mathbf{V}_{t+1}.
\end{aligned}$$

Hence, the one-step-ahead forecast problem is completely solved.

These Kalman recursions provide the necessary and efficient algorithms to compute prediction in an online manner. As for the initialization and parameter estimation problems as well as the proofs of these theorems, they can be found in Durbin and Koopman (2012), Tsay (2010, Chapter 11), Harvey (1989), and so on. In what follows, we shall discuss some applications of the state space model and Kalman recursions.

## 8.3 Local-Level Model and SARIMAX Models

### *8.3.1 Local-Level Model*

In this subsection, we consider a famous state space model, namely, the *local-level model* or *random walk plus noise model*. It takes the form

$$Y_t = \mu_t + \varepsilon_t,\ \varepsilon_t \sim \text{iidN}(0, \sigma_\varepsilon^2), \tag{8.14}$$

$$\mu_{t+1} = \mu_t + \eta_t,\ \eta_t \sim \text{iidN}(0, \sigma_\eta^2) \tag{8.15}$$

where $\{\varepsilon_t\}$ and $\{\eta_t\}$ are mutually uncorrelated and are independent of $\mu_1 \sim \text{N}(a_1, p_1)$. In some literature, $\mu_t$ is known as the trend of the series $Y_t$. Clearly, $Y_t$ is stationary and has no trend if $\sigma_\eta = 0$ and $\mu_t$ is actually a stochastic trend of the series $Y_t$ (also a random walk) if $\sigma_\eta \neq 0$. Moreover, $Y_t = \mu_t$ if $\sigma_\varepsilon = 0$. When $\sigma_\varepsilon \neq 0$, we first-difference the series $Y_t$ and arrive at

$$(1 - B)Y_t = \eta_{t-1} + \varepsilon_t - \varepsilon_{t-1}.$$

Let $\xi_t = \eta_{t-1} + \varepsilon_t - \varepsilon_{t-1}$. Obviously, $\xi_t$ is stationary and $\xi_t \sim \text{N}(0, 2\sigma_\varepsilon^2 + \sigma_\eta^2)$. Furthermore, the autocorrelations of $\xi_t$

$$\rho_k = \begin{cases} -\sigma_\varepsilon^2/(2\sigma_\varepsilon^2 + \sigma_\eta^2) \neq 0 & \text{if } k = 1, \\ 0 & \text{if } k > 1. \end{cases}$$

This illustrates that $\xi_t$ follows an MA(1) model (see Table 3.1), and therefore the local-level model (8.14)–(8.15) has an ARIMA(0,1,1) representation

$$(1 - B)Y_t = (1 + \theta B)\omega_t$$

where $\omega_t$ is a white noise series. This representation can be viewed as a special case of the ARIMAX model. And this is also an example to translating a state space model into an equivalent ARIMAX model.

Let $\sigma_\varepsilon = 2$ and $\sigma_\eta = 1$. Then it turns out that $\rho_1 = -2^2/(2 \times 2^2 + 1) = -4/9$. At this point, we can simulate a sample of size 300 from the local-level model with $\sigma_\varepsilon = 2$ and $\sigma_\eta = 1$. Its time series plot is shown in Fig. 8.1, which displays

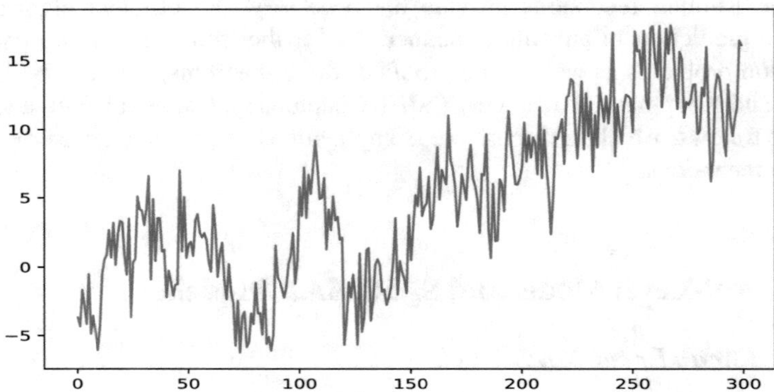

**Fig. 8.1** Time plot of the sample simulated from the local-level model with $\sigma_\varepsilon = 2$ and $\sigma_\eta = 1$
pyTSA_SimLLM

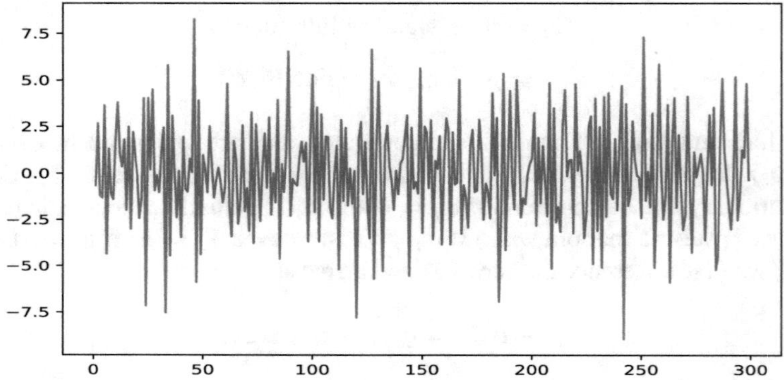

**Fig. 8.2** Time series plot of the differences of the sample from the local-level model
pyTSA_SimLLM

nonstationarity of the series. If we first-difference the sample and get the differenced series of the sample, then we can, respectively, draw the time plot and ACF plot of the differenced series shown in Figs. 8.2 and 8.3. We see that the time plot exhibits stationarity of the differenced series, the sample ACF value at lag 1 (viz., sample $\rho_1$) is around $-0.4$ (almost the same as the true value $-4/9$) and statistically significant, and at the same time, all the other ACF values are not significant.

```
>>> import numpy as np
>>> import matplotlib.pyplot as plt
>>> from PythonTsa.RandomWalk import RandomWalk_with_drift
>>> from PythonTsa.plot_acf_pacf import acf_pacf_fig
>>> np.random.seed(1379)
>>> rw0= RandomWalk_with_drift(drift=0.0, nsample=300, burnin=10)
>>> wn=np.random.normal(loc=0, scale=2.0, size=300)
>>> y=rw0+wn
>>> y.plot();plt.show()
```

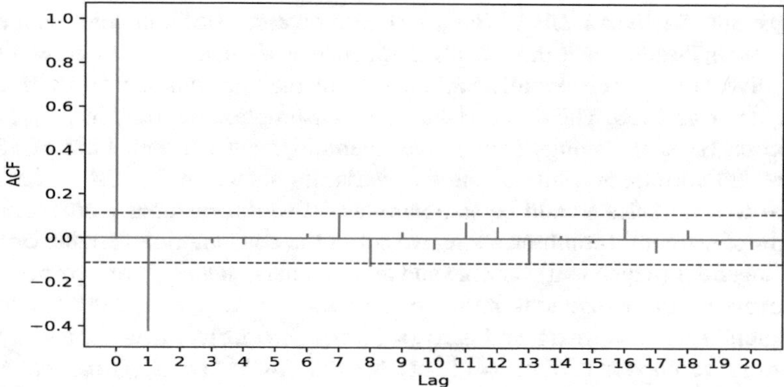


**Fig. 8.3** ACF plot of the differenced series of the sample from the local-level model

pyTSA_SimLLM

```
>>> dy=y.diff().dropna()
>>> dy.plot();plt.show()
>>> acf_pacf_fig(dy, both=False, lag=20)
>>> plt.show()
```

The local-level model is a simple case of the structural time series model (or unobserved component model). The interested reader can refer to Durbin and Koopman (2012) and Harvey (1989), among others.

### 8.3.2 SARIMAX Models

If we consider outer factors' impact on a time series, then we need to import exogenous regressors, namely, input variables. SARIMAX models are such a kind of models as to describe functionality of input variables. They have a few forms and we here introduce an often used form as follows:

$$\begin{cases} X_t = \boldsymbol{\beta}_t \mathbf{Y}_t + Z_t, & (8.16\text{a}) \\ \varphi(B)\Phi(B^s)(1-B)^d(1-B^s)^D Z_t = \theta(B)\Theta(B^s)\varepsilon_t, \ \varepsilon_t \sim \text{WN}(0, \sigma_\epsilon^2) & (8.16\text{b}) \end{cases}$$

where $X_t$ is a univariate time series considered, $\mathbf{Y}_t = (Y_{t1}, Y_{t2}, \cdots, Y_{tk})'$ is the input variables (exogenous regressors), $\boldsymbol{\beta}_t = (\beta_{t1}, \beta_{t2}, \cdots, \beta_{tk})$ is the coefficients corresponding to $\mathbf{Y}_t$, and $Z_t$ is the regressing error. And Eq. (8.16b) is a SARIMA$(p, d, q)$ $(P, D, Q)_s$ model presented in Chap. 5. Thus Eqs. (8.16a)–(8.16b) are a regression with SARIMA errors. Note that in many cases, $\boldsymbol{\beta}_t = \boldsymbol{\beta}$ is time-invariant. Now let us look at an example to implementing SARIMAX modeling with Python. The Python function used here is `SARIMAX()` in the following module `statsmodels.tsa.statespace.sarimax`.

*Example 8.4 (SARIMAX Model Building)* The dataset "USEconomicChange.csv" in the folder Ptsadata is a time series that consists of changes (viz., growth rates) of the five US macroeconomic variables from the first quarter of 1970 to the third quarter of 2016. The five variables are consumption (`cons`), income (`inc`), production (`prod`), savings (`sav`), and unemployment (`unem`). First of all, we observe the correlation plots of the five variables shown in Fig. 8.4. It turns out that the five variables should be stationary as a five-dimensional vector series. In the light of economic common sense, we select the consumption variable `cons` as the endogenous (dependent) variable and all the others, namely, `inc`, `prod`, `sav`, and `unem` as the exogenous regressors. In the function `SARIMAX()`, let the parameters `endog` = `cons` and `exog=` {`inc`, `prod`,`sav`, `unem`}, and by trial and error, we choose `order` = (1,0,1) for the ARIMA$(p, d, q)$ model. At the same time, all the other parameters are by default. Thus the ARMA model will be written as a state space Harvey representation. For more details on the Harvey representation, see Durbin and Koopman (2012) and Harvey (1989), among others. The estimated SARIMAX model is shown in the `SARIMAX Results` table in the following Python code. We see that the estimated coefficients of `inc` and `prod` are positive and the estimated coefficients of `sav` and `unem` negative. This is economically desirable. Furthermore, the ACF plot and $p$-value plot for Ljung-Box test of the residuals of the estimated SARIMAX model are, respectively, shown in Figs. 8.5 and 8.6. They clearly illustrate that the residual series can be viewed as a white noise and hence the estimated model should be adequate. However, we unexpectedly find that there exist autocorrelations (ARCH effect) in the squared residual series when we draw the ACF plot and $p$-value plot for Ljung-Box test of the squared residuals shown in Figs. 8.7 and 8.8. Therefore we need to continue to model the residuals of the estimated SARIMAX model. We fit a GARCH model to the residuals and arrive at the estimated GARCH(1,1) model shown in the `Zero Mean - GARCH Model Results` table in the following Python code. The $p$-value plots for Ljung-Box test of the residuals and squared residuals of the estimated GARCH(1,1) model are, respectively, shown in Figs. 8.9 and 8.10. They suggest that the residuals and squared residuals can be both seen as white noise series and there are no autocorrelations in them anymore. In conclusion, we build a complete and appropriate model for the given dataset as follows:

$$\begin{cases} cons_t = 0.253 + 0.731inc_t + 0.051prod_t - 0.046sav_t - 0.169unem_t + Z_t, \\ Z_t + 0.815Z_{t-1} = \eta_t + 0.729\eta_{t-1}, \\ \eta_t = \sigma_t\varepsilon_t,\ \varepsilon_t \sim \text{iidN}(0, 1), \\ \sigma_t^2 = 0.004 + 0.195\eta_{t-1}^2 + 0.776\sigma_{t-1}^2. \end{cases}$$

```
>>> import numpy as np
>>> import pandas as pd
>>> import matplotlib.pyplot as plt
>>> import statsmodels.api as sm
>>> from PythonTsa.plot_acf_pacf import acf_pacf_fig
```

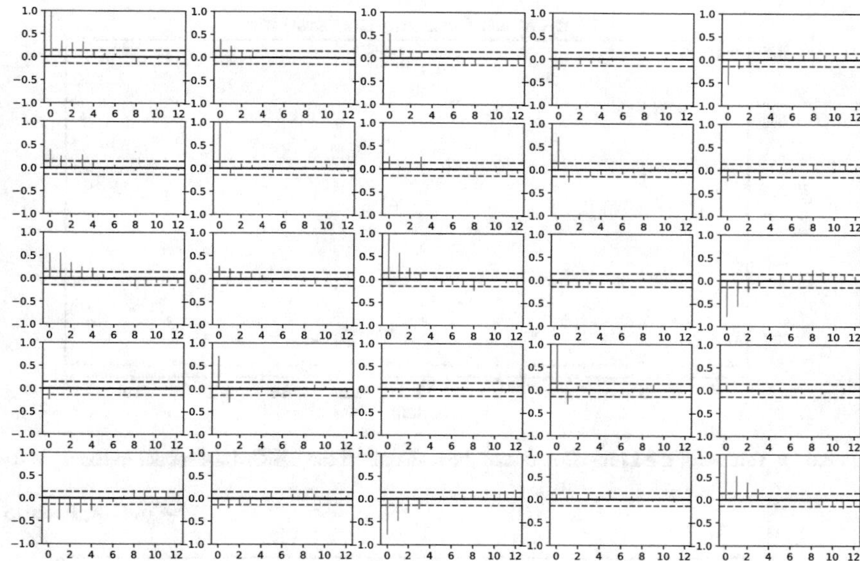

**Fig. 8.4** Correlation plots of the US macroeconomic change time series in Example 8.4

pyTSA_EconUS

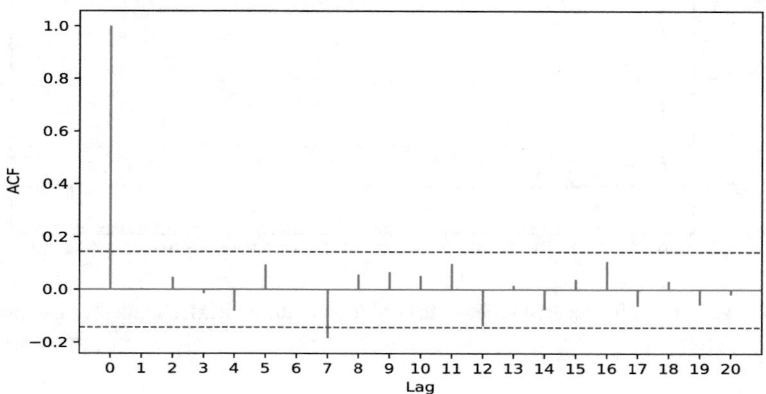


**Fig. 8.5** ACF plot of the residual series of the SARIMAX model in Example 8.4

pyTSA_EconUS

```
>>> from PythonTsa.LjungBoxtest import plot_LB_pvalue
>>> from PythonTsa.plot_multi_ACF import multi_ACFfig
>>> from statsmodels.tsa.statespace.sarimax import SARIMAX
>>> from arch import arch_model
>>> from PythonTsa.datadir import getdtapath
>>> dtapath=getdtapath()
>>> uscc=pd.read_csv(dtapath + 'USEconomicChange.csv')
```

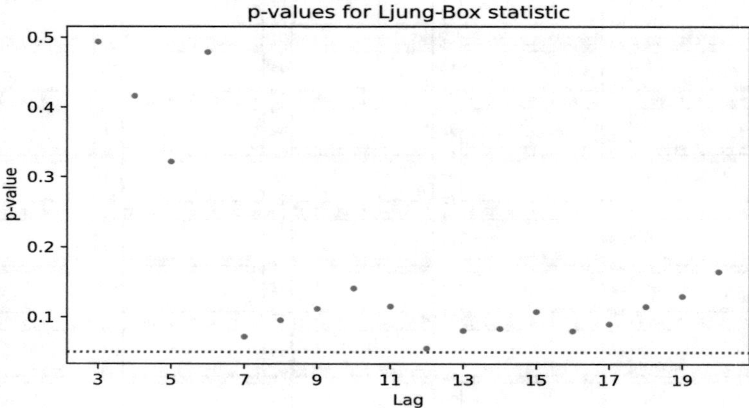


**Fig. 8.6** *p*-Values for the Ljung-Box test of the residuals of the SARIMAX model in Example 8.4

pyTSA_EconUS

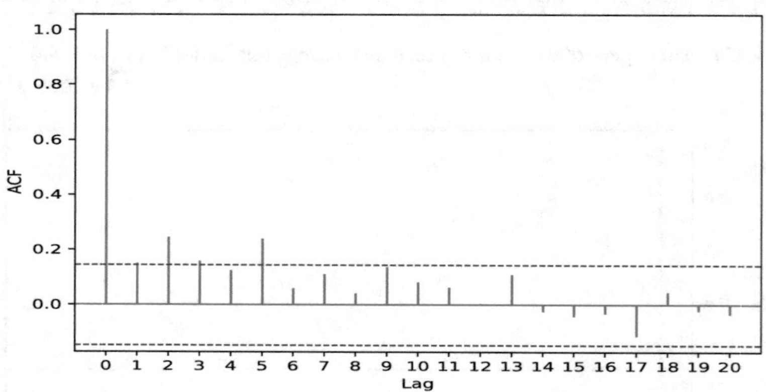


**Fig. 8.7** ACF plot of the squared series of the residuals of the SARIMAX model in Example 8.4

pyTSA_EconUS

```
>>> dates=pd.date_range('1970/3/31', periods=len(uscc), freq='Q')
>>> uscc.index=dates
>>> uscc=uscc.drop(columns=['Time'])
>>> uscc.rename(columns={'Consumption':'cons', 'Income':'inc',
'Production':'prod','Savings':'sav','Unemployment':'unem'},
inplace=True)
>>> multi_ACFfig(uscc, nlags=12)
>>> plt.show()
>>> X=uscc[['inc','prod','sav','unem']]
>>> Y=uscc['cons']
>>> X=sm.add_constant(X, prepend=False)
>>> sarimaxmod=SARIMAX(endog=Y,exog=X, order=(1,0,1))
```

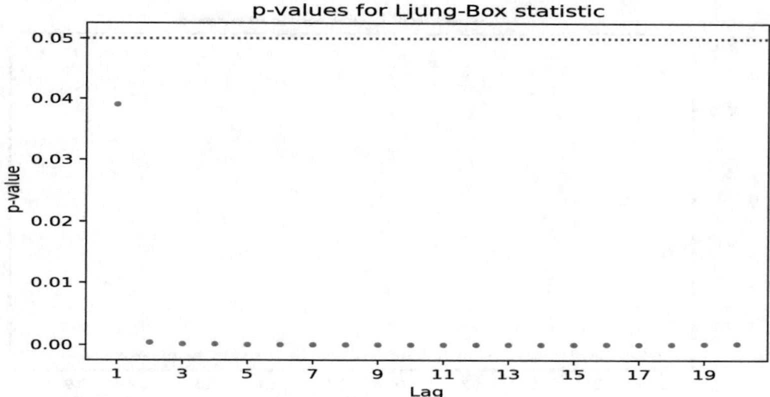


**Fig. 8.8** $p$-Values for the Ljung-Box test of the squared residuals of the SARIMAX model in Example 8.4

pyTSA_EconUS

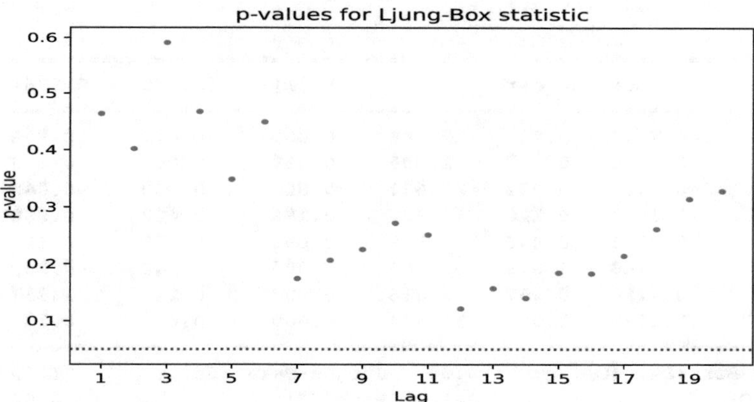


**Fig. 8.9** $p$-Values for the Ljung-Box test of the residuals of the GARCH model in Example 8.4

pyTSA_EconUS

```
>>> sarimaxfit=sarimaxmod.fit(disp=False)
>>> print(sarimaxfit.summary())
                          SARIMAX Results
===================================================================
Dep. Variable:                  cons   No. Observations:        187
Model:              SARIMAX(1, 0, 1)   Log Likelihood        -52.873
Date:               Sat, 04 Dec 2021   AIC                   121.746
Time:                       20:15:53   BIC                   147.595
Sample:                   03-31-1970   HQIC                  132.220
                        - 09-30-2016
Covariance Type:                 opg
```

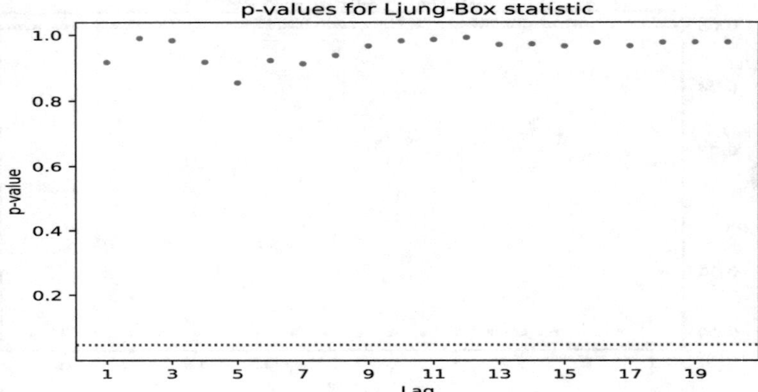


**Fig. 8.10** $p$-Values for the Ljung-Box test of the squared residuals of the GARCH model in Example 8.4

pyTSA_EconUS

```
==============================================================================
                 coef    std err          z      P>|z|      [0.025      0.975]
------------------------------------------------------------------------------
inc            0.7313      0.033     22.148      0.000       0.667       0.796
prod           0.0506      0.027      1.898      0.058      -0.002       0.103
sav           -0.0460      0.002    -23.691      0.000      -0.050      -0.042
unem          -0.1690      0.128     -1.322      0.186      -0.420       0.082
const          0.2534      0.038      6.589      0.000       0.178       0.329
ar.L1         -0.8148      0.218     -3.743      0.000      -1.242      -0.388
ma.L1          0.7290      0.257      2.836      0.005       0.225       1.233
sigma2         0.1030      0.009     11.474      0.000       0.085       0.121
==============================================================================
Ljung-Box (L1) (Q):              0.00   Jarque-Bera (JB):              21.79
Prob(Q):                         0.96   Prob(JB):                       0.00
Heteroskedasticity (H):          0.51   Skew:                           0.52
Prob(H) (two-sided):             0.01   Kurtosis:                       4.32
==============================================================================
Warnings:
[1] Covariance matrix calculated using the outer product
    of gradients (complex-step).
>>> sarimaxresid=sarimaxfit.resid
>>> acf_pacf_fig(sarimaxresid, both=False, lag=20)
>>> plt.show()
>>> plot_LB_pvalue(sarimaxresid, noestimatedcoef=2, nolags=20)
>>> plt.show()
# the residuals are of Model (8.16b) and so noestimatedcoef=2.
>>> acf_pacf_fig(sarimaxresid**2, both=False, lag=20)
>>> plt.show()
>>> plot_LB_pvalue(sarimaxresid**2, noestimatedcoef=0, nolags=20)
>>> plt.show()
>>> archmod = arch_model(sarimaxresid, mean='Zero').fit(disp='off')
```

```
# mean='Zero' means no mean equation in GARCH models.
>>> print(archmod.summary())
                     Zero Mean - GARCH Model Results
==================================================================
Dep. Variable:                None   R-squared:                0.000
Mean Model:              Zero Mean   Adj. R-squared:           0.005
Vol Model:                   GARCH   Log-Likelihood:        -35.8494
Distribution:               Normal   AIC:                    77.6988
Method:         Maximum Likelihood   BIC:                    87.3921
                                     No. Observations:           187
Date:             Sat, Dec 04 2021   Df Residuals:               184
Time:                     21:42:20   Df Model:                     3
                           Volatility Model
==================================================================
                coef   std err        t      P>|t|    95.0% Conf. Int.
------------------------------------------------------------------
omega   4.0515e-03 3.711e-03   1.092     0.275 [-3.222e-03,1.133e-02]
alpha[1]    0.1950 7.952e-02   2.453 1.418e-02   [3.917e-02,  0.351]
beta[1]     0.7756 7.608e-02  10.195 2.078e-24      [  0.627,  0.925]
==================================================================
Covariance estimator: robust
>>> archresid = archmod.std_resid
>>> plot_LB_pvalue(archresid, noestimatedcoef=0, nolags=20)
>>> plt.show()
>>> plot_LB_pvalue(archresid**2, noestimatedcoef=0, nolags=20)
>>> plt.show()
```

## 8.4 Markov Switching Models

The Markov switching model is a flexible class of nonlinear time series models. It has been popular, especially in economic and business cycle analysis since the publication of Hamilton (1989). This section will present the model in brief.

### 8.4.1 Definitions

The Markov switching model is widely used in econometrics and other disciplines. It involves multiple structures that characterize the time series variable behaviors in different regimes and permits switching between these structures. The Markov switching model provides such a kind of switching mechanism so that it is controlled by an unobservable state variable which follows a first-order Markov chain. It is sometimes also referred to as the *(Markov) regime switching model*. Frühwirth-Schnatter (2006) gives a detailed Bayesian analysis of this model. Kim and Nelson (1999) elaborate on a class of state space models with Markov switching. For a time series $X_t$, a general Markov switching model is of the form

$$\begin{aligned} X_t =&\mu_{S_t} + \mathbf{Y}_t\boldsymbol{\beta}_{S_t} + \varphi_{1,S_t}(X_{t-1} - \mu_{S_{t-1}} - \mathbf{Y}_{t-1}\boldsymbol{\beta}_{S_{t-1}}) + \cdots \\ &+ \varphi_{p,S_t}(X_{t-p} - \mu_{S_{t-p}} - \mathbf{Y}_{t-p}\boldsymbol{\beta}_{S_{t-p}}) + \varepsilon_t, \ \varepsilon_t \sim \text{WN}(0, \sigma^2_{S_t}) \end{aligned} \tag{8.17}$$

where $\mathbf{Y}_t = (Y_{t1}, Y_{t2}, \cdots, Y_{tk})$ are the input variables (exogenous regressors), $\boldsymbol{\beta}_{S_t}$ are the corresponding regression coefficient vectors, and $S_t$ is a Markov chain for regimes. Here a regime is viewed as a state that the Markov chain may take. Assume that all the possible values of state variable $S_t$ are $S = \{1 : K\}$. Then $S_t$ satisfies the Markov property

$$P(S_t = j|S_{t-1} = i, S_{t-2} = s_{t-2}, \cdots, S_1 = s_1) = P(S_t = j|S_{t-1} = i) = p_{ij}$$

where $j, i, s_{t-2}, \cdots, s_1 \in S$ and $p_{ij}$ is called the *one-step transition probability* from state $i$ to state $j$. And the state transition is governed by the (state) transition (probability) matrix $\mathbf{P} = [p_{ij}]$. Besides, the error $\varepsilon_t$ is also written as $\varepsilon_t = \sigma_{S_t}\eta_t$ where $\eta_t \sim \text{WN}(0, 1)$. And it is often assumed that the error $\varepsilon_t$ is normally distributed, namely, $\varepsilon_t \sim \text{iidN}(0, \sigma^2_{S_t})$.

If there are no exogenous regressors in Eq. (8.17), it is reduced to

$$X_t = \mu_{S_t} + \varphi_{1,S_t}(X_{t-1} - \mu_{S_{t-1}}) + \cdots + \varphi_{p,S_t}(X_{t-p} - \mu_{S_{t-p}}) + \varepsilon_t, \ \varepsilon_t \sim \text{WN}(0, \sigma^2_{S_t})$$

and called the *Markov switching autoregressive model*. Hamilton (1989) uses a special case of the Markov switching autoregressive model to study the business cycle dynamic of US GDP growth rates. And if there are no autoregressive terms in Eq. (8.17), it is reduced to

$$X_t = \mu_{S_t} + \mathbf{Y}_t\boldsymbol{\beta}_{S_t} + \varepsilon_t, \ \varepsilon_t \sim \text{WN}(0, \sigma^2_{S_t}) \tag{8.18}$$

and known as the *Markov switching (dynamic) regression model*. Moreover, in some applications, the autoregressive coefficients have no switching, and then the Markov switching model is reduced to

$$\begin{aligned} X_t =&\mu_{S_t} + \mathbf{Y}_t\boldsymbol{\beta}_{S_t} + \varphi_1(X_{t-1} - \mu_{S_{t-1}} - \mathbf{Y}_{t-1}\boldsymbol{\beta}_{S_{t-1}}) + \cdots \\ &+ \varphi_p(X_{t-p} - \mu_{S_{t-p}} - \mathbf{Y}_{t-p}\boldsymbol{\beta}_{S_{t-p}}) + \varepsilon_t, \ \varepsilon_t \sim \text{WN}(0, \sigma^2_{S_t}). \end{aligned} \tag{8.19}$$

Naturally, the variance of the error $\varepsilon_t$ may also have no switching: $\sigma^2_{S_t} = \sigma^2$. Nevertheless, for the Markov switching model in Eq. (8.17) to be meaningful, there must be a switching mechanism in at least one of the const term, the regression coefficients, the autoregressive coefficients, and the error variance. Furthermore, note that an important alternative to Eq. (8.17) is

$$X_t = \varphi_{0,S_t} + \varphi_{1,S_t}X_{t-1} + \cdots + \varphi_{p,S_t}X_{t-p} + \mathbf{Y}_t\boldsymbol{\beta}_{S_t} + \varepsilon_t, \ \varepsilon_t \sim \text{WN}(0, \sigma^2_{S_t}). \tag{8.20}$$

This is another form of the Markov switching model. For more information on it, see Tsay and Chen (2019), McCulloch and Tsay (1994), and so forth.

*Example 8.5 (A Two-State Markov Switching Model)* Consider the following two-state Markov switching model with the autoregressive order 1 and one exogenous regressor:

$$X_t = \mu_{S_t} + Y_t\beta_{S_t} + \varphi_{S_t}(X_{t-1} - \mu_{S_{t-1}} - Y_{t-1}\beta_{S_{t-1}}),\ \varepsilon_t \sim \text{WN}(0, \sigma_{S_t}^2).$$

If the two states are $S = \{0, 1\}$, then the state transition of the model is driven by the transition probabilities

$$P(S_t = 1|S_{t-1} = 0) = a,\quad P(S_t = 0|S_{t-1} = 1) = b$$

where $0 < a, b < 1$. And the transition matrix is

$$\mathbf{P} = \begin{bmatrix} 1-a & a \\ b & 1-b \end{bmatrix}.$$

In this case, we need to estimate the parameters $a$, $b$, $\mu_0, \mu_1, \beta_0, \beta_1, \varphi_0, \varphi_1, \sigma_0^2$, and $\sigma_1^2$ for building the two-state Markov switching model.

### 8.4.2 Examples

On theoretical discussion of the estimation issue of the Markov switching model, the reader can refer to Tsay and Chen (2019), Frühwirth-Schnatter (2006), and Kim and Nelson (1999), among others. As for carrying out estimation of the Markov switching model, there are two Python functions (classes): `MarkovRegression` and `MarkovAutoregression` available in the package `statsmodels`. The function `MarkovAutoregression` is used to estimate the model in Eq. (8.17) and `MarkovRegression` to the Markov switching regression model in Eq. (8.18). Note that in order to estimate the model in Eq. (8.20), we can view the lagged dependent variables $X_{t-1}, \cdots, X_{t-p}$ as a part of the exogenous regressors and then use the function `MarkovRegression`. What follows, we give a few examples to applications of the Markov switching model.

*Example 8.6 (Log Returns of Germany DAX Daily Index)* The dataset "DAXlogret" in the folder Ptsadata is the log return series of the Germany DAX daily index from October 24, 2014 to July 19, 2019 and denoted as `logret`. Its time series plot is shown in Fig. 8.11 and clearly displays heteroscedasticity. Thus we try to fit a Markov switching autoregressive model to the dataset. We find that we can select the order $p = 1$. The estimated model is shown in the table `Markov Switching Model Results` of the following Python code. Both

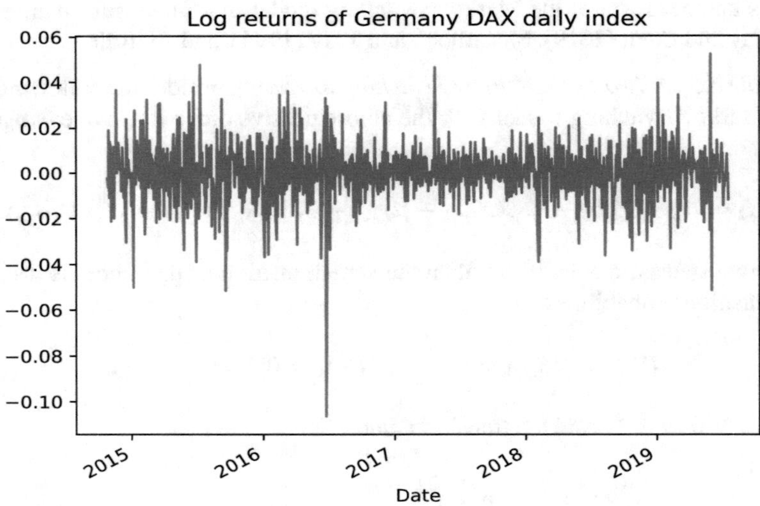


**Fig. 8.11** Time series plot of the log returns of Germany DAX daily index in Example 8.6
pyTSA_MarkovReturnsDAX

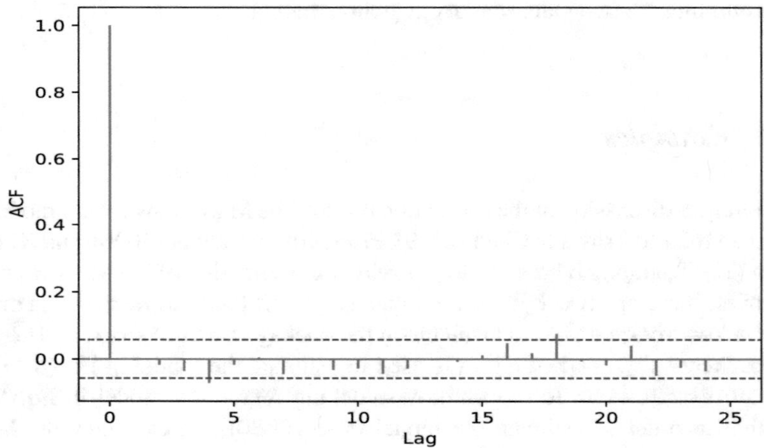


**Fig. 8.12** ACF plot of the residuals of the estimated Markov switching model in Example 8.6
pyTSA_MarkovReturnsDAX

ACF plot shown in Fig. 8.12 and *p*-value plot for the Ljung-Box test shown in Fig. 8.13 of the model residuals illustrate that the estimated model is appropriate.

```
>>> import numpy as np
>>> import pandas as pd
>>> import statsmodels.api as sm
>>> import matplotlib.pyplot as plt
```

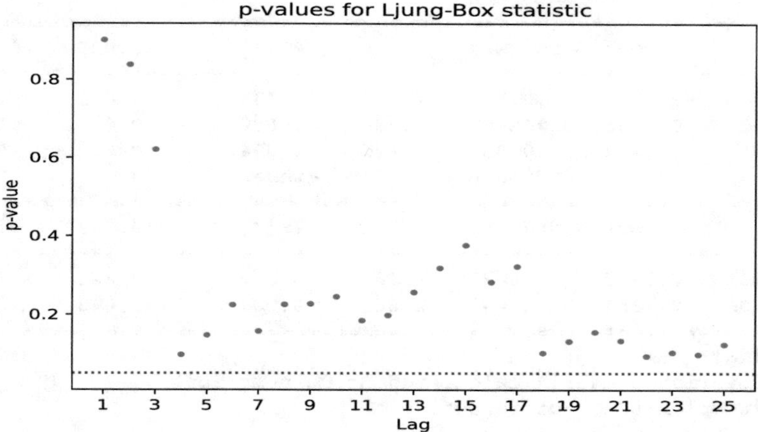


**Fig. 8.13** *p*-Value plot for the Ljung-Box test of the model residual series in Example 8.6
pyTSA_MarkovReturnsDAX

```
>>> from PythonTsa.plot_acf_pacf import acf_pacf_fig
>>> from PythonTsa.LjungBoxtest import plot_LB_pvalue
>>> from PythonTsa.datadir import getdtapath
>>> dtapath=getdtapath()
>>> daxlogret=pd.read_csv(dtapath + 'DAXlogret.csv',header=0)
>>> daxlogret.index=pd.DatetimeIndex(daxlogret.Date)
>>> logret=daxlogret.Logret
>>> logret.plot(title='Log returns of Germany DAX daily index')
>>> plt.show()
>>> mod= sm.tsa.MarkovAutoregression(logret, k_regimes=2,
     order=1, switching_variance=True)
>>> modfit=mod.fit()
>>> print(modfit.summary())

                  Markov Switching Model Results
================================================================
Dep. Variable:                 Logret   No. Observations:   1190
Model:           MarkovAutoregression   Log Likelihood  3692.208
Date:                Mon, 20 Dec 2021   AIC            -7368.417
Time:                        16:23:08   BIC            -7327.763
Sample:                             0   HQIC           -7353.096
                               - 1190
Covariance Type:               approx
                        Regime 0 parameters
================================================================
             coef    std err        z     P>|z|     [0.025    0.975]
----------------------------------------------------------------
const      0.0008      0.000    2.643     0.008      0.000     0.001
sigma2 5.438e-05    5.55e-06    9.791     0.000   4.35e-05  6.53e-05
ar.L1     -0.0253      0.040   -0.630     0.529     -0.104     0.053
                        Regime 1 parameters
```

```
==============================================================================
                 coef    std err          z      P>|z|      [0.025      0.975]
------------------------------------------------------------------------------
const         -0.0013      0.001     -1.483      0.138      -0.003       0.000
sigma2         0.0003   3.44e-05      8.724      0.000       0.000       0.000
ar.L1         -0.1172      0.055     -2.125      0.034      -0.225      -0.009
                        Regime transition parameters
==============================================================================
                 coef    std err          z      P>|z|      [0.025      0.975]
------------------------------------------------------------------------------
p[0->0]        0.9287      0.022     42.967      0.000       0.886       0.971
p[1->0]        0.1354      0.047      2.869      0.004       0.043       0.228
==============================================================================
Warnings:
[1] Covariance matrix calculated using numerical
    (complex-step) differentiation.
>>> modresid=modfit.resid
>>> acf_pacf_fig(modresid, both=False, lag=25); plt.show()
>>> plot_LB_pvalue(modresid, noestimatedcoef=0, nolags=25)
>>> plt.show()
```

*Example 8.7 (Federal Funds Rate, Output Gap, and Inflation Rate)* In this example, we are to analyze an American economic quarterly dataset that is built in the Python package `statsmodels`. It consists of three time series variables: `fedfunds`, `ogap`, and `inf`. Here `fedfunds` is the time series of federal funds rate, `ogap` the series of output gap (viz., GDP gap), and `inf` the series of inflation rate. The sample period for `fedfunds` and `ogap` is from July 1, 1954 to October 1, 2010, but `inf` is from July 1, 1955 to October 1, 2010. The time series plots of the three variables are shown in Fig. 8.14. Note here that `ogap` and `inf` are viewed as two exogenous variables which might have an influence on the endogenous variable `fedfunds`. By trial and error, we finally select the Markov switching model in Eq. (8.17) with order $p = 4$, the number of regimes 2, and the variance of the error $\varepsilon_t$ as well as the regression coefficients having no switching to fit to the dataset. The estimated model is shown in the table `Markov Switching Model Results` in the following Python code. We see that the estimates of most parameters are statistically significant, especially the estimates of the autoregressive coefficients at lag 4 whether the regime is 0 or 1. What is more, observing the ACF plot of the model residual series shown in Fig. 8.15 and the $p$-value plot for the Ljung-Box test of the residuals shown in Fig. 8.16, we conclude that the residual series can be seen as a white noise. (By the way, we have tried to estimate several other Markov switching models, but their residuals all appear serially correlated.) Therefore, we have built an adequate Markov switching model for the three variables. Lastly we can plot the smoothed probabilities of the high and low federal funds rate regimes shown in Fig. 8.17 and calculate the expected durations of the high and low regimes. It turns out that the high regime is expected to persist for about 3.6 quarters, whereas the low regime is expected to persist for about 38.4 quarters.

```
>>> import numpy as np
>>> import pandas as pd
```

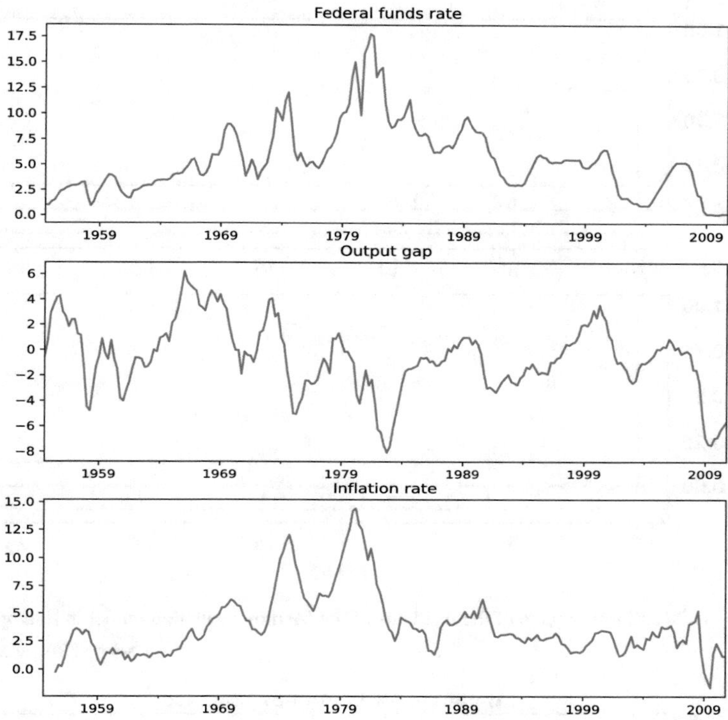


**Fig. 8.14** Time series plots of the three US economic variables in Example 8.7

pyTSA_MacroUS2

```
>>> import statsmodels.api as sm
>>> import matplotlib.pyplot as plt
>>> from PythonTsa.plot_acf_pacf import acf_pacf_fig
>>> from PythonTsa.LjungBoxtest import plot_LB_pvalue
>>> from statsmodels.tsa.regime_switching.tests.test
        _markov_regression import fedfunds, ogap, inf
>>> index=pd.date_range('1954-07-01', '2010-10-01', freq='QS')
>>> dta_fedfunds = pd.Series(fedfunds, index=index)
>>> dta_ogap = pd.Series(ogap, index=index)
>>> dta_inf = pd.Series(inf, index=index)
>>> dta_inf.head()
1954-07-01          NaN
1954-10-01          NaN
1955-01-01          NaN
1955-04-01          NaN
1955-07-01    -0.234724
>>> fig = plt.figure()
>>> dta_fedfunds.plot(ax=fig.add_subplot(311))
>>> plt.title("Federal funds rate")
>>> dta_ogap.plot(ax=fig.add_subplot(312))
```

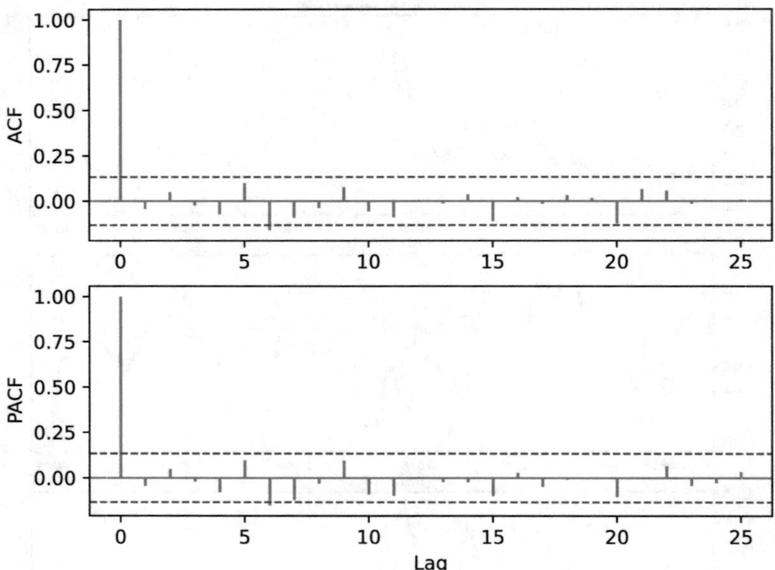


**Fig. 8.15** ACF and PACF plots of the residuals of the Markov switching model in Example 8.7
pyTSA_MacroUS2

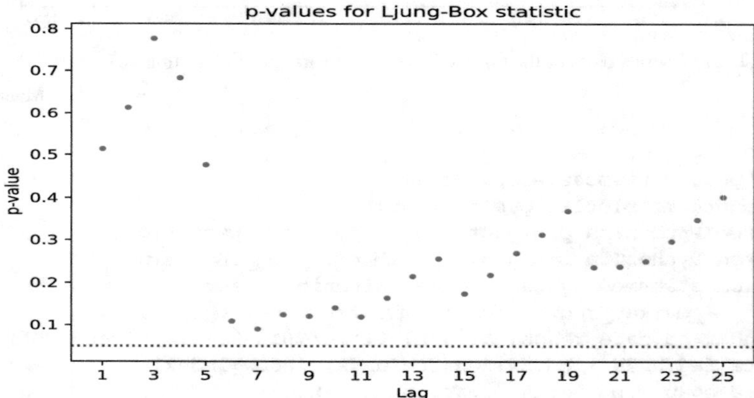


**Fig. 8.16** $p$-Value plot for the Ljung-Box test of the model residual series in Example 8.7
pyTSA_MacroUS2

```
>>> plt.title("Output gap")
>>> dta_inf.plot(ax=fig.add_subplot(313))
>>> plt.title("Inflation rate")
>>> plt.show()
>>> exogl = pd.concat((dta_ogap, dta_inf), axis=1).iloc[4:]
>>> mymod_fedfunds = sm.tsa.MarkovAutoregression(
```

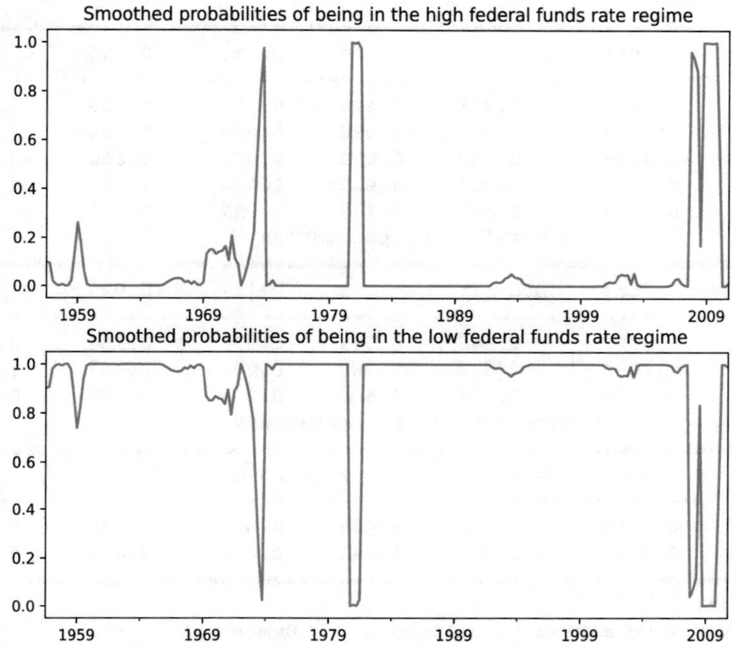


**Fig. 8.17** Smoothed probabilities of being in the high and low rate regimes in Example 8.7
pyTSA_MacroUS2

```
    dta_fedfunds.iloc[4:], k_regimes=2, order=4, exog=exog1)
>>> myres_fedfunds = mymod_fedfunds.fit()
>>> print(myres_fedfunds.summary())
                 Markov Switching Model Results
=================================================================
Dep. Variable:                    y  No. Observations:        218
Model:        MarkovAutoregression  Log Likelihood        -264.575
Date:             Thu, 16 Dec 2021  AIC                    559.149
Time:                     22:42:03  BIC                    609.917
Sample:                 07-01-1955  HQIC                   579.655
                      - 10-01-2010
Covariance Type:            approx
                     Regime 0 parameters
==================================================================
            coef     std err        z     P>|z|     [0.025     0.975]
------------------------------------------------------------------
const      0.5097      0.289    1.766     0.077     -0.056      1.075
ar.L1      0.1808      0.130    1.386     0.166     -0.075      0.437
ar.L2      0.2104      0.172    1.220     0.222     -0.128      0.548
ar.L3     -0.4561      0.206   -2.212     0.027     -0.860     -0.052
ar.L4     -0.5517      0.208   -2.654     0.008     -0.959     -0.144
                     Regime 1 parameters
```

```
==============================================================================
                 coef    std err          z      P>|z|      [0.025      0.975]
------------------------------------------------------------------------------
const          0.1389      0.335      0.415      0.678      -0.518       0.796
ar.L1          1.1467      0.072     15.930      0.000       1.006       1.288
ar.L2         -0.4909      0.097     -5.079      0.000      -0.680      -0.301
ar.L3          0.4459      0.093      4.813      0.000       0.264       0.627
ar.L4         -0.1438      0.065     -2.218      0.027      -0.271      -0.017
                         Non-switching parameters
==============================================================================
                 coef    std err          z      P>|z|      [0.025      0.975]
------------------------------------------------------------------------------
x1             0.2657      0.042      6.372      0.000       0.184       0.347
x2             1.2091      0.032     38.107      0.000       1.147       1.271
sigma2         0.5332      0.056      9.502      0.000       0.423       0.643
                        Regime transition parameters
==============================================================================
                 coef    std err          z      P>|z|      [0.025      0.975]
------------------------------------------------------------------------------
p[0->0]        0.7189      0.133      5.400      0.000       0.458       0.980
p[1->0]        0.0261      0.014      1.841      0.066      -0.002       0.054
==============================================================================
Warnings:
[1] Covariance matrix calculated using numerical
    (complex-step) differentiation.
>>> myresid=myres_fedfunds.resid
>>> acf_pacf_fig(myresid, both=True, lag=25); plt.show()
>>> plot_LB_pvalue(myresid, noestimatedcoef=0, nolags=25)
>>> plt.show()
>>> fig = plt.figure()
>>> ax1=fig.add_subplot(211)
>>> myres_fedfunds.smoothed_marginal_probabilities[0].plot(ax=ax1)
>>> plt.title('Smoothed probabilities of
             being in the high federal funds rate regime')
>>> ax2=fig.add_subplot(212)
>>> myres_fedfunds.smoothed_marginal_probabilities[1].plot(ax=ax2)
>>> plt.title('Smoothed probabilities of
              being in the low federal funds rate regime')
>>> plt.show()
>>> print(myres_fedfunds.expected_durations)
[ 3.55737378 38.37500431]
```

*Example 8.8 (Absolute Returns of S&P 500)* The dataset to be analyzed is also built in the Python package `statsmodels` and denoted as `areturns`. It is the absolute returns of S&P 500 and a weekly time series from May 4, 2004 to May 3, 2014. Its time series plot is shown in Fig. 8.18 and plainly exhibits heteroscedasticity. We select the Markov switching autoregressive model with order 4 to be fitted to the dataset. The estimated model is shown in the first `Markov Switching Model Results` table in the following Python code. We observe that the estimate for the variance of error $\varepsilon_t$ is significantly 0.6316

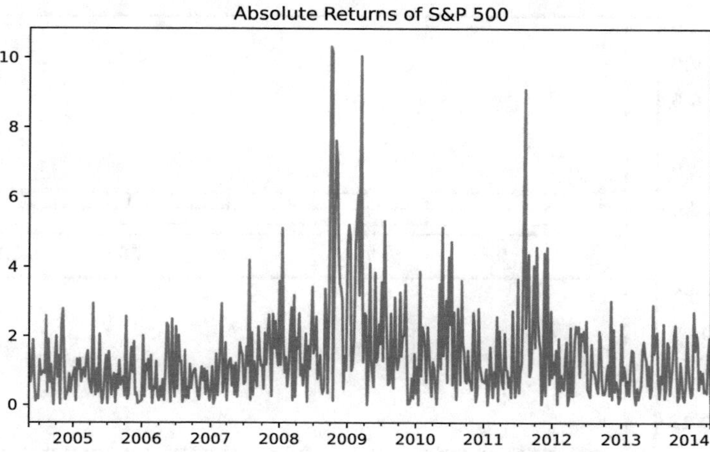


**Fig. 8.18** Time series plot of the absolute returns of S&P 500 in Example 8.8
pyTSA_MarkovReturnsSP500

(small) when the regime is 0, whereas significantly 2.4475 (big) when the regime is 1. This result illustrates that the switching is necessary. The ACF and PACF plots of the model residuals shown in Fig. 8.19 appear like ones of a white noise. However, the pity is that the $p$-value plot for the Ljung-Box test of the model residuals shown in Fig. 8.20 is undesirable. Now we take a logarithm of the time series data `areturns` and are to build a Markov switching autoregressive model for the log `areturns`. The estimated model is shown in the second `Markov Switching Model Results` table in the following Python code. In this case, the ACF and PACF plots of the model residuals shown in Fig. 8.21 are like ones of a white noise, and further the $p$-value plot for the Ljung-Box test of the model residuals shown in Fig. 8.22 is desirable (viz., all the $p$-values are greater or equal to 0.05). Thus the second estimated model is more acceptable.

```
>>> import numpy as np
>>> import pandas as pd
>>> import statsmodels.api as sm
>>> import matplotlib.pyplot as plt
>>> from PythonTsa.plot_acf_pacf import acf_pacf_fig
>>> from PythonTsa.LjungBoxtest import plot_LB_pvalue
>>> from statsmodels.tsa.regime_switching.tests.test
        _markov_regression import areturns
>>> index=pd.date_range("2004-05-04", "2014-5-03", freq="W")
>>> dta_areturns = pd.Series(areturns, index=index)
>>> dta_areturns.plot(title="Absolute Returns of S&P 500")
>>> plt.show()
>>> myMauto1= sm.tsa.MarkovAutoregression(dta_areturns,
    k_regimes=2, order=4, switching_variance=True)
>>> myfit1=myMauto1.fit()
>>> print(myfit1.summary())
```

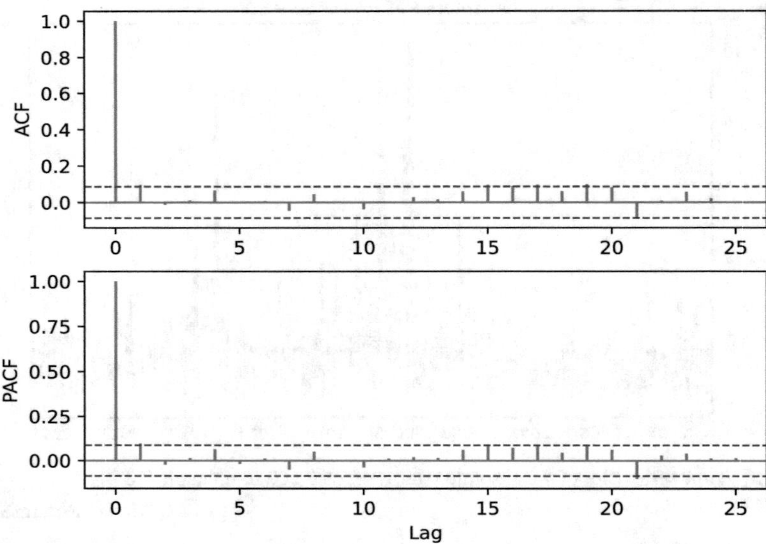


**Fig. 8.19** ACF and PACF plots of the model residuals of the absolute returns in Example 8.8
pyTSA_MarkovReturnsSP500

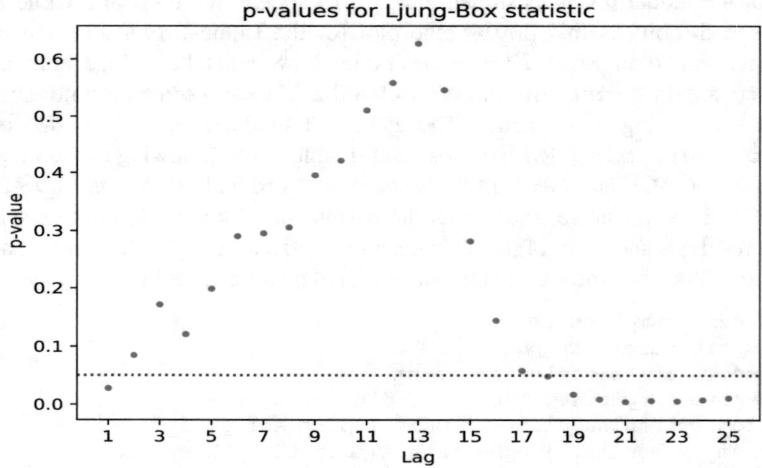


**Fig. 8.20** *p*-Values for the Ljung-Box test of the model residuals of `areturns` in Example 8.8
pyTSA_MarkovReturnsSP500

```
                     Markov Switching Model Results
================================================================================
Dep. Variable:                       y   No. Observations:                 517
Model:             MarkovAutoregression   Log Likelihood                -740.582
Date:                 Mon, 20 Dec 2021   AIC                           1509.164
```

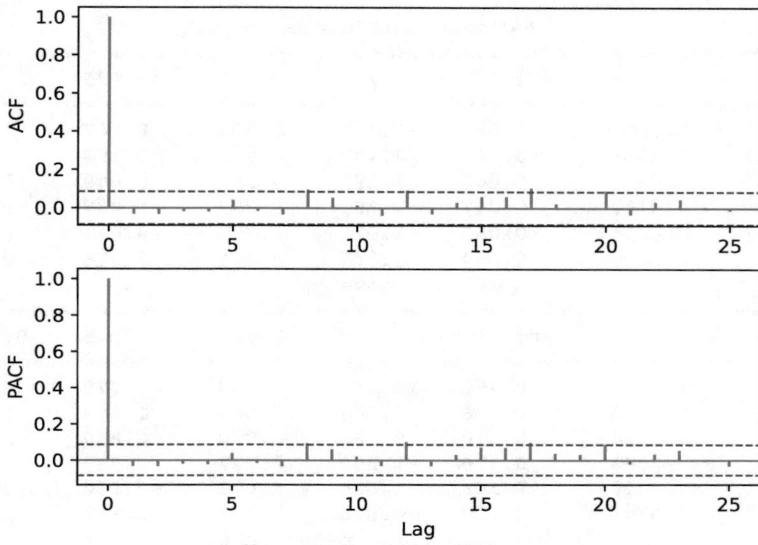


**Fig. 8.21** ACF and PACF plots of the model residuals of the log `areturns` in Example 8.8

pyTSA_MarkovReturnsSP500

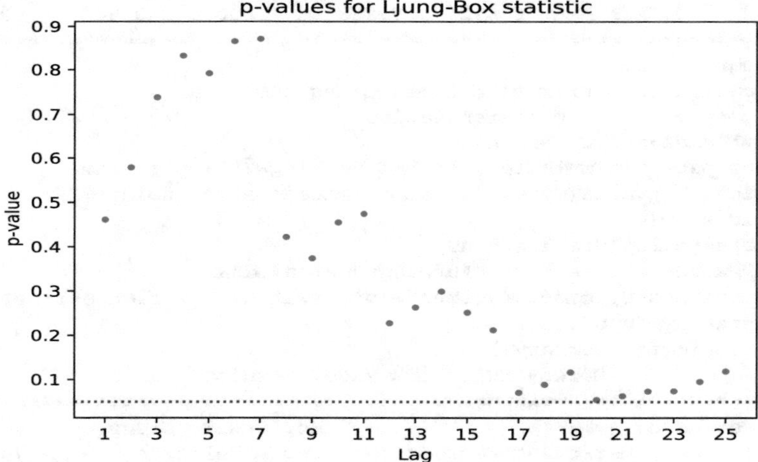


**Fig. 8.22** *p*-Values for the Ljung-Box test of the model residuals of the log `areturns` in Example 8.8

pyTSA_MarkovReturnsSP500

```
Time:                      04:01:16   BIC                     1568.636
Sample:                  05-09-2004   HQIC                    1532.467
                       - 04-27-2014
Covariance Type:             approx
```

```
                    Regime 0 parameters
=================================================================
              coef    std err          z      P>|z|    [0.025    0.975]
-----------------------------------------------------------------
const       1.0694      0.062     17.153      0.000     0.947     1.192
sigma2      0.6316      0.062     10.133      0.000     0.509     0.754
ar.L1       0.0538      0.048      1.127      0.260    -0.040     0.147
ar.L2       0.0142      0.047      0.306      0.760    -0.077     0.105
ar.L3       0.0564      0.055      1.026      0.305    -0.051     0.164
ar.L4       0.1808      0.053      3.387      0.001     0.076     0.285
                    Regime 1 parameters
=================================================================
              coef    std err          z      P>|z|    [0.025    0.975]
-----------------------------------------------------------------
const       4.4808      0.362     12.363      0.000     3.770     5.191
sigma2      2.4475      0.686      3.568      0.000     1.103     3.792
ar.L1       1.1134      0.130      8.586      0.000     0.859     1.368
ar.L2      -0.2053      0.170     -1.209      0.227    -0.538     0.128
ar.L3      -0.3090      0.169     -1.828      0.068    -0.640     0.022
ar.L4      -0.0948      0.110     -0.858      0.391    -0.311     0.122
                Regime transition parameters
=================================================================
              coef    std err          z      P>|z|    [0.025    0.975]
-----------------------------------------------------------------
p[0->0]     0.9461      0.017     57.150      0.000     0.914     0.979
p[1->0]     0.7261      0.092      7.888      0.000     0.546     0.907
=================================================================
Warnings:
[1] Covariance matrix calculated using numerical
    (complex-step) differentiation.
>>> myresid1=myfit1.resid
>>> acf_pacf_fig(myresid1, both=True, lag=25); plt.show()
>>> plot_LB_pvalue(myresid1, noestimatedcoef=0, nolags=25)
>>> plt.show()
>>> Ldta=np.log(dta_areturns)
>>> myMauto= sm.tsa.MarkovAutoregression(Ldta,
    k_regimes=2, order=4, trend='n', switching_variance=True)
>>> myfit=myMauto.fit()
>>> print(myfit.summary())
                Markov Switching Model Results
=================================================================
Dep. Variable:                    y   No. Observations:        517
Model:         MarkovAutoregression   Log Likelihood        -774.828
Date:              Tue, 21 Dec 2021   AIC                   1573.655
Time:                      06:15:31   BIC                   1624.632
Sample:                  05-09-2004   HQIC                  1593.630
                       - 04-27-2014
Covariance Type:             approx
                    Regime 0 parameters
=================================================================
              coef    std err          z      P>|z|    [0.025    0.975]
-----------------------------------------------------------------
sigma2      6.3165      1.774      3.560      0.000     2.839     9.794
ar.L1       0.5610      0.080      6.977      0.000     0.403     0.719
```

```
ar.L2        0.0465       0.154      0.301      0.763      -0.256      0.349
ar.L3        0.0727       0.188      0.387      0.699      -0.296      0.441
ar.L4       -0.3439       0.096     -3.590      0.000      -0.532     -0.156
                          Regime 1 parameters
=================================================================================
                 coef    std err          z      P>|z|      [0.025      0.975]
---------------------------------------------------------------------------------
sigma2       0.7327       0.062     11.830      0.000       0.611      0.854
ar.L1        0.0732       0.040      1.839      0.066      -0.005      0.151
ar.L2        0.0719       0.041      1.761      0.078      -0.008      0.152
ar.L3        0.0718       0.038      1.886      0.059      -0.003      0.146
ar.L4        0.1677       0.041      4.138      0.000       0.088      0.247
                     Regime transition parameters
=================================================================================
                  coef    std err          z      P>|z|      [0.025      0.975]
---------------------------------------------------------------------------------
p[0->0]    6.567e-07        nan        nan        nan         nan        nan
p[1->0]       0.1206      0.033      3.710      0.000       0.057      0.184
=================================================================================
Warnings:
[1] Covariance matrix calculated using numerical
    (complex-step) differentiation.
>>> myresid=myfit.resid
>>> acf_pacf_fig(myresid, both=True, lag=25); plt.show()
>>> plot_LB_pvalue(myresid, noestimatedcoef=0, nolags=25)
>>> plt.show()
```

In this chapter, we have addressed the linear state space model and its application as well as the Markov switching model. They are both useful in various domains. Of course, the handling here is introductory. For more information on state space methods including nonlinear and non-Gaussian state space models, see, for example, Tsay and Chen (2019); Douc et al. (2014); and Durbin and Koopman (2012). As to the Markov switching model, the reader can refer to Frühwirth-Schnatter (2006) and Kim and Nelson (1999), among others.

## Problems

**8.1** Verify that the state space models (8.4) and (8.5) specify the ARMA(1, 1) model (8.3) and so the former is equivalent to the latter.

**8.2** Find a state space form for the causal AR($p$) model as follows:

$$Y_t = \varphi_1 Y_{t-1} + \cdots + \varphi_p Y_{t-p} + \varepsilon_t.$$

**8.3** If the state space model (8.7)–(8.9) is time-invariant, that is, the parameters $\mathbf{c}_t,\ \mathbf{F}_t,\ \mathbf{R}_t,\ \mathbf{Q}_t,\ \mathbf{d}_t,\ \mathbf{Z}_t,$ *and* $\mathbf{H}_t$ do not change over time and derive the Kalman filtering (forecasting and smoothing) formulae.

**8.4** The dataset "realGdpConsInv" in the folder Ptsadata is from the Federal Reserve Bank of St. Louis, USA. It consists of the three economic variables `realgdp`, `realcons`, and `realinv`. Let `realcons` and `realinv` be the exogenous regressors. Then fit the SARIMAX model to the dataset. Lastly conduct an analysis of the model residuals.

**8.5** As for the variable `logret` in Example 8.6, let the lagged `logret` be an exogenous regressor. Then use the function `MarkovRegression` to fit the following Markov switching model

$$X_t = \mu_{S_t} + X_{t-1}\beta_{S_t} + \varepsilon_t, \ \varepsilon_t \sim \text{WN}(0, \sigma^2_{S_t})$$

to the dataset in Example 8.6. Lastly compare your results with the results of Example 8.6.

**8.6** Fit the Markov switching model with the number of regimes 2

$$X_t = \mu_{S_t} + X_{t-1}\beta_{S_t} + \varepsilon_t, \ \varepsilon_t \sim \text{WN}(0, \sigma^2)$$

to the time series `fedfunds` in Example 8.7, and then conduct an analysis of the model residuals. Besides, how about the case of the variance of the error $\varepsilon_t$ having switching.

**8.7** As for the three variables `fedfunds`, `ogap`, and `inf` in Example 8.7, let the lagged `fedfunds` be a part of the exogenous regressors, and then use the function `MarkovRegression` to build a Markov switching (auto)regression model like the following model

$$X_t = \mu_{S_t} + (X_{t-1}, Y_t, Z_t)\boldsymbol{\beta}_{S_t} + \varepsilon_t, \ \varepsilon_t \sim \text{WN}(0, \sigma^2)$$

with the number of regimes 2 where $X_{t-1}$ is the lagged `fedfunds`, $Y_t$ is `ogap`, and $Z_t$ is `inf`. Finally make a diagnostic check of the estimated model. Is it appropriate?

**8.8** Consider the Markov switching model

$$X_t = \mu_{S_t} + X_{t-1}\beta_{S_t} + \varepsilon_t, \ \varepsilon_t \sim \text{WN}(0, \sigma^2_{S_t}).$$

Use the Python function `MarkovRegression` to fit the model to the dataset in Example 8.8. Are you satisfied with your results?

# Chapter 9
# Nonstationarity and Cointegrations

This chapter deals with some advanced topics of time series analysis. We define the two concepts of stochastic trend and stochastic seasonality, introduce a few unit root and stationarity tests, as well as implement them with Python. We also elaborate on how to simulate a standard Brownian motion which is very useful in fields of finance and other disciplines. Finally, we concisely discuss Granger's representation theorem and vector error correction models.

## 9.1 Stochastic Trend and Stochastic Seasonality

In Sect. 2.4, we describe the features of a deterministic trend and features of deterministic seasonality in a time series. What is more, in practice, a large number of time series possess either deterministic seasonality (component) or a deterministic trend (component), and some of them may have both. Naturally, they are all nonstationary. On the other hand, a time series that has neither a deterministic trend nor deterministic seasonality is not necessarily stationary. A random walk is such an example. Thus, it is not easy to check if a time series without deterministic trend and seasonality is stationary. Besides, it must be noticed that for a time series with a deterministic trend, its mean function is nonconstant.

### *9.1.1 Deterministic Trend and Stochastic Trend*

What is a stochastic trend in a time series? Let us have a look at the following example before answering that question.

 
C. Huang, A. Petukhina, *Applied Time Series Analysis and Forecasting with Python*, Statistics and Computing,
https://doi.org/10.1007/978-3-031-13584-2_9

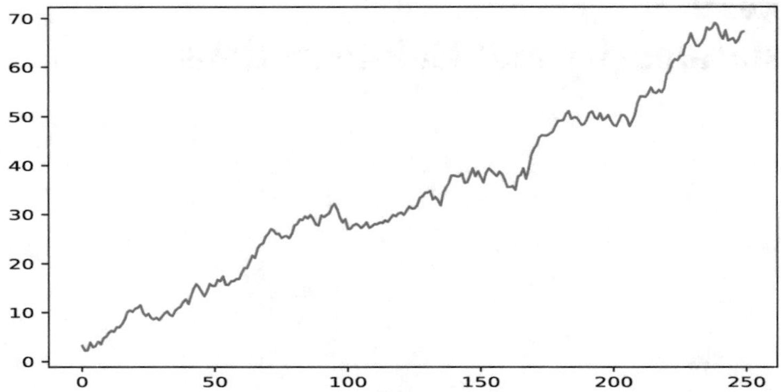


**Fig. 9.1** Time series plot of the process $X_t = 0.3 + 0.2t + \eta_t$ in Example 9.1

pyTSA_Trend

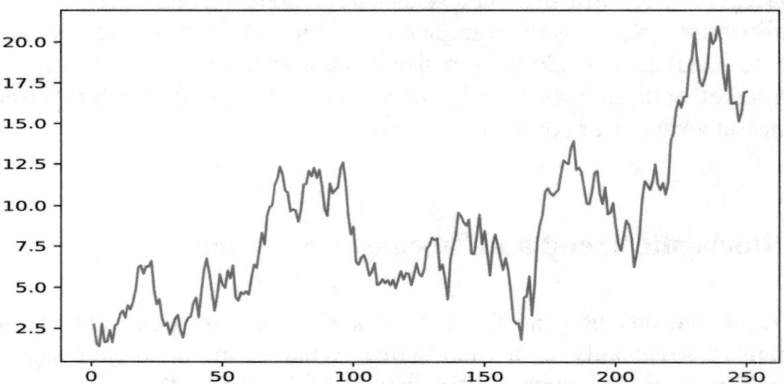


**Fig. 9.2** Time plot of the detrended series of the process $X_t = 0.3 + 0.2t + \eta_t$ in Example 9.1

pyTSA_Trend

*Example 9.1 (A Time Series with Trend Components)* Consider the time series $X_t = \alpha + \beta t + \eta_t$ where $\beta \neq 0$ and $\eta_t = \eta_{t-1} + \varepsilon_t$, $\varepsilon_t \sim \text{WN}(0, 1)$ is a random walk. Its time series plot is shown in Fig. 9.1 when $\alpha = 0.3$ *and* $\beta = 0.2$. Clearly there is a deterministic trend in the time series, and thus it is nonstationary. At the same time, the series has no seasonality. Furthermore, if we detrend the deterministic trend from the time series, we arrive at the series $\eta_t$ that is a random walk and also nonstationary, seeing Fig. 9.2. In other words, there are two components in the time series, which of both make it nonstationary. One component is well known to us, that is, the deterministic trend. Another is new and will be defined below. In addition, if we difference the deterministic-trend-removed time series $\eta_t$, then the differenced series is the white noise $\nabla \eta_t = \varepsilon_t$ and so stationary, seeing Figs. 9.3 and 9.4.

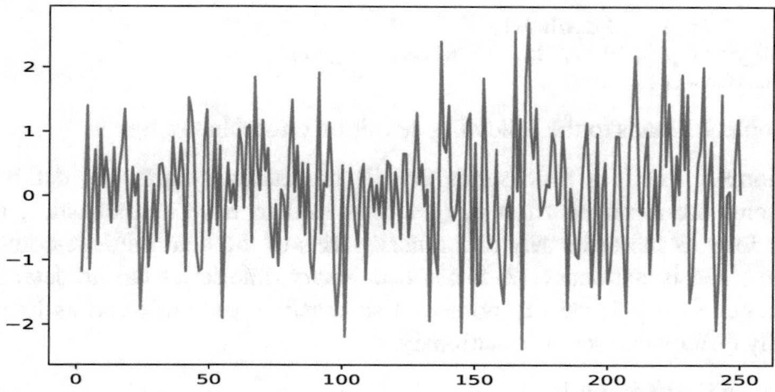


**Fig. 9.3** Time plot of the differenced series of the process $\eta_t$ in Example 9.1

pyTSA_Trend

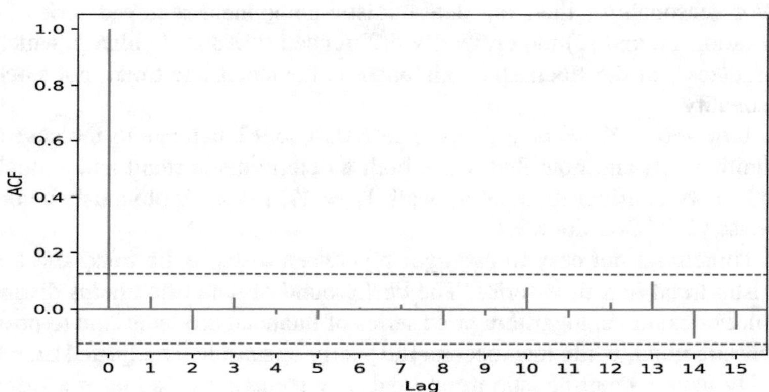


**Fig. 9.4** ACF plot of the differenced series of the process $\eta_t$ in Example 9.1

pyTSA_Trend

The following is the Python code used in this example.

```
>>> import numpy as np
>>> import matplotlib.pyplot as plt
>>> from PythonTsa.RandomWalk import RandomWalk_with_drift
>>> from PythonTsa.plot_acf_pacf import acf_pacf_fig
>>> np.random.seed(1373)
>>> rw0= RandomWalk_with_drift(drift=0.0, nsample=250, burnin=10)
# burnin: the number of observation at the beginning of the
# sample to drop. Used to reduce dependence on initial values.
>>> rw0.plot(); plt.show()
>>> t=np.arange(1,len(rw0)+1)
>>> Mydata=0.3+0.2*t+rw0
>>> Mydata.plot(); plt.show()
>>> drw=rw0.diff(1).dropna()
```

```
>>> drw.plot(); plt.show()
>>> acf_pacf_fig(drw, both=False, lag=15)
>>> plt.show()
```

Example 9.1 leads to the following definition on stochastic trends.

**Definition 9.1** (1) If a time series is still nonstationary after its deterministic components are removed from it, then it is said to have a stochastic (random) trend as long as the ordinarily differenced series of the deterministic-component-removed series is stationary. (2) If a nonstationary time series has no deterministic components, then it is said to possess a stochastic (random) trend as long as its ordinarily differenced series is stationary.

Remarks on Definition 9.1:

- A time series with a stochastic trend should satisfy two conditions: (1) it is nonstationary, and if it has deterministic components (deterministic trend and/or seasonality), then the deterministic-component-removed series is still nonstationary, and (2) the ordinarily differenced series is stationary, which also guarantees that the stochastic component is the stochastic trend, not stochastic seasonality.
- The time series $X_t = \alpha + \beta t + \eta_t$ in Example 9.1 belongs to the case (1) of Definition 9.1, and note that it has both a deterministic trend and a stochastic trend. If we consider the random walk $Y_t = Y_{t-1} + \varepsilon_t$, it obviously belongs to the case (2) of Definition 9.1.
- It is sometimes not easy to distinguish between a stochastic trend and a deterministic trend in a time series. The background of data often helps distinguish them. For example, logarithm price series of financial products tend to possess a stochastic trend, while macroeconomic yearly or seasonally adjusted time series usually have a deterministic trend (and may sometimes also have a stochastic trend). Two real examples are given in Example 9.2 and Problem 9.13.

*Example 9.2 (Stochastic Trend in Closing Price Series of IBM Stock)* The dataset "IBM.csv" in the folder Ptsadata is the IBM stock monthly price time series from January 1966 to March 2020. We now consider only the log series of closing prices. Its time series plot is shown in Fig. 9.5. The time plot illustrates nonstationarity of the time series and as well displays always-changing directions of the series. That is to say, the time plot suggests that the time series possesses a stochastic trend. In addition, the time plot of the differenced log series (viz., return series) shown in Fig. 9.6 is a typical time plot of a stationary series; ACF plot shown in Fig. 9.7 is like one of a white noise; and $p$-value plot for Ljung-Box test shown in Fig. 9.8 suggests that the differenced log series can be surely seen as a white noise.

```
>>> import numpy as np
>>> import pandas as pd
>>> import matplotlib.pyplot as plt
>>> from PythonTsa.plot_acf_pacf import acf_pacf_fig
>>> from PythonTsa.LjungBoxtest import plot_LB_pvalue
```

**Fig. 9.5** Time series plot of the log closing price series of IBM stock

pyTSA_TrendIBM

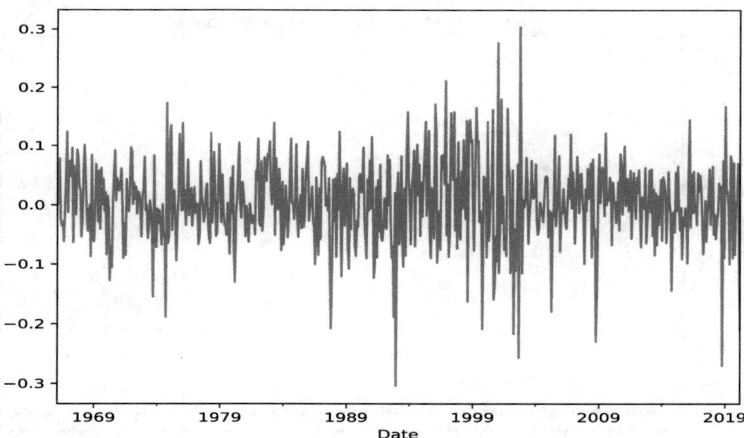


**Fig. 9.6** Time series plot of the differenced log closing price series of IBM stock

pyTSA_TrendIBM

```
>>> from PythonTsa.datadir import getdtapath
>>> dtapath=getdtapath()
>>> IBM=pd.read_csv(dtapath + 'IBM.csv', header=0)
>>> IBM['Date']=pd.to_datetime(IBM['Date'])
>>> IBM.index=IBM['Date']
>>> IBMclose=IBM.Close
>>> LIBMclose=np.log(IBMclose)
>>> LIBMclose.plot()
>>> plt.xticks(rotation=15)
>>> plt.show()
```

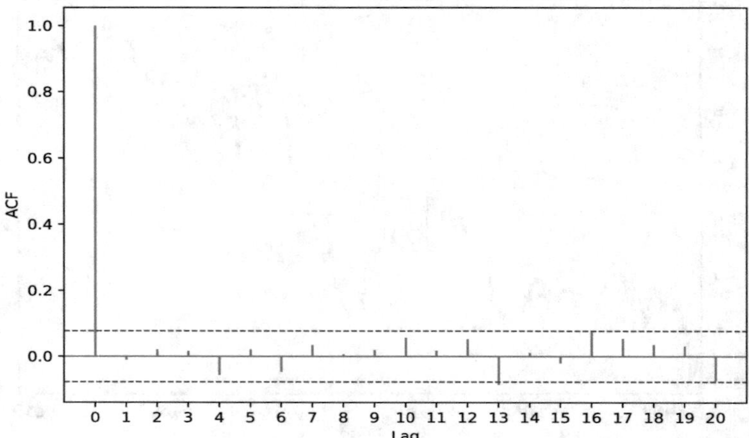


**Fig. 9.7** ACF plot of the differenced log closing price series of IBM stock

pyTSA_TrendIBM

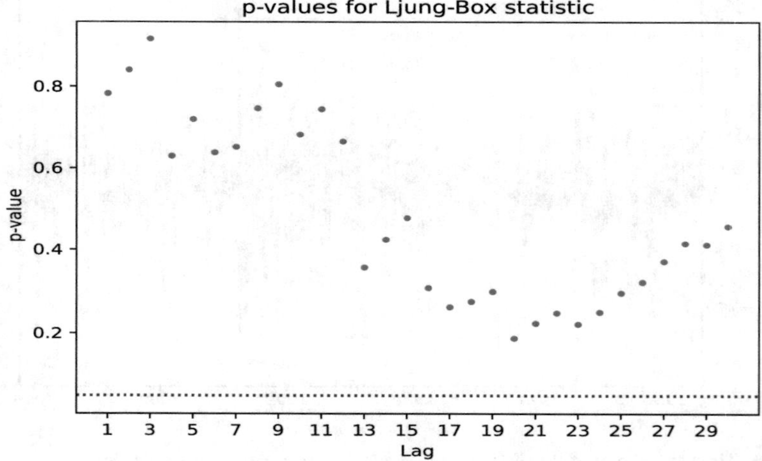


**Fig. 9.8** *p*-Value plot for Ljung-Box test of the differenced log closing price series of IBM stock

pyTSA_TrendIBM

```
>>> dLIBMclose=LIBMclose.diff(1).dropna()
>>> dLIBMclose.plot()
>>> plt.show()
>>> acf_pacf_fig(dLIBMclose, both=False, lag=20)
>>> plt.show()
>>> plot_LB_pvalue(dLIBMclose, noestimatedcoef=0, nolags=30)
>>> plt.show()
```

### 9.1.2 Deterministic Seasonality and Stochastic Seasonality

Now we turn to the notion of stochastic seasonality. Let us look at an example about seasonality in a time series.

*Example 9.3 (Stationary Seasonal AR(1) Processes and Seasonal Random Walk)* The ordinary random walk $X_t = X_{t-1} + \varepsilon_t$ is well known. The *seasonal random walk* is defined as $X_t = X_{t-p} + \varepsilon_t$ where $p > 1$ is a seasonal period and $\varepsilon_t \sim \text{WN}(0, \sigma^2)$. They are both nonstationary. In this example, we consider the seasonal AR(1) process generated by $X_t = \varphi X_{t-4} + \varepsilon_t$ where the seasonal period $p = 4$, that is, the time series, can be viewed as a quarterly series. Moreover, if $|\varphi| < 1$, then the series is a stationary SAR$(1)_4$ process, and if $\varphi = 1$, then the series is the seasonal random walk with seasonal period 4. In the cases of $\varphi = 0.2$, $\varphi = 0.6$, $\varphi = 0.8$, and $\varphi = 1$, their time plots are shown in Fig. 9.9, ACF plots in Fig. 9.10, and seasonal plots in Fig. 9.11. We observe that their stationarity becomes weaker and weaker as $\varphi$ approaches to 1. More interesting is that when $\varphi$ is smaller, every seasonal sample mean is almost equal, and every seasonal plot is almost the same pattern. In other words, the concatenate series of the seasonal subseries is still stationary (see Fig. 9.12). In fact, as long as $|\varphi| < 1$, the mean at every time point is the same due to its stationarity. On the contrary, when $\varphi = 1$, that is, $X_t$ is a seasonal random walk and nonstationary, although the means at different time points in the same season are identical as $X_t = X_{t-p} + \varepsilon_t$ means $\mathsf{E}X_t = \mathsf{E}X_{t-p}$ for every time point $t$, the means at the different seasons tend to be various. And the concatenate series of its seasonal subseries is also nonstationary (see Fig. 9.13). At this point, $X_t$ is said to have stochastic seasonality as defined below. To reproduce the plots in this example, run the following Python code.

```
>>> import numpy as np
>>> import matplotlib.pyplot as plt
>>> import pandas as pd
>>> import statsmodels.api as sm
>>> from PythonTsa.plot_acf_pacf import acf_pacf_fig
>>> from statsmodels.tsa.arima_process
        import arma_generate_sample
>>> sar1=np.array([1, 0, 0, 0, -0.2])
>>> sar2=np.array([1, 0, 0, 0, -0.6])
>>> sar3=np.array([1, 0, 0, 0, -0.8])
>>> sar4=np.array([1, 0, 0, 0, -1.0])
>>> np.random.seed(137)
>>> x1= arma_generate_sample(ar=sar1, ma=[1], nsample=200)
>>> x2= arma_generate_sample(ar=sar2, ma=[1], nsample=200)
>>> x3= arma_generate_sample(ar=sar3, ma=[1], nsample=200)
>>> x4= arma_generate_sample(ar=sar4, ma=[1], nsample=200)
>>> x1=pd.Series(x1); x2=pd.Series(x2)
>>> x3=pd.Series(x3); x4=pd.Series(x4)
>>> fig = plt.figure()
>>> x1.plot(marker='.',ax= fig.add_subplot(221))
>>> plt.title('$X_t=0.2X_{t-4}+\epsilon_t$')
>>> x2.plot(marker='.',ax= fig.add_subplot(222))
```

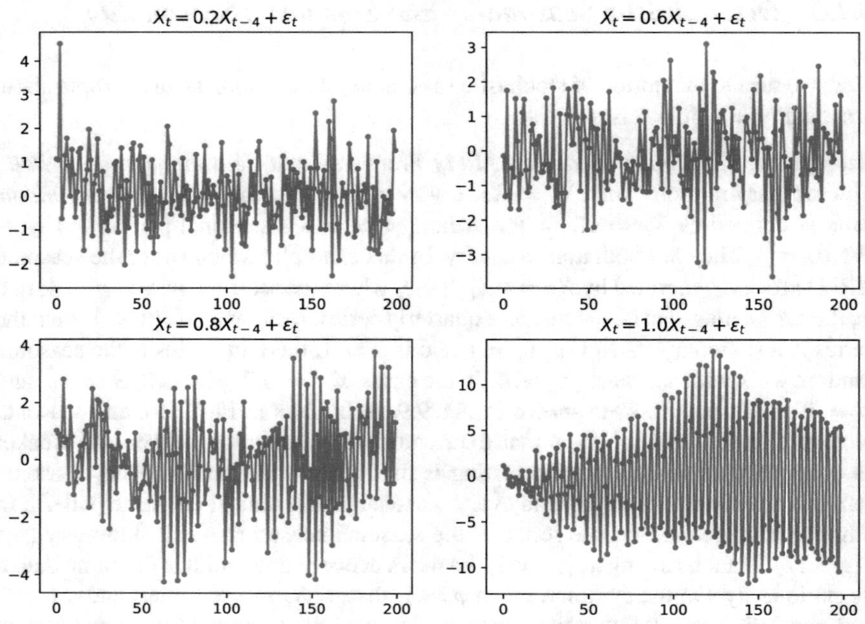

Fig. 9.9 Time plots of the seasonal AR processes and seasonal random walk in Example 9.3

pyTSA_Seasonality

```
>>> plt.title('$X_t=0.6X_{t-4}+\epsilon_t$')
>>> x3.plot(marker='.',ax= fig.add_subplot(223))
>>> plt.title('$X_t=0.8X_{t-4}+\epsilon_t$')
>>> x4.plot(marker='.',ax= fig.add_subplot(224))
>>> plt.title('$X_t=1.0X_{t-4}+\epsilon_t$')
>>> plt.show()
>>> fig = plt.figure()
>>> ax= fig.add_subplot(221)
>>> acf_pacf_fig(x1,both=False,lag=16)
>>> plt.title('$X_t=0.2X_{t-4}+\epsilon_t$')
>>> ax= fig.add_subplot(222)
>>> acf_pacf_fig(x2,both=False,lag=16)
>>> plt.title('$X_t=0.6X_{t-4}+\epsilon_t$')
>>> ax= fig.add_subplot(223)
>>> acf_pacf_fig(x3,both=False,lag=16)
>>> plt.title('$X_t=0.8X_{t-4}+\epsilon_t$')
>>> ax= fig.add_subplot(224)
>>> acf_pacf_fig(x4,both=False,lag=16)
>>> plt.title('$X_t=1.0X_{t-4}+\epsilon_t$')
>>> plt.show()
>>> from statsmodels.graphics.tsaplots import quarter_plot
>>> speriod=pd.date_range('2001-01', periods=len(x1),freq='Q')
>>> x1.index=speriod
>>> x2.index=speriod
```

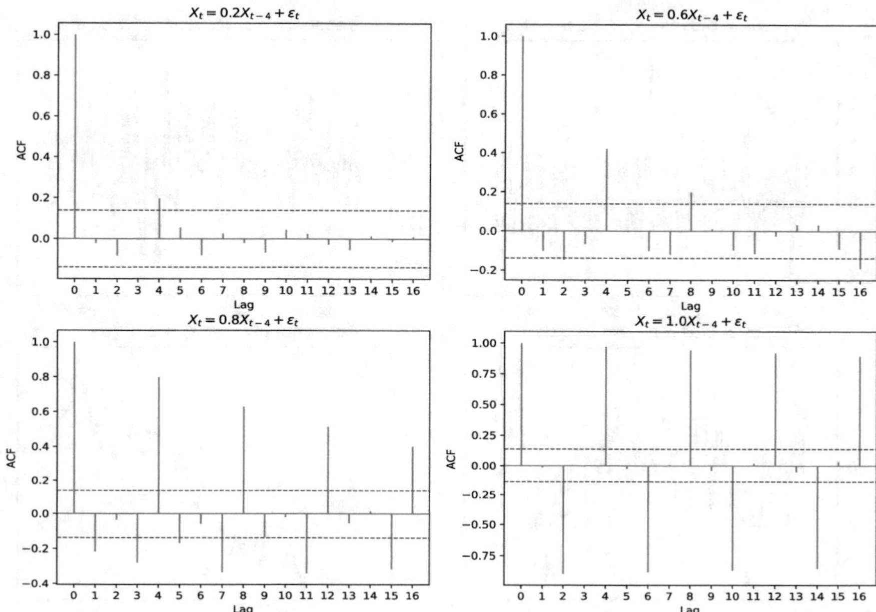


**Fig. 9.10** ACF plots of the seasonal AR processes and seasonal random walk in Example 9.3

pyTSA_Seasonality

```
>>> x3.index=speriod
>>> x4.index=speriod
>>> fig = plt.figure()
>>> quarter_plot(x1, ax= fig.add_subplot(221))
>>> plt.title('Seasonal plot for $X_t=0.2X_{t-4}+\epsilon_t$')
>>> quarter_plot(x2, ax= fig.add_subplot(222))
>>> plt.title('Seasonal plot for $X_t=0.6X_{t-4}+\epsilon_t$')
>>> quarter_plot(x3, ax= fig.add_subplot(223))
>>> plt.title('Seasonal plot for $X_t=0.8X_{t-4}+\epsilon_t$')
>>> quarter_plot(x4, ax= fig.add_subplot(224))
>>> plt.title('Seasonal plot for $X_t=1.0X_{t-4}+\epsilon_t$')
>>> plt.show()
>>> y=pd.DataFrame(index=range(0,int(len(x1)/4)),
                   columns=['0','1','2','3'])
>>> for i in range(0,4):
        for j in range(i, len(x1), 4):
            y.iat[int(j/4),i]=x1[j]
>>> z=pd.concat([y['0'],y['1'],y['2'],y['3']],ignore_index=True)
>>> fig = plt.figure()
>>> ax= fig.add_subplot(211)
>>> z.plot()
>>> plt.title('The concatenate series of the seasonal subseries
              for $X_t=0.2X_{t-4}+\epsilon_t$')
>>> ax= fig.add_subplot(212)
```

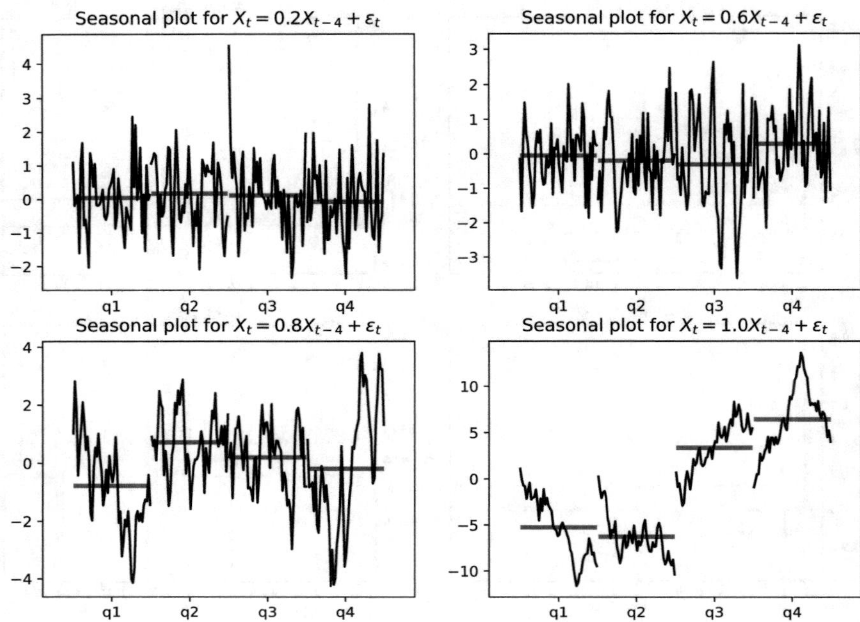


**Fig. 9.11** Seasonal plots of the seasonal AR processes and seasonal random walk in Example 9.3 pyTSA_Seasonality

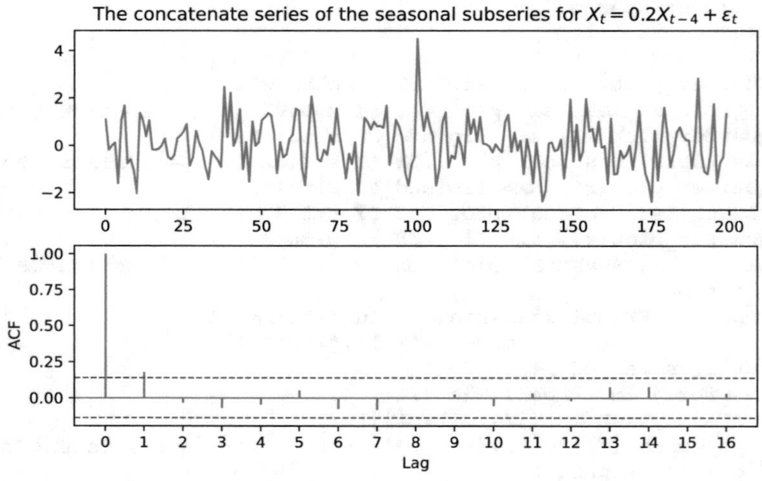


**Fig. 9.12** The concatenate series of the seasonal subseries for $X_t = 0.2X_{t-4} + \varepsilon_t$ in Example 9.3 pyTSA_Seasonality

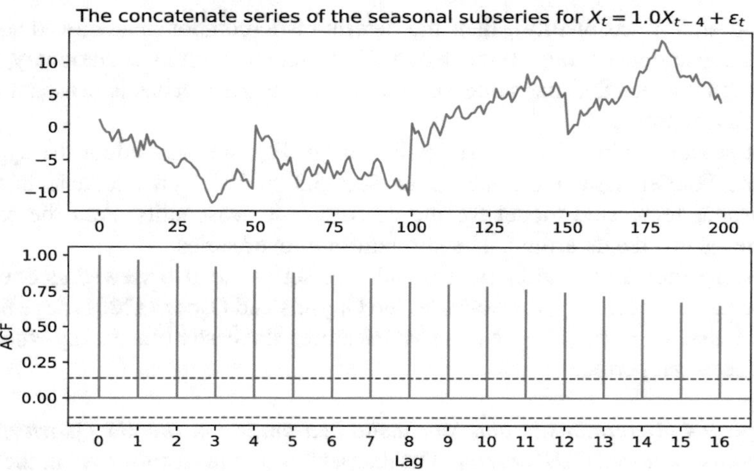


**Fig. 9.13** The concatenate series of the seasonal subseries for $X_t = 1.0X_{t-4} + \varepsilon_t$ in Example 9.3
pyTSA_Seasonality

```
>>> acf_pacf_fig(z,both=False,lag=16)
>>> plt.show()
>>> srw=pd.DataFrame(index=range(0,int(len(x1)/4)),
                     columns=['0','1','2','3'])
>>> for i in range(0,4):
        for j in range(i, len(x4), 4):
            srw.iat[int(j/4),i]=x4[j]
>>> csrw=pd.concat([srw['0'],srw['1'],srw['2'],srw['3']],
                   ignore_index=True)
>>> fig = plt.figure()
>>> ax= fig.add_subplot(211)
>>> csrw.plot()
>>> plt.title('The concatenate series of the seasonal subseries
              for $X_t=1.0X_{t-4}+\epsilon_t$')
>>> ax= fig.add_subplot(212)
>>> acf_pacf_fig(csrw,both=False,lag=16)
>>> plt.show()
```

**Definition 9.2** (1) If a seasonal time series is still nonstationary after its deterministic components are removed from it, then it is said to have stochastic (random) seasonality as long as the seasonally differenced series of the deterministic-component-removed series is stationary. (2) If a nonstationary seasonal time series has no deterministic components, then it is said to possess stochastic (random) seasonality as long as its seasonally differenced series is stationary.

Remarks about Definition 9.2:

- A seasonal time series with stochastic seasonality should satisfy two conditions: (1) it is nonstationary, and if it has deterministic components (deterministic trend and/or seasonality), then the deterministic-component-removed series is still nonstationary, and (2) the seasonally differenced series is stationary, which also guarantees that the stochastic component is the stochastic seasonality, not stochastic trend.

- If a seasonal series has deterministic seasonality, we may adopt the approach (viz., Fourier series) introduced in Sect. 5.3 to build an adequate harmonic seasonal regression model for the deterministic seasonality. And the residual series is just the deterministic-seasonality-removed series.
- In some literature, stationary seasonal time series are also viewed as ones with stochastic seasonality, for example, see Ghysels and Osborn (2001, page 8), they think that $X_t = \varphi X_{t-p} + \varepsilon_t,\ |\varphi| < 1$ features stochastic seasonality where $p$ is the seasonal period.

*Example 9.4 (Deterministic and Stochastic Seasonality in the US Quarterly Not-Seasonally-Adjusted GDP Series)* The dataset "usGDPnotAdjust.csv" in the folder Ptsadata is the US quarterly not-seasonally-adjusted GDP series (in millions of dollars) from the first quarter of 1947 to the third quarter of 2021. It is available from the Federal Reserve Bank of St. Louis. First of all, we take logarithm on the series for reducing the magnitude and volatility of it and arrive at the log series `lusgdp`. Its time series plot is shown in Fig. 9.14. Clearly it is nonstationary. Now we try to decompose the log series using the function `seasonal_decompose`. The decomposition result is shown in Fig. 9.15. The residual series `myresid` is simply the remainder of removing the deterministic trend and seasonality from the log series. Its time plot (bottom panel of Fig. 9.15) is not displayed as all the data are closely around 1. We separately draw the time plot shown in Fig. 9.16. We

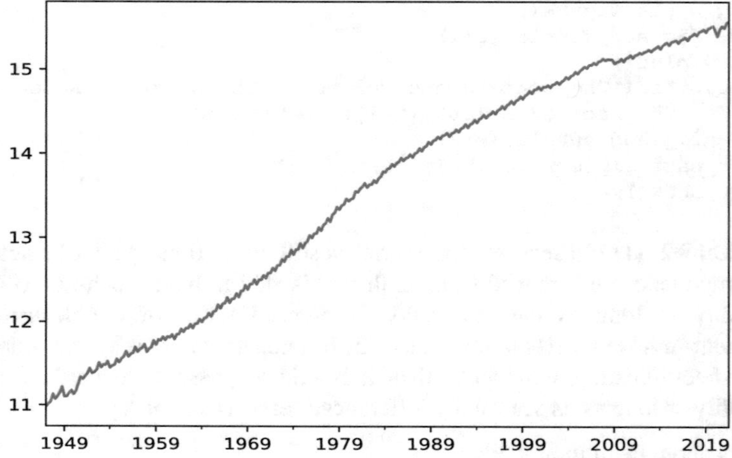


**Fig. 9.14** Time plot of the log series of US quarterly not-seasonally-adjusted GDP

pyTSA_SeasonalityUSGDP

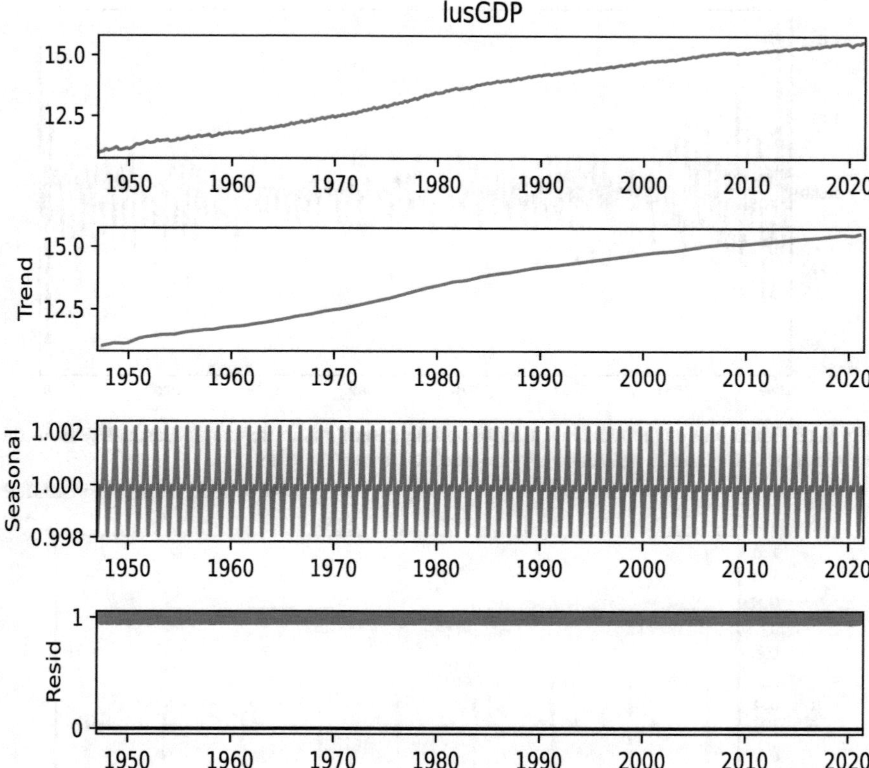


**Fig. 9.15** Decomposition of the log series of US quarterly not-seasonally-adjusted GDP

pyTSA_SeasonalityUSGDP

observe that there is still seasonality in the residual series although the deterministic seasonality has been removed. At the same time, the ACF plot of the residual series shown in Fig. 9.17 illustrates that it is nonstationary. Next, we seasonally difference the residual series and obtain the seasonally differenced series `dmyresid`. There is no seasonality in the differenced series anymore, and it is stationary, seeing Figs. 9.18 and 9.19. Besides, we can give more evidence to support the assertion that the differenced sequence is stationary. Actually, we are able to make a KPSS stationarity test (see Sect. 9.3.3 for more details) of the seasonally differenced series. The $p$-value for the KPSS test is greater than 0.1, which strongly support our assertion. To summarize, in addition to the deterministic trend and seasonality, there is stochastic seasonality in the log series of the US quarterly not-seasonally-adjusted GDP.

```
>>> import numpy as np
>>> import pandas as pd
>>> import matplotlib.pyplot as plt
```

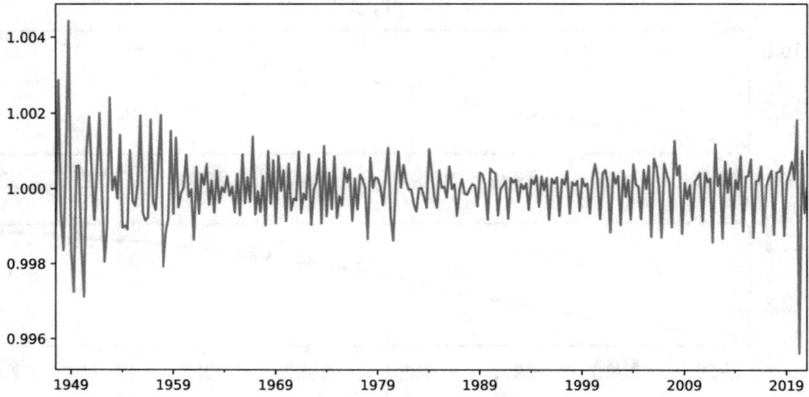


**Fig. 9.16** Time plot of the residuals after removing deterministic trend and seasonality from the log series of US quarterly not-seasonally-adjusted GDP in Example 9.4

pyTSA_SeasonalityUSGDP

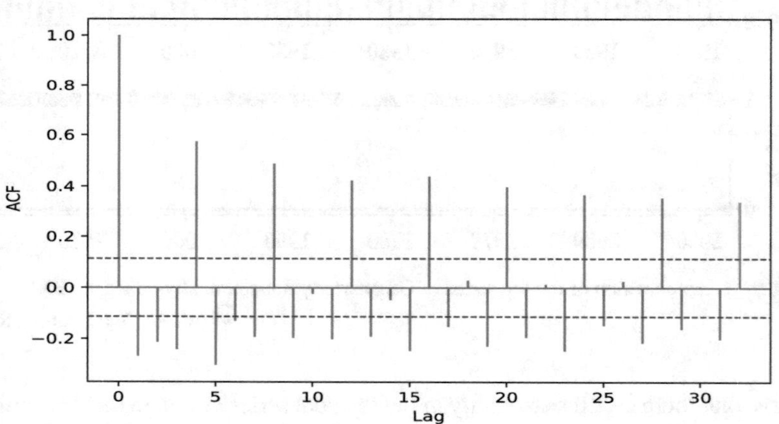


**Fig. 9.17** ACF plot of the residuals after removing deterministic trend and seasonality from the log series of US quarterly not-seasonally-adjusted GDP in Example 9.4

pyTSA_SeasonalityUSGDP

```
>>> from statsmodels.tsa.seasonal import seasonal_decompose
>>> from statsmodels.tsa.statespace.tools import diff
>>> from PythonTsa.plot_acf_pacf import acf_pacf_fig
>>> from statsmodels.tsa.stattools import kpss
>>> from PythonTsa.datadir import getdtapath
>>> dtapath=getdtapath()
>>> usgdp=pd.read_csv(dtapath + 'usGDPnotAdjust.csv',header=0)
>>> timeindx=pd.date_range('1947-01',periods=len(usgdp),freq='QS')
>>> usgdp.index=timeindx
>>> usgdp=usgdp['NA000334Q']
```

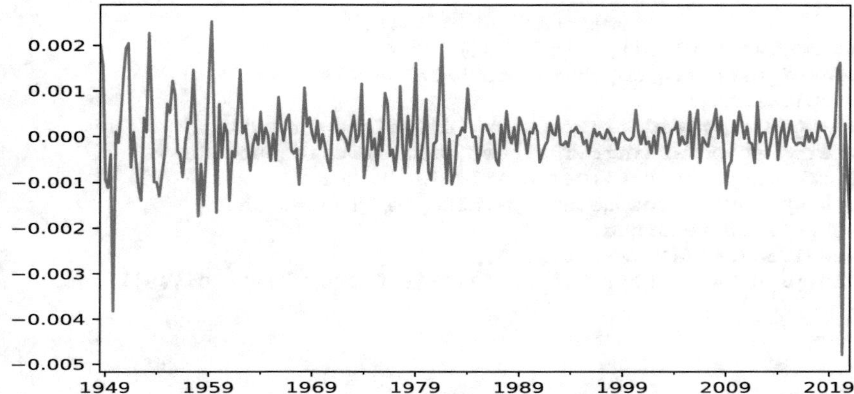


**Fig. 9.18** Time plot of the seasonally differenced residuals after removing deterministic trend and seasonality from the log series of US quarterly not-seasonally-adjusted GDP in Example 9.4

pyTSA_SeasonalityUSGDP

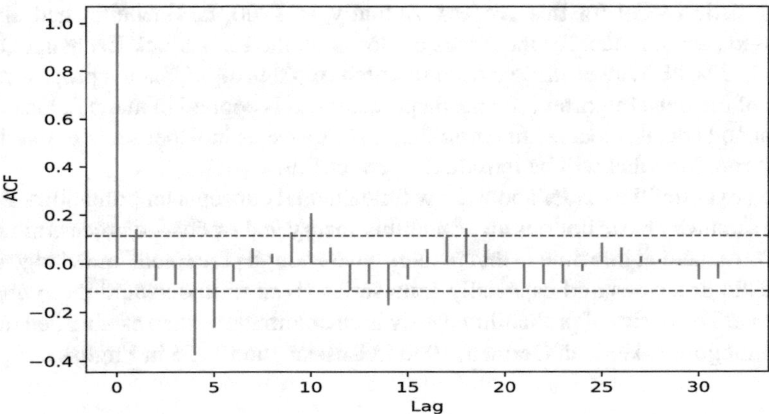


**Fig. 9.19** ACF plot of the seasonally differenced residuals after removing deterministic trend and seasonality from the log series of US quarterly not-seasonally-adjusted GDP in Example 9.4

pyTSA_SeasonalityUSGDP

```
>>> lusgdp=np.log(usgdp)
>>> lusgdp=lusgdp.rename('lusGDP')
>>> lusgdp.plot(); plt.show()
>>> xdec=seasonal_decompose(lusgdp, model='multi')
>>> xdec.plot(); plt.show()
>>> myresid=xdec.resid.dropna()
>>> myresid.plot(); plt.show()
>>> acf_pacf_fig(myresid, both=False, lag=32)
>>> plt.show()
>>> dmyresid=diff(myresid, k_diff=0, k_seasonal_diff=1,
```

```
                    seasonal_periods=4)
>>> dmyresid.plot(); plt.show()
>>> acf_pacf_fig(dmyresid, both=False, lag=32)
>>> plt.show()
>>> kpss(dmyresid, regression='c', nlags='auto')
InterpolationWarning: The test statistic is outside
of the range of p-values available in the
look-up table. The actual p-value is greater than
the p-value returned.
(0.06178378477917526, 0.1, 5,
{'10%': 0.347, '5%': 0.463, '2.5%': 0.574, '1%': 0.739})
```

## 9.2 Brownian Motions and Simulation

In 1828, with a microscope, the Scottish botanist R. Brown observed the irregular motion process of pollen particles in a liquid. In 1905, A. Einstein provided a probabilistic model for this process. Actually, in 1900, L. Bachelier had already proposed such a model for the prices of stocks on the Paris Stock Exchange. Later, in the 1920s, N. Wiener made a deep research into this topic. Such a process model is one of the most important stochastic processes. It is applied in many fields of both natural and social sciences. In particular, this process is indispensable to the theory of unit root tests that will be introduced in Sect. 9.3.

We next brief the reader about a few fundamental concepts on probability theory. The reason why basic notions on probability theory and stochastic process are rigorously presented at this time is that readers have enough perceptual knowledge about stochastic processes and especially time series so as to understand these abstract concepts. The basics of probability theory is an axiomatic system established mainly by Kolmogorov (1933, in German, 1936 in Russian, and 1956 in English).

### 9.2.1 Probability Space

A probability space is an abstract structure that is designed for describing stochastic phenomenon. Suppose that $\Omega$ is a given nonempty set, the elements of which we shall call *elementary events*. Then we are able to give some basic concepts from general probability theory.

**Definition 9.3** Let $\mathscr{F}$ be a collection of subsets of a given nonempty set $\Omega$. Then $\mathscr{F}$ is known as a $\sigma$-algebra (or Borel field) on $\Omega$ if it has the following properties:

(i) The empty set $\emptyset \in \mathscr{F}$.
(ii) If $A \in \mathscr{F}$, then $A^c = \Omega \setminus A \in \mathscr{F}$.
(iii) If $A_i \in \mathscr{F}, i = 1, 2, \cdots$, then $\bigcup_{i=1}^{\infty} A_i \in \mathscr{F}$.

At this point, the pair $(\Omega, \mathscr{F})$ is called a measurable space, an element of $\mathscr{F}$ is referred to as a (random) event, and $\mathscr{F}$ is also known as the field of events.

Note that (1) due to properties (i) and (ii), we have $\Omega \in \mathscr{F}$; (2) it can be proved that the new set resulted from some operations of random events is still in $\mathscr{F}$, for example, if any $A, B, C \in \mathscr{F}$, then $A \bigcap B = AB \in \mathscr{F}, A \bigcup B = A + B \in \mathscr{F}, A \bigcap (B \bigcup C) = A(B + C) = AB + AC \in \mathscr{F}, A \setminus B = AB^c \in \mathscr{F}$.

**Definition 9.4** A probability measure on a measurable space $(\Omega, \mathscr{F})$ is a function $P$ with domain $\mathscr{F}$ such that:

(i) For any $A \in \mathscr{F}$, $P(A) \geq 0$.
(ii) $P(\Omega) = 1$.
(iii) If $A_i \in \mathscr{F}, i = 1, 2, \cdots$ and are pairwise disjoint, that is, for any $i \neq j$, $A_i A_j = \emptyset$, then $P(\bigcup_{i=1}^{\infty} A_i) = \sum_{i=1}^{\infty} P(A_i)$.

At this time, the triple $(\Omega, \mathscr{F}, P)$ is said to be a probability space. For any event $A \in \mathscr{F}$, the value $P(A)$ is the probability that the event $A$ occurs. If $P(A) = 1$, we say that the event $A$ occurs with probability 1 or almost surely (a.s.).

Evidently for the empty event $\emptyset$, the probability that it occurs is $P(\emptyset) = 0$. Furthermore, theoretically, for every event $A \in \mathscr{F}$, the probability that it occurs can be computed out.

**Definition 9.5** Let $(\Omega, \mathscr{F}, P)$ be a probability space. A real single-valued function $X(\omega)$ defined on $\Omega$ is called a random variable if for each real number $x$, $\{\omega; X(\omega) \leq x\} \in \mathscr{F}$ and further $F(x) = P(\{\omega; X(\omega) \leq x\})$ is said to be the cumulative distribution function (cdf) of random variable $X(\omega)$.

Thus, a random variable is actually a function from $\Omega$ into the real numbers. However, the random variable $X(\omega)$ is often simply written as $X$. We also say that the random variable $X(\omega)$ is $\mathscr{F}$-measurable since $\{\omega; X(\omega) \leq x\} \in \mathscr{F}$ for each real number $x$. It can be proved that the cdf of a random variable completely determines the probability law of the random variable.

**Definition 9.6** A stochastic (or random) process is a parameterized collection of random variables $\{X_t(\omega); t \in T\}$ defined on a probability space $(\Omega, \mathscr{F}, P)$ where $T$ is an ordered set called a parameter space.

Remarks about Definition 9.6:

- The parameter space $T$ may be any set but $\Omega$, for example, the halfline $[0, \infty)$, any interval $[a, b]$, the integers, the nonnegative integers, and so on. If $T$ is discrete such as the integers, $\{X_t(\omega); t \in T\}$ is often called a discrete-time stochastic process or time series. On the other hand, if $T$ is continuous such as the interval $[0, 1]$, $\{X_t(\omega); t \in T\}$ is a continuous-time stochastic process.
- For each fixed $t \in T$, $X_t(\omega)$ is a random variable.
- For each fixed $\omega_0 \in \Omega$, $X_t(\omega_0)$ may be considered a function of the parameter $t \in T$ and denoted by $X(t, \omega_0)$, which is called a (sample) path or realization of the stochastic process $X_t(\omega)$.

- We may also regard the process $X_t(\omega)$ as a function of two variables $(t, \omega)$ from $T \times \Omega = \{(t, \omega); t \in T, \omega \in \Omega\}$ into the real numbers. This is a natural point of view in stochastic analysis.
- For simplicity, a stochastic process $\{X_t(\omega); t \in T\}$ is often written as $\{X_t; t \in T\}$ or $\{X(t); t \in T\}$.

In the next subsection, we introduce a very important continuous-time stochastic process: Brownian motion. It is a fundamental process in stochastic analysis.

### 9.2.2 Brownian Motions

As we know, the random walk and its first differences play a crucial role in time series analysis. And the Brownian motion is the counterpart of the random walk in continuous-time processes. It constitutes a fundamental building block in stochastic analysis. The following is the definition for a standard Brownian motion.

**Definition 9.7** A continuous-time stochastic process $\{B(t); 0 \le t \le 1\}$ is referred to as a standard Brownian motion (SBM) if it satisfies:

(1) $B(0) = 0$ a.s.
(2) For any integer $n \ge 3$ and any set of time points $0 \le t_1 < t_2 < \cdots < t_n \le 1$, the increments $\{B(t_i) - B(t_{i-1}); i = 2 : n\}$ are independent.
(3) For $0 \le s < t \le 1$, $B(t) - B(s) \sim \mathrm{N}(0, t - s)$.

The following is a few remarks on Definition 9.7:

- In the light of Kolmogorov's continuity theorem, it can be proved that there always exists a standard Brownian motion $B(t)$ which is continuous as a function of time $t$ with probability 1. See, for example, (Øksendal 2003, pp13–14).
- Although the standard Brownian motion $B(t)$ is continuous with probability 1, it is nowhere differentiable.
- The last two properties indicate that the Brownian motion has stationary independent increments.
- The Brownian motion is also called a Wiener process.
- Clearly $B(t) \sim \mathrm{N}(0, t)$ for any $t \in (0, 1]$.
- In $K$-dimensional case, there is a similar definition of $K$-dimensional SBM $\{\mathbf{B}(t) = (B_1(t), \cdots, B_K(t))'; 0 \le t \le 1\}$. At this point, for $0 \le s < t \le 1$, $\mathbf{B}(t) - \mathbf{B}(s) \sim \mathrm{N}[\mathbf{0}, (t - s)\mathbf{I}_K]$ where $\mathbf{I}_K$ is an identity matrix.

It is N. Wiener who showed in 1923 that there exists a stochastic process with continuous sample paths satisfying the properties in Definition 9.7. Later the process has come to be known as a Brownian motion or Wiener process.

To simulate a Brownian motion, we need to discretize it. The following theorem provides a guaranty.

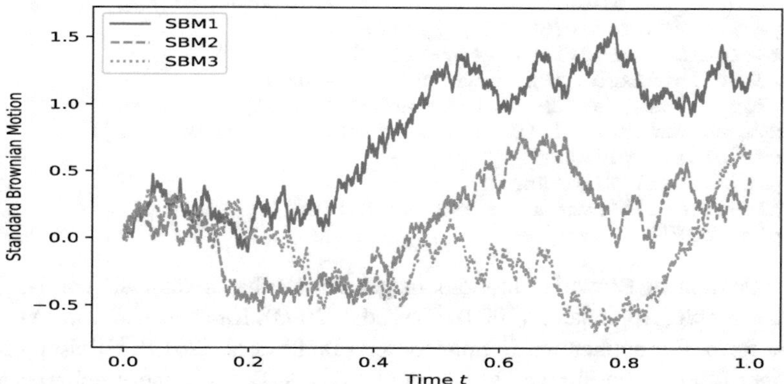


**Fig. 9.20** Three simulated sample paths of the standard Brownian motion

pyTSA_SimBM

**Theorem 9.1 (Donsker's Theorem)** *Assume that* $\{Z_i; i = 1 : n\}$ *is a sequence of iid random variables with mean 0 and variance 1. For any given* $t \in [0, 1]$*, we have*

$$S_n(t) = \frac{1}{\sqrt{n}} \sum_{i=1}^{[nt]} Z_i \xrightarrow{d} B(t),$$

*that is,* $S_n(t)$ *converges in distribution to a standard Brownian motion as* $n \to \infty$*, where* $[nt]$ *is the greatest integer part of* $nt$ *and if* $[nt] = 0$*, define* $S_n(t) = 0$.

Remarks about Theorem 9.1:

- Theorem 9.1 is a simple form of the functional central limit theorem. See, for example (Billingsley 1999).
- Sometimes we assume that $\{Z_i; i = 1 : n\}$ are independent standard normal random variables.
- Convergence in distribution (denoted as $\xrightarrow{d}$ or $\Longrightarrow$ or $\xrightarrow{L}$ in some literature) is also called weak convergence.
- If $\{\mathbf{Z}_i; i = 1 : n\}$ is a series of iid $K$-dimensional random vectors with mean vector $\mathbf{0}$ and covariance matrix $\mathbf{I}_K$, then the partial sum converges in distribution to a $K$-dimensional standard Brownian motion $\mathbf{B}(t)$.

In the `PythonTsa` package, the function `simulSBM(n, seed, fig)` generates a sample path of standard Brownian motion. For example, running the following code produces three simulated sample paths of the standard Brownian motion shown in Fig. 9.20.

```
>>> import pandas as pd
>>> import matplotlib.pyplot as plt
>>> from PythonTsa.SimulSBM import simulSBM
>>> x=simulSBM(seed=1357, fig=False)
```

```
>>> y=simulSBM(seed=357, fig=False)
>>> z=simulSBM(seed=3571, fig=False)
# if seed is fixed, the same SBM is reproduced.
# if fig = True, a SBM plot is automatically generated.
>>> xyz=pd.DataFrame({'SBM1': x, 'SBM2': y, 'SBM3': z})
>>> xyz.plot(style=['-', '--', ':'])
>>> plt.xlabel('Time $t$')
>>> plt.ylabel('Standard Brownian Motion')
>>> plt.show()
```

Application of Brownian motions in general stochastic analysis can be found in, for example, Applebaum (2009), Øksendal (2003), Karatzas and Shreve (1998), and so forth. For application in finance, see Franke et al. (2019), Hilpisch (2015), Klebaner (2005), and Shreve (2004), among others. Besides, these references also help understand the substance of the next section.

## 9.3 Stationarity, Nonstationarity, and Unit Root Tests

As we know, most time series in the real world are nonstationary, and there are different types of nonstationarity. In this section, we first introduce two kinds of nonstationarity and then discuss unit root as well as stationarity test problems.

### *9.3.1 Trend Stationarity and Difference Stationarity*

There exist two kinds of nonstationary time series, which are very important for characterizing economic and financial variables. We first give definitions for them below.

**Definition 9.8** (1) If a time series $Y_t$ is of the form $Y_t = T_t + X_t$ where $T_t$ is its deterministic trend and $X_t$ is stationary, then $Y_t$ is said to be trend-stationary. (2) If a time series has a stochastic trend (component) and its differenced series is stationary, then it is said to be difference-stationary.

Note that both trend-stationary and difference-stationary series are all nonstationary. They are often encountered in the fields of economics and finance and so on. Now we use an example to better understand Definition 9.8.

*Example 9.5 (A Typical Trend-Stationary Model and Difference-Stationary Model)* (1) Consider the time series $X_t = \alpha + \beta t + \varepsilon_t$ where $\beta \neq 0$ and $\varepsilon_t \sim \text{WN}(0, \sigma^2)$ is a white noise. Clearly the deterministic-trend-removed series is stationary and so it is trend-stationary. If we were to difference $X_t$ and obtain the differenced series $\nabla X_t = \beta + \varepsilon_t - \varepsilon_{t-1}$, then although the differenced series is stationary, it is not desirable since it is noninvertible. (2) Have a look at the series $Y_t = \beta + Y_{t-1} + \varepsilon_t$. It is a random walk plus constant $\beta$ and known as a *random walk with drift* $\beta$. Its

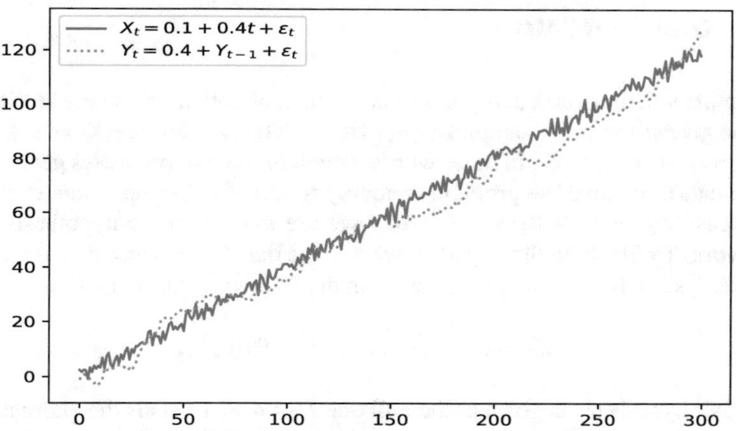


**Fig. 9.21** Time series plots of the series $X_t = 0.1 + 0.4t + \varepsilon_t$ and random walk with drift 0.4

pyTSA_Stationarity

differenced series $\nabla Y_t = \beta + \varepsilon_t$ is obviously stationary, and thus it is difference-stationary. In addition, it is easy to obtain another expression of the random walk with drift: $Y_t = \beta t + Y_0 + \sum_{i=1}^{t} \varepsilon_i$. That is, the random walk with drift seems to have a deterministic trend. However, the deterministic-trend-removed series $Z_t = Y_0 + \sum_{i=1}^{t} \varepsilon_i$ is still nonstationary. In summary, the properties of the trend-stationary model are significantly different from the ones of the difference-stationary model, and hence treatment methods for the trend-stationary model are also very distinct from the ones for the difference-stationary model.

The time series plots of both $X_t = 0.1 + 0.4t + \varepsilon_t$ with $\varepsilon_t \sim \text{WN}(0, 2^2)$ and the random walk with drift 0.4 are shown in Fig. 9.21. We observe that the two time series plots appear similar to each other and the random walk with drift 0.4 seems also to have a linear trend with slope 0.4. This gives rise to a problem: which of both models should be chosen to fit to this sample if we are given the sample that is used to draw the time plots in Fig. 9.21 and do not know which of both time series process generate the sample?

```
>>> import numpy as np
>>> import pandas as pd
>>> import matplotlib.pyplot as plt
>>> from PythonTsa.RandomWalk import RandomWalk_with_drift
>>> t=pd.Series(range(300),dtype='float64')
>>> Tt=0.1+0.4*t
>>> np.random.seed(13711)
>>> wn=np.random.normal(loc=0, scale=2, size=300)
>>> X=Tt+wn
>>> Y=RandomWalk_with_drift(drift=0.4, nsample=300, burnin=10)
>>> XY=pd.DataFrame({'$X_t=0.1+0.4t+\epsilon_t$':X,
                     '$Y_t=0.4 +Y_{t-1}+ \epsilon_t$':Y})
>>> XY.plot(style=['-', ':']); plt.show()
```

### 9.3.2 Unit Root Tests

In Example 9.5, the problem is put forward: which of both models may be elected to fit to the given time series sample$\{X_{1:T}\}$? Here the two models are $X_t = \alpha + \beta t + \varepsilon_t$ and $X_t = \beta + X_{t-1} + \varepsilon_t$. In other words, which of the two processes generates the sample data? To solve the problem is simply to test if the sample data come from the process $X_t = \beta + X_{t-1} + \varepsilon_t$. Now we are to construct a hypothesis testing framework. To illustrate the key idea, we assume that $X_0 = 0$ and $\beta = 0$ are given. Consider testing for the AR coefficient $\alpha$ in the following AR(1) model

$$X_t = \alpha X_{t-1} + \varepsilon_t, \ \varepsilon_t \sim \text{iidN}(0, 1). \tag{9.1}$$

And a pair hypothesis is given as the null one $H_0 : \alpha = 1$ versus the alternative one $H_1 : \alpha < 1$. Under the null hypothesis, the model in Eq. (9.1) is a random walk. If using the sample data $\{X_{1:T}\}$ lets the test pass, then it could be accepted that the sample $\{X_{1:T}\}$ follows a random walk. Such a test is called random walk or *unit root tests* since the AR polynomial is $\varphi(z) = 1 - z$ and $z = 1$ is its root, and the null hypothesis is simply that there is a unit root in the series $X_t$. It is not hard to arrive at the ordinary least squares estimate (OLSE) $\hat{\alpha}$ of $\alpha$ in Eq. (9.1)

$$\hat{\alpha} = \frac{\sum_{t=1}^{T} X_t X_{t-1}}{\sum_{t=1}^{T} X_{t-1}^2}.$$

Under the null hypothesis $H_0 : \alpha = 1$, we also have

$$T(\hat{\alpha} - 1) = \frac{\frac{1}{T}\sum_{t=1}^{T} X_{t-1}\varepsilon_t}{\frac{1}{T^2}\sum_{t=1}^{T} X_{t-1}^2}. \tag{9.2}$$

To understand the asymptotic properties of the test statistic in Eq. (9.2), it is necessary to first study the asymptotic properties of both the numerator and denominator. For the numerator in Eq. (9.2), we have

$$X_t^2 = X_{t-1}^2 + 2X_{t-1}\varepsilon_t + \varepsilon_t^2, \text{ namely, } X_{t-1}\varepsilon_t = \frac{1}{2}(X_t^2 - X_{t-1}^2 - \varepsilon_t^2).$$

Therefore,

$$\frac{1}{T}\sum_{t=1}^{T} X_{t-1}\varepsilon_t = \frac{1}{2}\left(\frac{X_T^2}{T} - \frac{1}{T}\sum_{t=1}^{T}\varepsilon_t^2\right).$$

By the strong law of large numbers, the second term

$$\frac{1}{T}\sum_{t=1}^{T}\varepsilon_t^2 \to \mathsf{E}(\varepsilon_t^2) = \mathrm{Var}(\varepsilon_t) = 1 \; a.s..$$

On the other hand, $X_t = \sum_{i=1}^{t}\varepsilon_i \sim \mathrm{N}(0, t)$ and $B(1) \sim \mathrm{N}(0, 1)$. Thus,

$$\frac{X_T}{\sqrt{T}} \sim \mathrm{N}(0, 1) \text{ and } \frac{X_T^2}{T} \xrightarrow{d} B^2(1).$$

Now using Itô's rule, we obtain

$$\frac{1}{T}\sum_{t=1}^{T}X_{t-1}\varepsilon_t \xrightarrow{d} \frac{1}{2}[B^2(1) - 1] = \int_0^1 B(t)dB(t). \tag{9.3}$$

As for the denominator in Eq. (9.2), notice $X_t = \sum_{i=1}^{t}\varepsilon_i$, and according to Theorem 9.1, we have

$$\frac{X_t}{\sqrt{T}} = \frac{1}{\sqrt{T}}\sum_{i=1}^{t}\varepsilon_i \xrightarrow{d} B(t)$$

and

$$\frac{1}{T^2}\sum_{t=1}^{T}X_{t-1}^2 = \sum_{t=1}^{T}\left(\frac{X_{t-1}}{\sqrt{T}}\right)^2\frac{1}{T} \xrightarrow{d} \int_0^1 B^2(t)dt. \tag{9.4}$$

In summary, substituting (9.3) and (9.4) into (9.2), we finally arrive at

$$T(\hat{\alpha} - 1) \xrightarrow{d} \frac{\int_0^1 B(t)dB(t)}{\int_0^1 B^2(t)dt} \quad \text{as } T \to \infty.$$

If the test statistic is $\tau = (\hat{\alpha} - 1)/\mathrm{se}(\hat{\alpha})$, then it can also be proved that

$$\tau \xrightarrow{d} \frac{\int_0^1 B(t)dB(t)}{\sqrt{\int_0^1 B^2(t)dt}} \quad \text{as } T \to \infty.$$

These limiting random variables are usually called the unit root test statistics or Dickey-Fuller statistics. As early as 1979, Dickey and Fuller (1979) made a research into them. The corresponding limiting distributions do not have closed forms. Critical values of these test statistics and corresponding numerical percentiles have been tabulated via simulation. Additionally, these results have been generalized to such as the augmented Dickey-Fuller (ADF) unit root test. Suppose that the time series $X_t$ follows the AR($p$) model $\varphi(B)X_t = \varepsilon_t$ where

$$\varphi(z) = 1 - \varphi_1 z - \cdots - \varphi_p z^p.$$

If the AR polynomial has a single unit root, then it can be written as

$$\varphi(z) = \phi(z)(1 - z) \text{ where } \phi(z) = 1 - \phi_1 z - \cdots - \phi_{p-1} z^{p-1}.$$

And the AR($p$) model becomes $\phi(B)\nabla X_t = \varepsilon_t$, namely,

$$X_t = X_{t-1} + \phi_1 \nabla X_{t-1} + \cdots + \phi_{p-1} \nabla X_{t-p+1} + \varepsilon_t.$$

Therefore, testing for a unit root in $\varphi(z)$ is equivalent to testing $\rho = 0$ in the following model

$$\nabla X_t = \rho X_{t-1} + \phi_1 \nabla X_{t-1} + \cdots + \phi_{p-1} \nabla X_{t-p+1} + \varepsilon_t. \tag{9.5}$$

Under the null hypothesis $H_0 : \rho = 0$, it can be proved that the test statistic $\tau = \hat{\rho}/\text{se}(\hat{\rho})$ where $\hat{\rho}$ is the OLSE of $\rho$ has the same limiting distribution as in the random walk case. There is no constant term (viz., intercept) in the model (9.5). However, some time series need AR models with the constant term or a deterministic trend in order to fit to them. Besides the model (9.5), there are still two models that are often employed in the unit root test:

$$\nabla X_t = c + \rho X_{t-1} + \phi_1 \nabla X_{t-1} + \cdots + \phi_{p-1} \nabla X_{t-p+1} + \varepsilon_t$$

and

$$\nabla X_t = c + bt + \rho X_{t-1} + \phi_1 \nabla X_{t-1} + \cdots + \phi_{p-1} \nabla X_{t-p+1} + \varepsilon_t.$$

Phillips and Perron (1988) take account of both the autocorrelation and conditional heteroscedasticity in the innovation $\varepsilon_t$ and modify the test statistics to perform unit root tests. Their approach is now known as Phillips-Perron (PP) unit root test. PP and ADF tests are asymptotically equivalent. For more details on these tests, see Tsay (2014), Hamilton (1994), Phillips and Perron (1988), and Dickey and Fuller (1979), among others.

*Example 9.6 (Unit Root Processes)* If an AR model has a unit root, then the time series represented by the AR model is called a *unit root process*. The random walk with drift $X_t = \beta + X_{t-1} + \varepsilon_t$ is a well-known unit root process. Note that there are numerous unit root processes. The following AR(5) model is another example

$$X_t = 0.2X_{t-1} + 0.6X_{t-2} + 0.2X_{t-5} + \varepsilon_t.$$

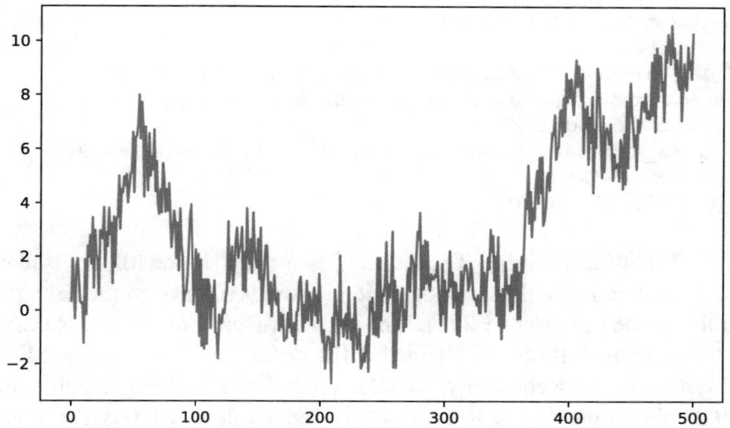


**Fig. 9.22** Time series plot of the AR(5) model with a unit root in Example 9.6

pyTSA_UnitRoot

Its AR polynomial is $\varphi(z) = 1 - 0.2z - 0.6z^2 - 0.2z^5$, and plainly $z = 1$ is its unit root. Its other roots are all greater than 1 in modulus, seeing the solution in the following Python code. This time series is thus a unit root process. Figure 9.22 shows its time series plot. Evidently there exists a stochastic trend in it and it is also a difference-stationary series. Now let

$$\varphi(z) = 1-0.2z-0.6z^2-0.2z^5 = (1-\phi_1 z-\phi_2 z^2-\phi_3 z^3-\phi_4 z^4)(1-z) = \phi(z)(1-z).$$

Then it is easy to obtain $\phi(z) = 1+0.8z+0.2z^2+0.2z^3+0.2z^4$, which has no unit roots. In general, for an AR($p$) model with a single unit root $\varphi(B)X_t = \varepsilon$, if $\varphi(z)$ is written as

$$\varphi(z) = 1 - \varphi_1 z - \cdots - \varphi_p z^p = (1 - \phi_1 z - \cdots - \phi_{p-1} z^{p-1})(1 - z) = \phi(z)(1 - z),$$

it is not hard to verify that for $1 \le k \le p - 1$,

$$\phi_k = \sum_{i=1}^{k} \varphi_i - 1 = - \sum_{i=k+1}^{p} \varphi_i. \tag{9.6}$$

The proof is left as an exercise for the reader.

```
>>> import numpy as np
>>> import pandas as pd
>>> import matplotlib.pyplot as plt
>>> from statsmodels.tsa.arima_process
        import arma_generate_sample
>>> p=[-0.2, 0.0, 0.0, -0.6, -0.2, 1]
```

```
>>> roots=np.roots(p)
>>> abs(roots)
array([1.35580506, 1.35580506, 1.64925479, 1.64925479, 1.    ])
>>> ar=np.array([1, -0.2, -0.6, 0.0, 0.0, -0.2])
>>> np.random.seed(12347)
>>> x= arma_generate_sample(ar=ar, ma=[1], nsample=500)
>>> x=pd.Series(x)
>>> x.plot(); plt.show()
```

*Example 9.7 (Unit Root Tests)* The dataset "us-q-rgdp" in the folder Ptsadata is the series of US quarterly real gross domestic product (GDP) from the first quarter of 1947 to the second quarter of 2021. The data are from the Federal Reserve Bank at St Louis and in billions of chained 2012 dollars and seasonally adjusted. In order to reduce heteroscedasticity, we take a logarithm on the data before analysis. Figure 9.23 shows the time series plot of the logged data. We observe that there is a deterministic almost-linear upward trend in the series. Hence we let the parameter `regression='ct'` in the ADF test function `adfuller()`, which means that there is a deterministic trend (viz., a constant and linear time trend) $c + bt$ in the regression model (for more details on `adfuller()`, see Sect. 9.3.3). The $p$-value for the ADF test is 0.9649, and so we have strong evidence to support the null hypothesis: there is a unit root in the series. The deterministic trend component can be extracted from the series through linear regression. The regression results exhibit in the following table "`OLS Regression Results`", and the fitted deterministic trend is also shown in Fig. 9.23. Explicitly, the residual series is just the remainder of removing the fitted trend part from the data. From its time plot shown in Fig. 9.24, we see that there is no deterministic trend in the residual series anymore. Note that there is still a unit root in the residual series since the $p$-value for the ADF test is 0.5214. In other words, removing the deterministic trend does not remove the unit root. Moreover, if we directly difference the logged data, it turns out that the differenced series should be stationary, seeing the time plot and ACF plot of the differenced series shown in Figs. 9.25 and 9.26 and noting the $p$-value for the ADF test is about 0.0000. We conclude that the time series of US quarterly real GDP is a difference-stationary series with a deterministic trend. The following Python code is for analysis of this example.

```
>>> import numpy as np
>>> import pandas as pd
>>> import statsmodels.api as sm
>>> from PythonTsa.plot_acf_pacf import acf_pacf_fig
>>> import matplotlib.pyplot as plt
>>> from statsmodels.tsa.stattools import adfuller, kpss
>>> from PythonTsa.datadir import getdtapath
>>> dtapath=getdtapath()
>>> usrgdp=pd.read_csv(dtapath + 'us-q-rgdp.csv', header=0)
>>> usrgdp.index=usrgdp['DATE']
>>> usrgdp=usrgdp.GDPC1
>>> lusrgdp=np.log(usrgdp)
```

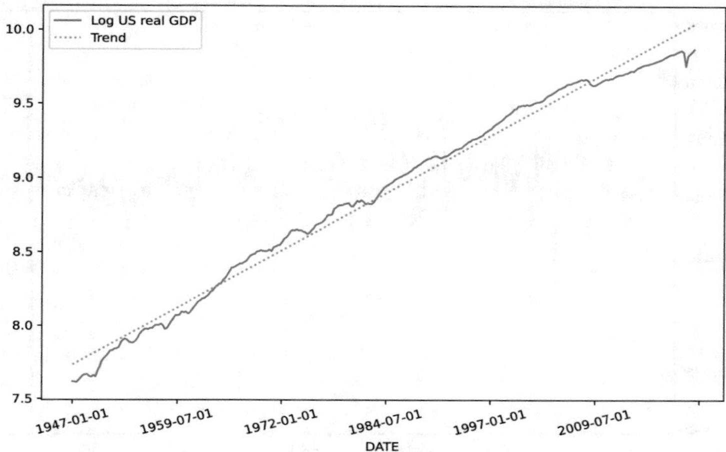


**Fig. 9.23** Time series plot of the US quarterly GDP in logarithm in Example 9.7

pyTSA_UnitRootTest

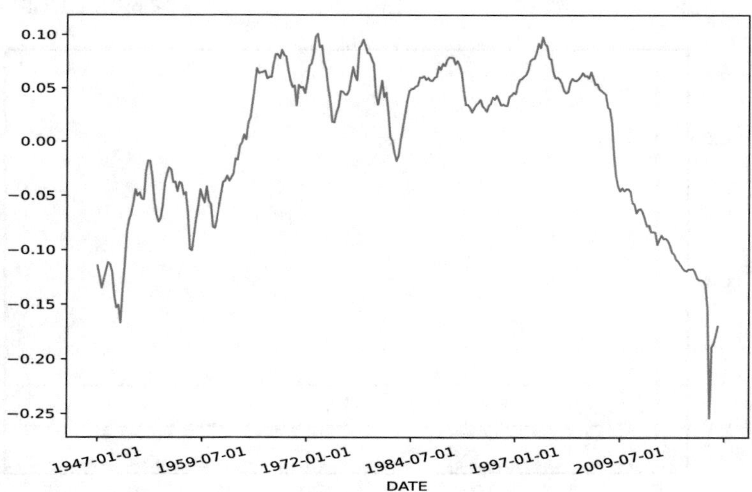


**Fig. 9.24** Time series plot of the deterministic-trend-removed series in Example 9.7

pyTSA_UnitRootTest

```
>>> lusrgdp.name='lrgdp'
>>> adfuller(lusrgdp, regression='ct')
(-0.8094, 0.9649, 0, 297,
{'1%': -3.9896, '5%': -3.4254, '10%': -3.1358}, -1719.2924)
# This result is edited.
>>> kpss(lusrgdp, regression='ct', nlags='auto')
InterpolationWarning: The test statistic is outside of the range
 of p-values available in the look-up table. The actual p-value
```

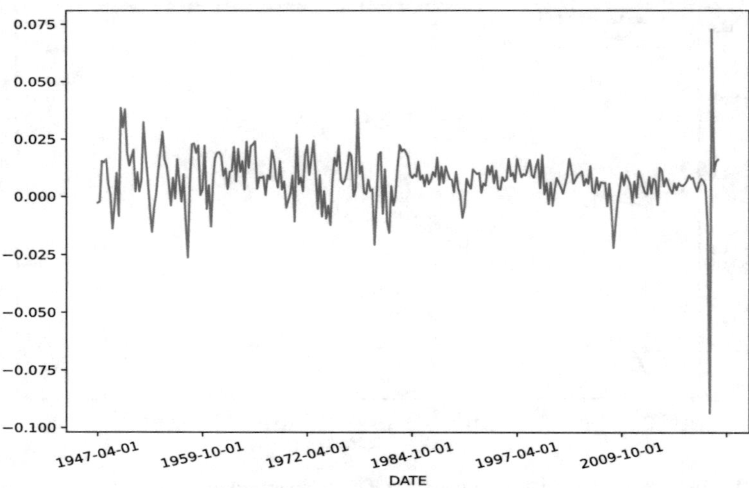


**Fig. 9.25** Time plot of the first differences of the logged series in Example 9.7

pyTSA_UnitRootTest

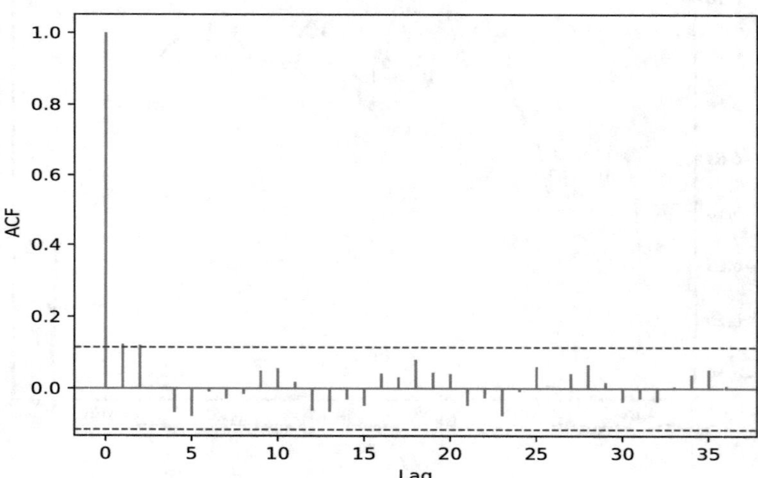


**Fig. 9.26** ACF plot of the first differences of the logged series in Example 9.7

pyTSA_UnitRootTest

```
 is smaller than the p-value returned.
(0.5538,0.01,10,{'10%':0.119,'5%':0.146,'2.5%':0.176,'1%':0.216})
# This result is edited.
>>> t=pd.Series(range(len(lusrgdp)),dtype='float64')
>>> t.index=lusrgdp.index
>>> t.name='trend'
>>> ct=sm.add_constant(t, prepend=False)
```

```
>>> modfit=sm.OLS(lusrgdp, ct).fit()
>>> print(modfit.summary())
                            OLS Regression Results
==============================================================================
Dep. Variable:                  lrgdp   R-squared:                       0.988
Model:                            OLS   Adj. R-squared:                  0.988
Method:                 Least Squares   F-statistic:                 2.467e+04
Date:                Thu, 14 Oct 2021   Prob (F-statistic):          4.07e-287
Time:                        08:17:08   Log-Likelihood:                 356.15
No. Observations:                 298   AIC:                            -708.3
Df Residuals:                     296   BIC:                            -700.9
Df Model:                           1
Covariance Type:            nonrobust
==============================================================================
                 coef    std err          t      P>|t|      [0.025      0.975]
------------------------------------------------------------------------------
trend          0.0078   4.95e-05    157.070      0.000       0.008       0.008
const          7.7328      0.008    910.597      0.000       7.716       7.749
==============================================================================
Omnibus:                       27.097   Durbin-Watson:                   0.025
Prob(Omnibus):                  0.000   Jarque-Bera (JB):               29.127
Skew:                          -0.725   Prob(JB):                     4.73e-07
Kurtosis:                       2.508   Cond. No.                         342.
==============================================================================
Notes:
[1] Standard Errors assume that the covariance matrix
    of the errors is correctly specified.
>>> myfitted=modfit.fittedvalues
>>> dnf=pd.DataFrame({'Log US real GDP':lusrgdp,'Trend':myfitted})
>>> dnf.plot(style=['-', ':'])
>>> plt.xticks(rotation=15)
>>> plt.show()
>>> myresid=modfit.resid
>>> myresid.plot()
>>> plt.xticks(rotation=15)
>>> plt.show()
>>> adfuller(myresid, regression='c')
(-1.5244, 0.5214, 2, 295,
{'1%': -3.4527, '5%': -2.8714, '10%': -2.5720}, -1716.7014)
# This result is edited.
>>> kpss(myresid, regression='c', nlags='auto')
(0.5538, 0.0296, 10,
{'10%': 0.347, '5%': 0.463, '2.5%': 0.574, '1%': 0.739})
# This result is edited.
>>> dlusrgdp=lusrgdp.diff().dropna()
>>> dlusrgdp.plot()
>>> plt.xticks(rotation=15)
>>> plt.show()
>>> acf_pacf_fig(dlusrgdp, both=False, lag=36)
>>> plt.show()
>>> adfuller(dlusrgdp, regression='c')
(-10.1855, 6.5257e-18, 1, 295,
{'1%': -3.4527, '5%': -2.8714, '10%': -2.5720}, -1711.3256)
# This result is edited.
```

```
>>> kpss(dlusrgdp, regression='c', nlags='auto')
(0.4930, 0.0432, 4,
{'10%': 0.347, '5%': 0.463, '2.5%': 0.574, '1%': 0.739})
# This result is edited.
```

### *9.3.3 Stationarity Tests*

Checking if a time series is stationary or has not been a tough and important problem in analysis of time series. Kwiatkowski et al. (KPSS, 1992) propose an approach to directly testing the null hypothesis of stationarity against the alternative of a unit root. Let $\{X_{1:T}\}$ be the observed series for which stationarity test is wanted. Assume that the time series can be expressed by

$$X_t = \beta' D_t + \sum_{i=1}^{t} \varepsilon_i + \eta_t \tag{9.7}$$

$$R_t = \sum_{i=1}^{t} \varepsilon_i = R_{t-1} + \varepsilon_t, \ \varepsilon_t \sim \text{iid}(0, \sigma_\varepsilon^2) \tag{9.8}$$

where $\beta' = (c, 0)$ or $(c, b)$, $D_t' = (1, t)$, that is, $\beta' D_t = c$ or $c + bt$ $(b \neq 0)$ is a constant or deterministic trend, and $\eta_t$ is stationary and may be conditionally heteroscedastic. Evidently, if the variance $\sigma_\varepsilon^2 = 0$, then the random walk $R_t = R_{t-1} + \varepsilon_t$ is actually zero, and so the time series $X_t$ in Eq. (9.7) is stationary (when $\beta' D_t = c$) or trend-stationary (when $\beta' D_t = c + bt$). The null hypothesis that $X_t$ is stationary or trend-stationary is formulated as $H_0 : \sigma_\varepsilon^2 = 0$. Under the null hypothesis $H_0 : \sigma_\varepsilon^2 = 0$, the KPSS stationarity test statistic is

$$KPSS_T = \frac{\sum_{t=1}^{T} S_t^2}{T^2 \hat{\sigma}_\eta^2}$$

where $S_t = \sum_{i=1}^{t} \hat{\eta}_i$, $\hat{\eta}_i$ is the residual series of the regression model in Eq. (9.7) and $\hat{\sigma}_\eta^2$ is a consistent estimate of the long-run variance of $\eta_t$. Using the Bartlett window $w(s, l) = 1 - s/(1 + l)$ as an optimal weighting function to estimate the long-run variance, it can be obtained that

$$\hat{\sigma}_\eta^2 = T^{-1} \sum_{t=1}^{T} \hat{\eta}_t^2 + 2T^{-1} \sum_{s=1}^{l} \left[ (1 - \frac{s}{1+l}) \sum_{t=1+s}^{T} \hat{\eta}_t \hat{\eta}_{t-s} \right].$$

Moreover, it can be proved that if $\beta' D_t = c$, then

$$KPSS_T \xrightarrow{d} \int_0^1 V^2(t)dt$$

where $V(t) = B(t) - tB(1)$ is known as the *first-level Brownian bridge*, and if $\beta' D_t = c + bt$, then

$$KPSS_T \xrightarrow{d} \int_0^1 V_2^2(t)dt$$

where

$$V_2(t) = B(t) + t(2-3t)B(1) + 6t(t-1)\int_0^1 B(s)ds$$

is called the *second-level Brownian bridge*. Critical values for these test statistics and numerical percentiles are derived through simulation. See Kwiatkowski et al. (1992) and Hobijn et al. (2004) for more details.

Now we turn to the problem of how to test stationarity of a time series with Python. First of all, there is a KPSS stationarity test function `kpss()` in the Python package `statsmodels`. Roughly speaking, if the parameter `regression` in `kpss()` is equal to `'c'` and the $p$-value for the KPSS test is greater than 0.05, then the series should be (level) stationary, and if the parameter `regression` is equal to `'ct'` and the $p$-value for the KPSS test is greater than 0.05, then the series should be trend-stationary. Otherwise, if the null hypothesis is rejected, that is, the $p$-value for the KPSS test is much smaller than 0.05, then the series should be nonstationary.

The ADF and PP tests are used to determine the existence of a unit root in the time series and thus help understand if the series is stationary or not. If the null hypothesis fails to be rejected, these tests provide evidence that the series should have a unit root or be nonstationary. In the package `statsmodels`, the function `adfuller()` is used to make an ADF unit root test. If the parameter `regression` in `adfuller()` is equal to `'nc'` or `'c'` and the $p$-value for the ADF test is much smaller than 0.05, then the series should be stationary, and if the parameter `regression` is equal to `'ct'` or `'ctt'` and the $p$-value for the ADF test is much smaller than 0.05, then the series should be trend-stationary. Besides, the PP unit root test can be found in the Python package `arch`.

*Example 9.8 (Example 9.7 Continued)* From the Python code in Example 9.7, we see that the $p$-value of the KPSS test for the logged series of US quarterly real GDP is smaller than 0.01 and hence the logged series should be nonstationary. Furthermore, the conclusion is the same as one by the ADF test. For the deterministic-trend-removed or residual series, both the KPSS test (the $p$-value is 0.0296 at this point) and ADF test (the $p$-value is 0.5214) obtain the identical result that the deterministic-trend-removed series still is nonstationary or has a unit root. Unfortunately, the KPSS test for stationarity of the differenced series fails since the $p$-value is 0.0432 at this point. Of course, it is well known that all statistical tests occasionally make mistakes.

In the previous chapters, there have been many examples of using the KPSS stationarity test. Here we do not want to take more examples.

## 9.4 Cointegrations and Granger's Representation Theorem

In an economic system, time series-level variables are usually nonstationary. However, some linear combinations of these variables may be stationary. This phenomenon is also found in other fields. The cointegration theory studies the effects of these combinations and the relations among the variables. Note that in this section, for simplicity, we call a stochastic process generated by a VAR model a VAR process regardless of whether it is stationary or not. At this point, we also say that the stochastic process follows the VAR model.

### 9.4.1 Spurious Regressions and $I(d)$ Processes

Regression analysis is widely used in data science and various other disciplines. However, care must be taken when some or all of the variables in the regression are nonstationary since the regressing results may or may not hold. This phenomenon is studied using simulation approach by Granger and Newbold (1974). Now we first give the notion of $I(d)$ processes.

**Definition 9.9** Suppose that $\boldsymbol{\varepsilon}_t$ is a $K$-dimensional white noise $\text{WN}(\mathbf{0}, \boldsymbol{\Sigma})$.

(1) A causal $K$-dimensional time series $\mathbf{X}_t = \boldsymbol{\Psi}(B)\boldsymbol{\varepsilon}_t = \sum_{h=0}^{\infty} \boldsymbol{\Psi}_h \boldsymbol{\varepsilon}_{t-h}$ is said to be integrated of order 0 (denoted as $I(0)$) or an $I(0)$ process if $\boldsymbol{\Psi}(1) = \sum_{h=0}^{\infty} \boldsymbol{\Psi}_h \neq \mathbf{0}$ where $\boldsymbol{\Psi}(B) = \sum_{h=0}^{\infty} \boldsymbol{\Psi}_h B^h$ and $B$ is the backshift operator.
(2) A $K$-dimensional time series $\mathbf{X}_t$ is called integrated of order $d$ (written as $I(d)$) if $(1-B)^d\mathbf{X}_t$ is $I(0)$ but $(1-B)^{d-1}\mathbf{X}_t$ is nonstationary.

Note the following remarks on Definition 9.9:

- Clearly, an $I(0)$ process is stationary. However, a stationary process is not necessarily $I(0)$. See Example 9.9 below.
- The $K$-dimensional white noise $\boldsymbol{\varepsilon}_t$ is $I(0)$.
- To restrict ourselves to a real integrated process, the condition that $\sum_{h=0}^{\infty} \boldsymbol{\Psi}_h \neq 0$ is imposed on the $I(0)$ process. For details, see Example 9.9 again.
- If $\mathbf{X}_t$ is $I(0)$, then $\boldsymbol{v} + \mathbf{X}_t$ is also considered $I(0)$ where $\boldsymbol{v}$ is a constant vector.
- That a univariate process is $I(d)$ implies that it has $d$ unit roots.

*Example 9.9 (Stationary Processes May or May Not Be $I(0)$)* Suppose that $X_t$ is a causal AR(1) process, that is, $X_t = \varphi X_{t-1} + \varepsilon_t$ with $|\varphi| < 1$ and $\varepsilon_t \sim \text{WN}(0, 1)$. Then $X_t$ has the MA($\infty$) representation $X_t = \psi(B)\varepsilon_t = \sum_{h=0}^{\infty} \psi_h \varepsilon_{t-h} = \sum_{h=0}^{\infty} \varphi^h \varepsilon_{t-h}$ and $\sum_{h=0}^{\infty} \varphi^h = 1/(1-\varphi) \neq 0$. According to Definition 9.9, $X_t$ is $I(0)$.

Let $Y_t = \nabla X_t = X_t - X_{t-1}$. Then $Y_t$ is stationary but not $I(0)$. In fact,

$$Y_t = \sum_{h=0}^{\infty} \varphi^h \varepsilon_{t-h} - \sum_{h=0}^{\infty} \varphi^h \varepsilon_{t-1-h} = \varepsilon_t + \sum_{h=1}^{\infty} (\varphi^h - \varphi^{h-1}) \varepsilon_{t-h} = \sum_{h=0}^{\infty} \hat{\varphi}_h \varepsilon_{t-h}$$

and

$$\sum_{h=0}^{\infty} \hat{\varphi}_h = 1 + \sum_{h=1}^{\infty} (\varphi^h - \varphi^{h-1}) = 0.$$

Actually, now that $X_t$ is stationary, it should not be differenced. In other words, an $I(d)$ process with $d \geq 1$ must be nonstationary.

Let $Z_t = \sum_{h=1}^{t} a_h$ where $a_h$ is any $I(0)$ process, for example, the white noise $\varepsilon_h$. That is, $Z_t = Z_{t-1} + a_t$ and is nonstationary. At the same time, the differences of $Z_t$ are $\nabla Z_t = a_t$. Thus, $Z_t$ is a real integrated process, namely, $I(1)$ process.

*Example 9.10 (Spurious Regression)* Assume that $X_t$ and $Y_t$ are two independent random walks as follows:

$$X_t = X_{t-1} + \varepsilon_t, \varepsilon_t \sim \text{iidN}(0, 1) \text{ and } Y_t = Y_{t-1} + \eta_t, \eta_t \sim \text{iidN}(0, 1).$$

Both $X_t$ and $Y_t$ are naturally $I(1)$ but noncointegrated (see Definition 9.10). We now choose a sample of size 300 drawn from $X_t$ and another sampled from $Y_t$ and show them in Fig. 9.27. They looked positively correlated, and then we regress $Y_t$ on $X_t$. The regressing results are in the first table "`OLS Regression Results`" of the following Python code. In the table we see that the estimated slope coefficient is 0.6639 with a large t-statistic of 17.342 (viz., significant), and the regression $R^2$ is moderate at 0.502. However, $X_t$ and $Y_t$ are independent; hence the regression $\hat{Y}_t = 2.3623 + 0.6639\hat{X}_t$ is spurious. Moreover, the time and ACF plots of the regression residuals shown in Figs. 9.28 and 9.29 clearly illustrate that the residual series is nonstationary. Using the ADF test, we obtain the same result since the $p$-value for the ADF test is 0.1201. These facts also explain that the regression equation is not acceptable. If we difference $X_t$ and $Y_t$ and obtain stationary $\nabla X_t$ and $\nabla Y_t$, then regressing $\nabla Y_t$ on $\nabla X_t$ can reveal the correct relationship between the two series. See the second table "`OLS Regression Results`" in the following Python code.

```
>>> import numpy as np
>>> import pandas as pd
>>> import matplotlib.pyplot as plt
```

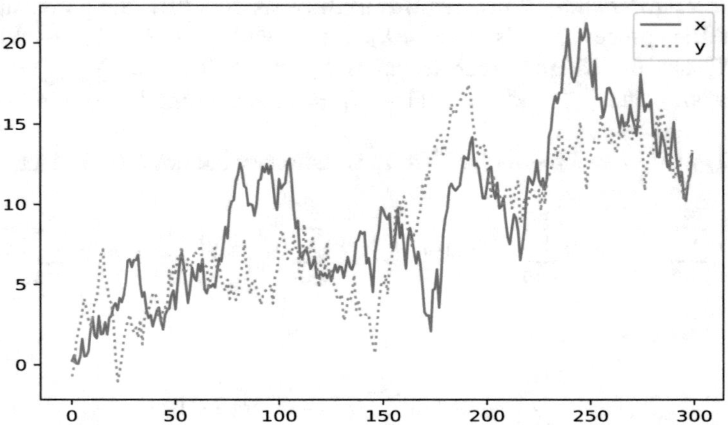


**Fig. 9.27** Simulated sample paths of two independent $I(1)$ processes

pyTSA_SpuriousRegression

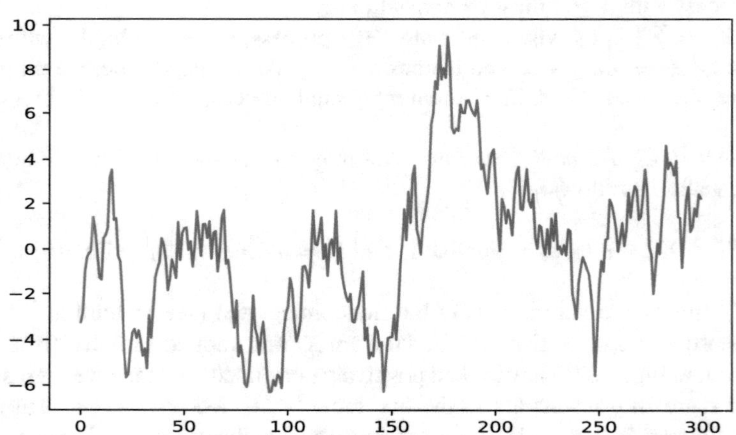

**Fig. 9.28** Time plot of the regression residuals for the simulated sample in Example 9.10

pyTSA_SpuriousRegression

```
>>> from statsmodels.tsa.stattools import adfuller, kpss
>>> from statsmodels.tsa.arima_process import arma_generate_sample
>>> import statsmodels.api as sm
>>> import statsmodels.formula.api as smf
>>> from PythonTsa.plot_acf_pacf import acf_pacf_fig
>>> ar=np.array([1, -1.0])
>>> np.random.seed(1373)
>>> x= arma_generate_sample(ar=ar, ma=[1], nsample=300)
>>> y= arma_generate_sample(ar=ar, ma=[1], nsample=300)
>>> x=pd.Series(x); y=pd.Series(y)
>>> xy=pd.DataFrame({'x': x, 'y': y})
```

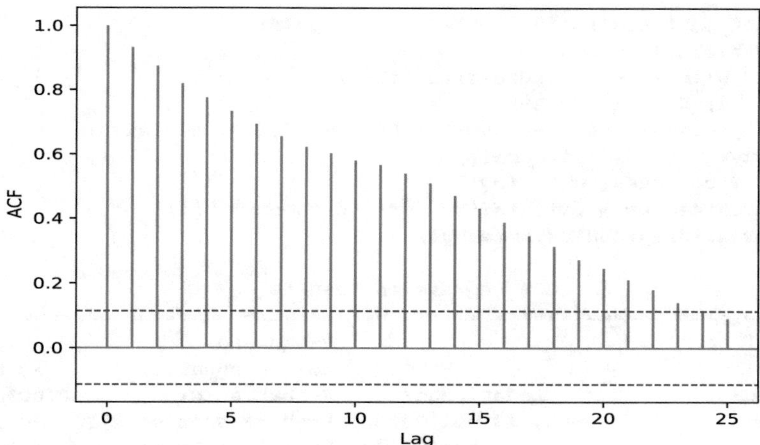


**Fig. 9.29** ACF plot of the regression residuals for the simulated sample in Example 9.10

pyTSA_SpuriousRegression

```
>>> xy.plot(style=['-', ':']); plt.show()
>>> regresults = smf.ols('y ~ x', data=xy).fit()
>>> print(regresults.summary())
                OLS Regression Results
==============================================================
Dep. Variable:                  y   R-squared:           0.502
Model:                        OLS   Adj. R-squared:      0.501
Method:             Least Squares   F-statistic:         300.7
Date:            Sun, 19 Jul 2020   Prob (F-statistic): 4.60e-47
Time:                    09:39:37   Log-Likelihood:    -772.86
No. Observations:             300   AIC:                 1550.
Df Residuals:                 298   BIC:                 1557.
Df Model:                       1
Covariance Type:        nonrobust
==============================================================
              coef  std err       t    P>|t|   [0.025   0.975]
--------------------------------------------------------------
Intercept   2.3623    0.400   5.900    0.000    1.574    3.150
x           0.6639    0.038  17.342    0.000    0.589    0.739
==============================================================
Omnibus:                4.360   Durbin-Watson:           0.131
Prob(Omnibus):          0.113   Jarque-Bera (JB):        4.230
Skew:                   0.290   Prob(JB):                0.121
Kurtosis:               3.035   Cond. No.                 22.9
==============================================================
Warnings:
[1] Standard Errors assume that the covariance matrix of
    the errors is correctly specified.
>>> resid=regresults.resid
>>> resid.plot(); plt.show()
```

```
>>> acf_pacf_fig(resid, both=False, lag=25)
>>> plt.show()
>>> adfuller(resid, regression='ctt')
(-3.4724, 0.1201, 0, 299,
{'1%': -4.4103,'5%': -3.8523, '10%': -3.5656}, 878.2873)
>>> dxy=xy.diff(1).dropna()
>>> dxy.columns=['dx', 'dy']
>>> dregresults = smf.ols('dy ~dx', data=dxy).fit()
>>> print(dregresults.summary())

                    OLS Regression Results
==================================================================
Dep. Variable:                    dy   R-squared:            0.000
Model:                           OLS   Adj. R-squared:      -0.003
Method:                Least Squares   F-statistic:       0.009652
Date:               Sun, 19 Jul 2020   Prob (F-statistic):   0.922
Time:                       13:52:03   Log-Likelihood:     -409.56
No. Observations:                299   AIC:                  823.1
Df Residuals:                    297   BIC:                  830.5
Df Model:                          1
Covariance Type:           nonrobust
==================================================================
                coef    std err      t    P>|t|   [0.025    0.975]
------------------------------------------------------------------
Intercept     0.0473      0.055  0.855    0.393   -0.062     0.156
dx           -0.0056      0.057 -0.098    0.922   -0.117     0.106
==================================================================
Omnibus:                   3.083   Durbin-Watson:            2.192
Prob(Omnibus):             0.214   Jarque-Bera (JB):         3.159
Skew:                     -0.237   Prob(JB):                 0.206
Kurtosis:                  2.833   Cond. No.                  1.05
==================================================================
Warnings:
[1] Standard Errors assume that the covariance matrix of
    the errors is correctly specified.
```

As a rule of thumb, Granger and Newbold (1974) suggest that one should be cautious if the $R^2$ is greater than the Durbin-Watson statistic. Phillips (1986) provides a theoretical basis for this finding and points out that regression with $I(1)$ data makes sense only when the data are cointegrated. What is more, we shall give a real case for the spurious regression in Example 9.15, and we shall briefly present the concept and theory of cointegration in the following subsections.

### 9.4.2 *Cointegrations*

An individual economic time series variable can wander extensively. However, economic and market forces tend to keep many of these variables together and form a kind of equilibrium relationship. This leads to the concept of cointegration.

**Definition 9.10** Let $\mathbf{X}_t$ be a $K$-dimensional $I(1)$ process. If there is a vector $\boldsymbol{\beta} \neq \mathbf{0}$ such that $\boldsymbol{\beta}'\mathbf{X}_t$ is stationary by a suitable choice of $\boldsymbol{\beta}'\mathbf{X}_0$, then $\mathbf{X}_t$ is called cointegrated, $\boldsymbol{\beta}$ is the cointegrating vector, and $\boldsymbol{\beta}'\mathbf{X}_t$ is the cointegrating process (or series). The number of linearly independent cointegrating vectors is known as the cointegrating rank, and the space spanned by the cointegrating vectors is the cointegration space.

Remarks on Definition 9.10:

- In general, an $I(d)$ process is referred to as cointegrated if there exists a vector $\boldsymbol{\beta} \neq \mathbf{0}$ such that $\boldsymbol{\beta}'\mathbf{X}_t$ is $I(h)$ with $h < d$. Nevertheless, in real applications, the most common case is $d = 1$ and $h = 0$.
- Since $\mathbf{X}_t$ is $I(1)$ (and so is nonstationary), $\boldsymbol{\beta}'\mathbf{X}_t$ is required to be stationary only.
- If $\mathbf{X}_t$ is $I(1)$ and $\mathbf{A}$ is a matrix of full rank, then $\mathbf{A}\mathbf{X}_t$ is also $I(1)$. Furthermore, if $\boldsymbol{\beta}$ is the cointegrating vector, then $(\mathbf{A}^{-1})'\boldsymbol{\beta}$ is the cointegrating vector of $\mathbf{A}\mathbf{X}_t$.
- Even if one of the components of $\mathbf{X}_t$ is $I(1)$ and the rest of the components are $I(0)$, the multiple process $\mathbf{X}_t$ is still $I(1)$. In this way, stationarity of a single component in $\mathbf{X}_t$ becomes a special case of cointegration.

*Example 9.11 (A Two-Dimensional Cointegrated Process)* Consider the following two-dimensional VAR(1) process

$$\begin{bmatrix} X_{t1} \\ X_{t2} \end{bmatrix} = \begin{bmatrix} 1 & 0 \\ -\varphi & 0 \end{bmatrix} \begin{bmatrix} X_{t-1,1} \\ X_{t-1,2} \end{bmatrix} + \begin{bmatrix} \varepsilon_{t1} \\ \varepsilon_{t2} \end{bmatrix}$$

where $(\varepsilon_{t1}, \varepsilon_{t2})' \sim \text{WN}(\mathbf{0}, \mathbf{I}_2)$ and $\varphi$ is any real number. $\mathbf{X}_t = (X_{t1}, X_{t2})'$ is clearly nonstationary since the component $X_{t1}$ is a random walk $X_{t1} = X_{t-1,1} + \varepsilon_{t1}$. Differencing $\mathbf{X}_t$, we have

$$\nabla\mathbf{X}_t = \begin{bmatrix} \varepsilon_{t1} \\ \varepsilon_{t2} \end{bmatrix} + \begin{bmatrix} 0 & 0 \\ -\varphi & -1 \end{bmatrix} \begin{bmatrix} \varepsilon_{t-1,1} \\ \varepsilon_{t-1,2} \end{bmatrix} \text{ and } \mathbf{I}_2 + \begin{bmatrix} 0 & 0 \\ -\varphi & -1 \end{bmatrix} = \begin{bmatrix} 1 & 0 \\ -\varphi & 0 \end{bmatrix} \neq \mathbf{0}.$$

Therefore, $\nabla\mathbf{X}_t$ is $I(0)$ and $\mathbf{X}_t$ is $I(1)$. Now let $\boldsymbol{\beta} = (\varphi, 1)'$, and then the combination

$$\boldsymbol{\beta}'\mathbf{X}_t = \varphi X_{t1} + X_{t2} = \varphi X_{t1} - \varphi X_{t-1,1} + \varepsilon_{t2} = \varphi\varepsilon_{t1} + \varepsilon_{t2}$$

is explicitly stationary. Thus, the two-dimensional process $\mathbf{X}_t$ is cointegrated with cointegrating vector $\boldsymbol{\beta} = (\varphi, 1)'$.

*Example 9.12 (A Three-Dimensional Cointegrated Process)* Let $\boldsymbol{\varepsilon}_t = (\varepsilon_{t1}, \varepsilon_{t2}, \varepsilon_{t3})' \sim \text{WN}(\mathbf{0}, \mathbf{I}_3)$ and

$$\mathbf{X}_t = \begin{bmatrix} X_{t1} \\ X_{t2} \\ X_{t3} \end{bmatrix} = \begin{bmatrix} \varepsilon_{t2} + \sum_{h=1}^{t} \varepsilon_{h1} \\ \varepsilon_{t3} + \frac{1}{2} \sum_{h=1}^{t} \varepsilon_{h1} \\ \varepsilon_{t3} \end{bmatrix}.$$

First of all, $\mathbf{X}_t$ is nonstationary since the first two components consist of the common term $\sum_{h=1}^{t} \varepsilon_{h1}$ which is nonstationary (see Example 9.9). Second, the difference

$$\nabla X_t = \begin{bmatrix} \varepsilon_{t1} + \varepsilon_{t2} - \varepsilon_{t-1,2} \\ \frac{1}{2}\varepsilon_{t1} + \varepsilon_{t3} - \varepsilon_{t-1,3} \\ \varepsilon_{t3} - \varepsilon_{t-1,3} \end{bmatrix} = \begin{bmatrix} 1 & 1 & 0 \\ \frac{1}{2} & 0 & 1 \\ 0 & 0 & 1 \end{bmatrix} \begin{bmatrix} \varepsilon_{t1} \\ \varepsilon_{t2} \\ \varepsilon_{t3} \end{bmatrix} + \begin{bmatrix} 0 & -1 & 0 \\ 0 & 0 & -1 \\ 0 & 0 & -1 \end{bmatrix} \begin{bmatrix} \varepsilon_{t-1,1} \\ \varepsilon_{t-1,2} \\ \varepsilon_{t-1,3} \end{bmatrix},$$

and

$$\begin{bmatrix} 1 & 1 & 0 \\ \frac{1}{2} & 0 & 1 \\ 0 & 0 & 1 \end{bmatrix} + \begin{bmatrix} 0 & -1 & 0 \\ 0 & 0 & -1 \\ 0 & 0 & -1 \end{bmatrix} = \begin{bmatrix} 1 & 0 & 0 \\ \frac{1}{2} & 0 & 0 \\ 0 & 0 & 0 \end{bmatrix} \neq \mathbf{0}.$$

Hence, $\nabla \mathbf{X}_t$ is $I(0)$ and $\mathbf{X}_t$ is $I(1)$. Furthermore,

$$(1, -2, 0)\mathbf{X}_t = X_{t1} - 2X_{t2} = \varepsilon_{t2} - 2\varepsilon_{t3}$$

is stationary since $(\varepsilon_{t1}, \varepsilon_{t2}, \varepsilon_{t3})' \sim \text{WN}(\mathbf{0}, \mathbf{I}_3)$, and $\boldsymbol{\beta} = (1, -2, 0)'$ is a cointegrating vector for $\mathbf{X}_t$. Since $\mathbf{X}_t$ has no deterministic components and is nonstationary, and the first difference $\nabla \mathbf{X}_t$ is stationary, it must possess a stochastic trend. On the other hand, obviously, it is the common part $\sum_{h=1}^{t} \varepsilon_{h1}$ that makes the process nonstationary. We call the part $\sum_{h=1}^{t} \varepsilon_{h1}$ a *common stochastic trend* of the process $\mathbf{X}_t$.

### *9.4.3 Granger's Representation Theorem*

Now we consider the cointegration problem within the vector autoregressive framework. Recall the $K$-dimensional VAR($p$) model

$$\mathbf{X}_t = \boldsymbol{\Phi}_1 \mathbf{X}_{t-1} + \cdots + \boldsymbol{\Phi}_p \mathbf{X}_{t-p} + \boldsymbol{\varepsilon}_t, \text{ namely, } \boldsymbol{\Phi}(B)\mathbf{X}_t = \boldsymbol{\varepsilon}_t \tag{9.9}$$

where $\boldsymbol{\Phi}(z) = I_K - \boldsymbol{\Phi}_1 z - \cdots - \boldsymbol{\Phi}_p z^p$ is the VAR matrix polynomial. Here the time series $\mathbf{X}_t$ generated by VAR($p$) model (9.9) is simply called a VAR($p$) process, which may be nonstationary. Let

$$\boldsymbol{\Phi} = -\boldsymbol{\Phi}(1) = \boldsymbol{\Phi}_1 + \cdots + \boldsymbol{\Phi}_p - I_K \tag{9.10}$$

and

$$\boldsymbol{\Gamma} = -\frac{d\boldsymbol{\Phi}(z)}{dz}\bigg|_{z=1} + \boldsymbol{\Phi} = 2\boldsymbol{\Phi}_1 + 3\boldsymbol{\Phi}_2 + \cdots + (1+p)\boldsymbol{\Phi}_p - I_K. \tag{9.11}$$

With these notations, we give a well-known condition under which a stationary VAR($p$) process is $I(0)$.

**Theorem 9.2** *If* $\mathbf{X}_t$ *is a stationary VAR(p) process and* $\det[\boldsymbol{\Phi}(z)] \neq 0$ *for all complex numbers* $\{z; |z| < 1\}$, *then* $\mathbf{X}_t$ *can be given an initial distribution such that it becomes* $I(0)$ *if and only if* $\boldsymbol{\Phi}$ *is of full rank. In this case, there is no unit root for* $\mathbf{X}_t$ *and*

$$\mathbf{X}_t = \boldsymbol{\Psi}(B)\boldsymbol{\varepsilon}_t = \sum_{h=0}^{\infty} \boldsymbol{\Psi}_h \boldsymbol{\varepsilon}_{t-h}$$

*where* $\boldsymbol{\Psi}(z) = \sum_{h=0}^{\infty} \boldsymbol{\Psi}_h z^h = [\boldsymbol{\Phi}(z)]^{-1}$ *converges when* $|z| < 1+\delta$ *for some* $\delta > 0$.

Note that from (9.10), it is easy to see that the full rank condition of $\boldsymbol{\Phi}$ simply means that $\det[\boldsymbol{\Phi}(1)] \neq 0$. On the other hand, if $\boldsymbol{\Phi}$ is of reduced rank, then we have a cointegrated system.

What follows, we first give the definition of an orthogonal complement of matrix $\boldsymbol{\alpha}$ and then present Granger's representation theorem.

**Definition 9.11** Let $\boldsymbol{\alpha}$ be a $K \times r$ matrix of rank $r$ $(r < K)$. Then an orthogonal complement, denoted as $\boldsymbol{\alpha}_{\perp}$, of $\boldsymbol{\alpha}$ is a $K \times (K-r)$ matrix of rank $(K-r)$ so that $\boldsymbol{\alpha}'\boldsymbol{\alpha}_{\perp} = \mathbf{0}$. If $\boldsymbol{\alpha}$ is a nonsingular square matrix $(r = K)$, define $\boldsymbol{\alpha}_{\perp} = \mathbf{0}$.

**Theorem 9.3 (Granger's Representation Theorem)** *If* $\mathbf{X}_t$ *is a VAR(p) process and* $\det[\boldsymbol{\Phi}(z)] \neq 0$ *for all complex numbers* $\{z; |z| < 1\}$, *then* $\mathbf{X}_t$ *is* $I(1)$ *if and only if* $\boldsymbol{\Phi} = \boldsymbol{\alpha}\boldsymbol{\beta}'$ *where* $\boldsymbol{\alpha}$ *and* $\boldsymbol{\beta}$ *are* $K \times r$ *matrices which are of full rank* $r$ $(r < K)$ *and* $\boldsymbol{\alpha}_{\perp}'\boldsymbol{\Gamma}\boldsymbol{\beta}_{\perp}$ *is of full rank. In this case,* $\nabla\mathbf{X}_t$ *and* $\boldsymbol{\beta}'\mathbf{X}_t$ *can be given initial distributions so that they become* $I(0)$. *Moreover,* $\mathbf{X}_t$ *can be expressed as*

$$\mathbf{X}_t = \boldsymbol{\beta}_{\perp}(\boldsymbol{\alpha}_{\perp}'\boldsymbol{\Gamma}\boldsymbol{\beta}_{\perp})^{-1}\boldsymbol{\alpha}_{\perp}'\sum_{h=1}^{t}\boldsymbol{\varepsilon}_h + \boldsymbol{\Psi}(B)\boldsymbol{\varepsilon}_t, \tag{9.12}$$

*and satisfies the vector error correction equation (model)*

$$\nabla \mathbf{X}_t = \boldsymbol{\alpha}\boldsymbol{\beta}'\mathbf{X}_{t-1} + \sum_{h=1}^{p-1} \boldsymbol{\Gamma}_h \nabla \mathbf{X}_{t-h} + \boldsymbol{\varepsilon}_t \tag{9.13}$$

*where* $\boldsymbol{\Gamma}_h = -(\boldsymbol{\Phi}_{h+1} + \cdots + \boldsymbol{\Phi}_p), \; h = 1, 2, \cdots, p-1.$

Remarks on Granger's representation theorem:

- It gives the conditions under which an $I(1)$ VAR process is cointegrated. In this case, $\mathbf{X}_t$ is sometimes called cointegrated of rank $r$ or with cointegration rank $r$. In addition, $\boldsymbol{\beta}'\mathbf{X}_t$ is referred to as the cointegrating series.
- It decomposes the $I(1)$ VAR process into $I(1)$ and $I(0)$ components by (9.12) and obtains the common stochastic trend $\boldsymbol{\alpha}'_{\perp} \sum_{h=1}^{t} \boldsymbol{\varepsilon}_h$.
- If $\boldsymbol{\beta} = (\beta_1, \beta_2, \cdots, \beta_r)$, then every component $\beta_i$ $(i = 1, 2, \cdots, r)$ of $\boldsymbol{\beta}$ is a cointegrating vector, and in algebra, they are linearly independent. In this case, $\boldsymbol{\beta}$ is called a cointegration (cointegrating) matrix, and $\boldsymbol{\alpha}$ is the loading matrix.
- It derives the vector error correction model (VECM) (9.13) for the $I(1)$ VAR process with cointegration rank $r$.
- Sometimes the matrix $\boldsymbol{\Phi}$ is called the long-run impact matrix, and $\boldsymbol{\Gamma}_h$ are the short-run impact matrices.
- If the rank $r = 0$, then $\nabla \mathbf{X}_t$ is a causal VAR$(p-1)$ process, and $\mathbf{X}_t$ is noncointegrated; if $r = K$, then $\mathbf{X}_t$ is a causal VAR$(p)$ process due to Theorem 9.2.

*Example 9.13 (Application of Granger's Representation Theorem)* Consider the following two-dimensional VAR(1) process

$$\mathbf{X}_t = \boldsymbol{\Phi}_1 \mathbf{X}_{t-1} + \boldsymbol{\varepsilon}_t = \begin{bmatrix} 1-\alpha & 2\alpha \\ 0 & 1 \end{bmatrix} \begin{bmatrix} X_{t-1,1} \\ X_{t-1,2} \end{bmatrix} + \begin{bmatrix} \varepsilon_{t,1} \\ \varepsilon_{t,2} \end{bmatrix}$$

where $0 < \alpha < 1$ and $\boldsymbol{\varepsilon}_t = (\varepsilon_{t1}, \varepsilon_{t2})' \sim \text{WN}(\mathbf{0}, \mathbf{I}_2)$. At this point, the VAR matrix polynomial is $\boldsymbol{\Phi}(z) = I_2 - \boldsymbol{\Phi}_1 z$. Therefore, $\det[\boldsymbol{\Phi}(z)] = (1-z)(1+\alpha z - z)$ and

$$\boldsymbol{\Phi} = -\boldsymbol{\Phi}(1) = \begin{bmatrix} -\alpha & 2\alpha \\ 0 & 0 \end{bmatrix}, \; \boldsymbol{\Gamma} = 2\boldsymbol{\Phi}_1 - I_2 = \begin{bmatrix} 1-2\alpha & 4\alpha \\ 0 & 1 \end{bmatrix}.$$

Let $\det[\boldsymbol{\Phi}(z)] = (1-z)(1+\alpha z - z) = 0$. Due to $0 < \alpha < 1$, we have two roots $z = 1$ and $z = (1-\alpha)^{-1} > 1$. In other words, $\det[\boldsymbol{\Phi}(z)] \neq 0$ for all complex numbers $\{z; |z| < 1\}$. Let $\boldsymbol{\alpha} = (-\alpha, 0)'$ and $\boldsymbol{\beta} = (1, -2)'$, and then $\boldsymbol{\Phi} = \boldsymbol{\alpha}\boldsymbol{\beta}'$, $\boldsymbol{\alpha}_{\perp} = (0, 1)'$, $\boldsymbol{\beta}_{\perp} = (2, 1)'$. Moreover, $\boldsymbol{\alpha}'_{\perp}\boldsymbol{\Gamma}\boldsymbol{\beta}_{\perp} = 1$. Obviously, the condition of Granger's representation theorem is satisfied and thus $\mathbf{X}_t$ is $I(1)$.

Now consider the cointegration problem of $\mathbf{X}_t$. It is not hard to arrive at

$$\nabla \mathbf{X}_t = \begin{bmatrix} X_{t1} - X_{t-1,1} \\ X_{t2} - X_{t-1,2} \end{bmatrix} = \begin{bmatrix} -\alpha(X_{t-1,1} - 2X_{t-1,2}) + \varepsilon_{t1} \\ \varepsilon_{t2} \end{bmatrix}. \tag{9.14}$$

Hence, it follows that $X_{t1} - 2X_{t2} = (1-\alpha)(X_{t-1,1} - 2X_{t-1,2}) + \varepsilon_{t1} - 2\varepsilon_{t2}$. That is to say, $Y_t = (1-\alpha)Y_{t-1} + \eta_t$ is an AR(1) model where $Y_t = X_{t1} - 2X_{t2}$ and $\eta_t = \varepsilon_{t1} - 2\varepsilon_{t2}$. What is more, $Y_t = (1-\alpha)Y_{t-1} + \eta_t$ is causal since $0 < 1-\alpha < 1$. As a result, $Y_t = X_{t1} - 2X_{t2}$ has the MA($\infty$) representation

$$X_{t1} - 2X_{t2} = \sum_{h=0}^{\infty} (1-\alpha)^h (\varepsilon_{t-h,1} - 2\varepsilon_{t-h,2}). \tag{9.15}$$

Thus $\boldsymbol{\beta}'\mathbf{X}_t = X_{t1} - 2X_{t2}$ is $I(0)$ and $\mathbf{X}_t$ is cointegrated. In addition, substituting (9.15) into (9.14), it turns out that

$$\begin{aligned} \nabla \mathbf{X}_t &= \begin{bmatrix} -\alpha \sum_{h=0}^{\infty} (1-\alpha)^h (\varepsilon_{t-h-1,1} - 2\varepsilon_{t-h-1,2}) + \varepsilon_{t1} \\ \varepsilon_{t2} \end{bmatrix} \\ &= \begin{bmatrix} \varepsilon_{t1} \\ \varepsilon_{t2} \end{bmatrix} + \begin{bmatrix} -\alpha & 2\alpha \\ 0 & 0 \end{bmatrix} \begin{bmatrix} \varepsilon_{t-1,1} \\ \varepsilon_{t-1,2} \end{bmatrix} + \begin{bmatrix} -\alpha(1-\alpha) & 2\alpha(1-\alpha) \\ 0 & 0 \end{bmatrix} \begin{bmatrix} \varepsilon_{t-2,1} \\ \varepsilon_{t-2,2} \end{bmatrix} + \cdots \end{aligned}$$

It is not difficult to see that the sum of the matrix coefficients $\sum_{h=0}^{\infty} \boldsymbol{\Psi}_h \neq \mathbf{0}$. Consequently, $\nabla \mathbf{X}_t$ is also $I(0)$.

Clearly, the second component of $\mathbf{X}_t$ is a random walk $X_{t2} = X_{t-1,2} + \varepsilon_{t2}$. That is, $X_{t2} = \sum_{h=1}^{t} \varepsilon_{h2} + X_{02}$. Noting (9.15), we obtain

$$\begin{aligned} X_{t1} &= 2X_{t2} + \sum_{h=0}^{\infty} (1-\alpha)^h (\varepsilon_{t-h,1} - 2\varepsilon_{t-h,2}) \\ &= 2\sum_{h=1}^{t} \varepsilon_{h2} + \sum_{h=0}^{\infty} (1-\alpha)^h (\varepsilon_{t-h,1} - 2\varepsilon_{t-h,2}) + 2X_{02}. \end{aligned}$$

Furthermore, by letting $\mathbf{X}_0 = \mathbf{0}$, we arrive at

$$\begin{aligned} \mathbf{X}_t &= \begin{bmatrix} 2\sum_{h=1}^{t} \varepsilon_{h2} + \sum_{h=0}^{\infty} (1-\alpha)^h (\varepsilon_{t-h,1} - 2\varepsilon_{t-h,2}) \\ \sum_{h=1}^{t} \varepsilon_{h2} \end{bmatrix} \\ &= \begin{bmatrix} 0 & 2 \\ 0 & 1 \end{bmatrix} \begin{bmatrix} \sum_{h=1}^{t} \varepsilon_{h1} \\ \sum_{h=1}^{t} \varepsilon_{h2} \end{bmatrix} + \sum_{h=0}^{\infty} \begin{bmatrix} (1-\alpha)^h & -2(1-\alpha)^h \\ 0 & 0 \end{bmatrix} \begin{bmatrix} \varepsilon_{t-h,1} \\ \varepsilon_{t-h,2} \end{bmatrix}. \end{aligned}$$

That is, $\mathbf{X}_t = \boldsymbol{\beta}_\perp \boldsymbol{\alpha}_\perp' \sum_{h=1}^{t} \boldsymbol{\varepsilon}_h + \sum_{h=0}^{\infty} \boldsymbol{\Psi}_h \boldsymbol{\varepsilon}_{t-h}$, which is just Granger's representation for $\mathbf{X}_t$. And the common stochastic trend is $\boldsymbol{\alpha}_\perp' \sum_{h=1}^{t} \boldsymbol{\varepsilon}_h = \sum_{h=1}^{t} \varepsilon_{h2}$. At last, from (9.14), it easily follows that

$$\begin{aligned}\nabla \mathbf{X}_t &= \begin{bmatrix} -\alpha(X_{t-1,1} - 2X_{t-1,2}) + \varepsilon_{t1} \\ \varepsilon_{t2} \end{bmatrix} \\ &= \begin{bmatrix} -\alpha & 2\alpha \\ 0 & 0 \end{bmatrix} \mathbf{X}_{t-1} + \boldsymbol{\varepsilon}_t \\ &= \boldsymbol{\alpha}\boldsymbol{\beta}' \mathbf{X}_{t-1} + \boldsymbol{\varepsilon}_t,\end{aligned}$$

which is simply the vector error correction model for $\mathbf{X}_t$.

### 9.4.4 *Estimation of Vector Error Correction Models*

Now we consider how to apply Theorem 9.2 and estimate VECM (9.13) given $K$-dimensional time series data. The following steps are taken for specifying and estimating a VECM:

1. Checking if the data $\mathbf{X}_t$ is $I(1)$. For $\mathbf{X}_t$ being $I(1)$, it is required that $\mathbf{X}_t$ is nonstationary and the difference $\nabla \mathbf{X}_t$ is stationary. In practice, we simply verify that the original series is nonstationary and the first-differenced series is stationary.
2. Testing for cointegration. We can use the function `coint()` in the module `statsmodels.tsa.stattools` to test for cointegration of the data. The function `coint()` actually uses the augmented Engle-Granger two-step cointegration test procedure. Note that the null hypothesis is that there is no cointegration.
3. Determining order $p$ and cointegration rank $r$ for VECM (9.13). We can, respectively, use the functions `select_order` and `select_coint_rank` in the module `statsmodels.tsa.vector_ar.vecm` to select order $p$ and cointegration rank $r$.
4. Estimating parameters. At this step, we make use of the class `VECM` in the module `statsmodels.tsa.vector_ar.vecm` to estimate all the parameters in VECM (9.13).
5. Diagnostically checking the residuals. The portmanteau test statistic for residual autocorrelation is now asymptotically $\chi^2(mK^2 - K^2(p-1) - Kr)$-distributed. Here note that $K^2(p-1) + Kr$ is actually the number of estimated parameters in $\boldsymbol{\alpha}$ and $\boldsymbol{\Gamma}_h,\ h = 1, \cdots, p-1$. See Brüggemann et al. (2006) and Lütkepohl (2005, pp345–347), among others.

6. If the data-generating process model includes deterministic terms as it sometimes does in practice, we can then set up the VECM for the observed data as

$$\nabla \mathbf{X}_t = \boldsymbol{\alpha} \left[ \boldsymbol{\beta}' \ \boldsymbol{\eta}' \right] \begin{bmatrix} \mathbf{X}_{t-1} \\ \mathbf{D}_{t-1}^{co} \end{bmatrix} + \sum_{h=1}^{p-1} \boldsymbol{\Gamma}_h \nabla \mathbf{X}_{t-h} + \mathbf{C}\mathbf{D}_t + \boldsymbol{\varepsilon}_t \tag{9.16}$$

where $\mathbf{D}_{t-1}^{co}$ contains all the deterministic terms which are inside the cointegration relation (or restricted to the cointegration relation), $\mathbf{D}_t$ contains all remaining deterministic terms, and $\mathbf{C}$ is a corresponding coefficient matrix to be estimated. Note that it is assumed that a specific deterministic term appears only once, either in $\mathbf{D}_{t-1}^{co}$ or in $\mathbf{D}_t$. For more information, see, for example, Lütkepohl (2005) and Johansen (1995).

*Example 9.14 (Cointegration of Prices of WTI Crude Oil and Brent Crude Oil)* The dataset "WTI-Brent" in the folder Ptsadata is from the US Energy Information Administration. It is the monthly spot price (dollars per barrel) time series from May 1987 to July 2020 of the West Texas Intermediate (WTI) crude oil and Brent crude oil. We take a logarithm on the price time series for decreasing volatility in it. The original time series and logged time series plots are, respectively, shown in Figs. 9.30 and 9.31. Obviously both of them are nonstationary, seeing the correlation plots shown in Fig. 9.32. If we take a first difference on the logged time series, then the first-differenced logged time series becomes stationary, seeing the time plot shown in Fig. 9.33 and correlation plots shown in Fig. 9.34. In other words, the logged time series can be viewed as $I(1)$.

Now we can try to build an appropriate VECM for the logged time series. Using the function `coint`, we obtain the $p$-value for the cointegration test $1.0182 \times 10^{-5}$. That is, the logged time series is cointegrated. We can select the order $p = 2$ by the function `select_order` and the cointegration rank $r = 1$ by the function `select_coint_rank`. Furthermore, through the class `VECM` and method `fit`, we gain an estimated VECM for the logged time series, seeing the result table in the following code. Finally, we make a diagnostic check of the VECM residual series. Comparing the correlation plots shown in Fig. 9.32 of the logged time series with ones shown in Fig. 9.35 of the VECM residual series, we see that Fig. 9.32 is typical correlation plots of a two-dimensional nonstationary series; however, Fig. 9.35 is correlation plots of a two-dimensional white noise. What is more, the $p$-value plot for portmanteau test of the VECM residual series shown in Fig. 9.36 indicates that all the $p$-values are greater than 0.05 and so the estimated VECM is adequate.

```
>>> import numpy as np
>>> import pandas as pd
>>> import matplotlib.pyplot as plt
>>> from statsmodels.tsa.stattools import coint, adfuller, kpss
>>> from statsmodels.tsa.vector_ar.vecm
    import VECM, coint_johansen, select_coint_rank, select_order
>>> from PythonTsa.plot_multi_ACF import multi_ACFfig
>>> from PythonTsa.plot_multi_Q_pvalue import MultiQpvalue_plot
```

**Fig. 9.30** Time series plots of prices of WTI and Brent crude oil

pyTSA_CointegrationOil

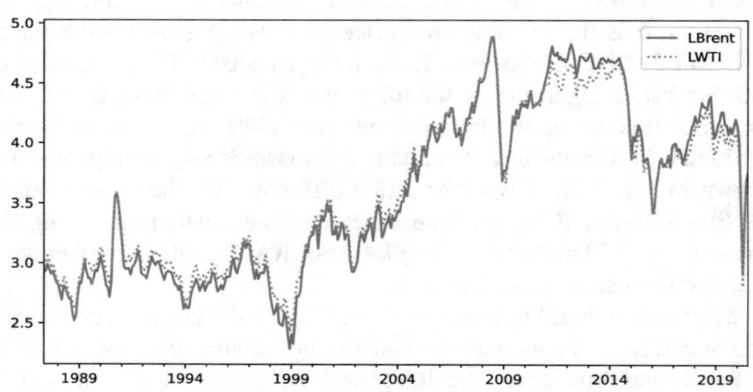


**Fig. 9.31** Time series plots of logged prices of WTI and Brent crude oil

pyTSA_CointegrationOil

```
>>> from PythonTsa.datadir import getdtapath
>>> dtapath=getdtapath()
>>> bwoil=pd.read_csv(dtapath + 'WTI-Brent.csv')
# By default, header=0 in pd.read_csv
>>> dates=pd.date_range('1987-05', periods=len(bwoil),freq='M')
>>> bwoil.index=dates
>>> bwoil.plot(style=['-', ':']); plt.show()
>>> logbw=np.log(bwoil)
>>> logbw.columns=['LBrent', 'LWTI']
>>> logbw.plot(style=['-', ':']); plt.show()
>>> multi_ACFfig(logbw, nlags=18)
>>> plt.show()
# the 2 results below have been edited.
>>> adfuller(logbw.LWTI, regression='c')
(-1.666959, 0.448211, 4, 394, {'1%': -3.4471, '5%': -2.8689,
                               '10%': -2.5707}, -723.0741)
```

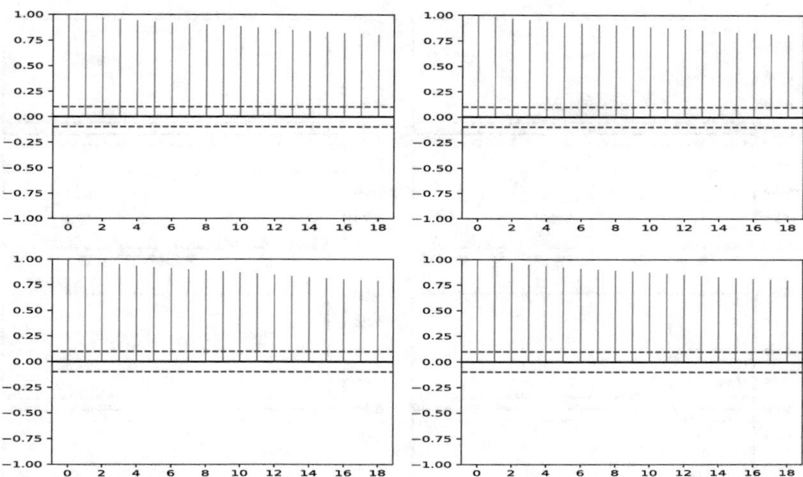

**Fig. 9.32** Correlation plots of logged prices of WTI and Brent crude oil
pyTSA_CointegrationOil

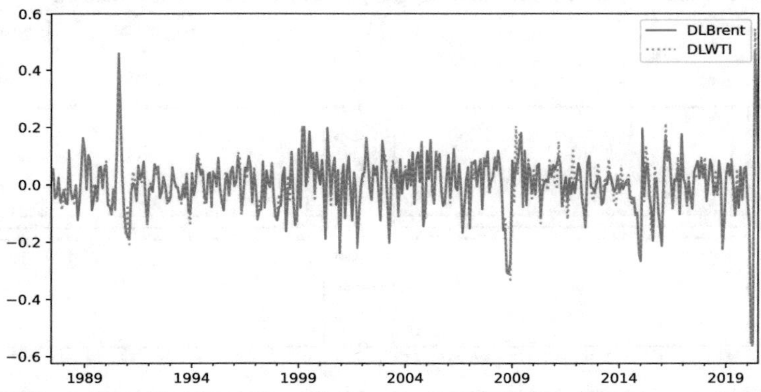


**Fig. 9.33** Time series plots of differenced logged prices of WTI and Brent crude oil
pyTSA_CointegrationOil

```
>>> adfuller(logbw.LBrent, regression='c')
(-1.5777887, 0.494741, 4, 394, {'1%': -3.4471, '5%': -2.8689,
                                '10%': -2.5707}, -693.6955)
>>> dlogbw=logbw.diff(1).dropna()
>>> dlogbw.columns=['DLBrent', 'DLWTI']
>>> dlogbw.plot(style=['-', ':']); plt.show()
>>> multi_ACFfig(dlogbw, nlags=18)
>>> plt.show()
>>> coint(logbw.LWTI, logbw.LBrent, trend='c')
(-5.62099238958, 1.0182223032005589e-05,
array([-3.92416899, -3.35152508, -3.05512345]))
```

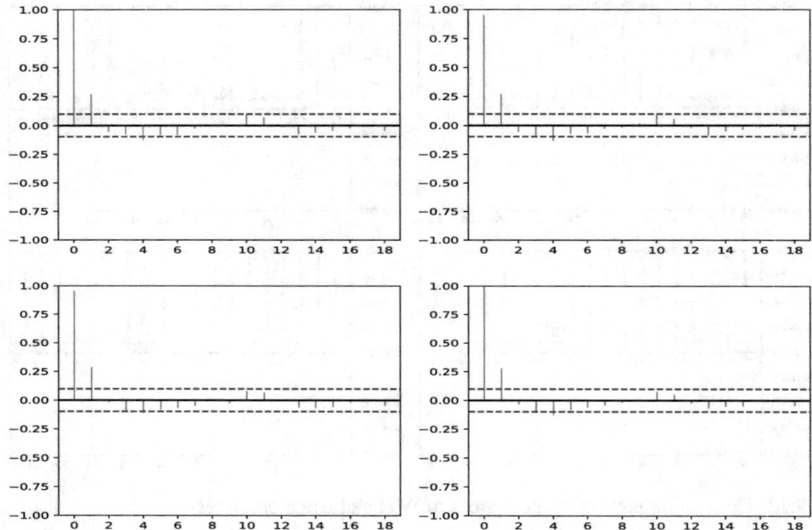


**Fig. 9.34** Correlation plots of differenced logged prices of WTI and Brent crude oil
pyTSA_CointegrationOil

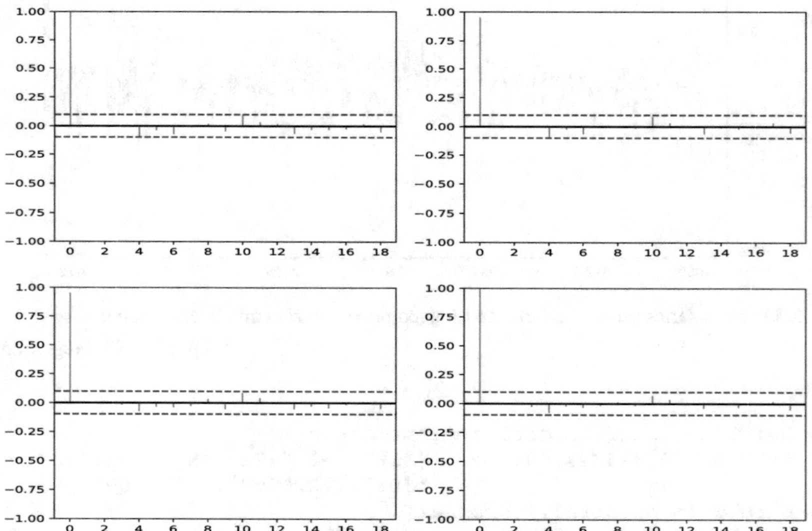


**Fig. 9.35** Correlation plots of VECM residuals of logged prices of WTI and Brent crude oil
pyTSA_CointegrationOil

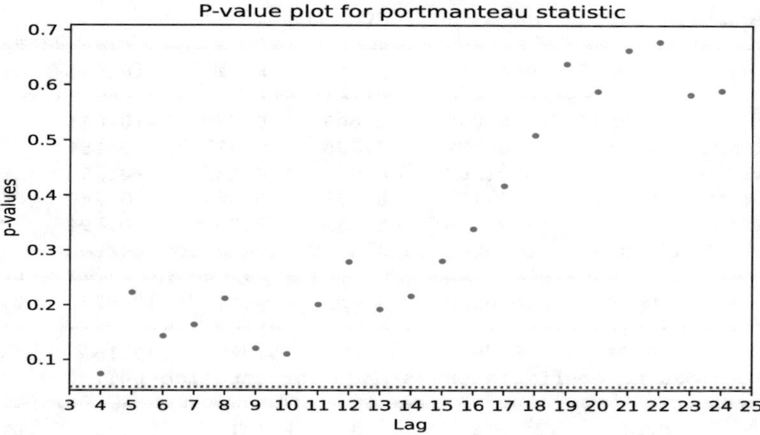


**Fig. 9.36** *p*-Values for portmanteau test of VECM residuals of logged prices of WTI and Brent oil

pyTSA_CointegrationOil

```
>>> selord=select_order(data=logbw,maxlags=10,deterministic='co')
>>> selord.selected_orders
{'aic': 2, 'bic': 1, 'hqic': 2, 'fpe': 2}
>>> scr=select_coint_rank(logbw, det_order=0, k_ar_diff=2)
>>> print(scr.summary())
Johansen cointegration test using trace test statistic
with 5% significance level
=====================================
r_0 r_1 test statistic critical value
-------------------------------------
  0   2          28.20          15.49
  1   2          3.031          3.841
-------------------------------------
>>> scr.rank
1
>>> vecmod=VECM(endog=logbw, k_ar_diff=2, coint_rank=1,
            deterministic='co')
>>> vecmfit=vecmod.fit()
>>> print(vecmfit.summary())
Det. terms outside the coint. relation & lagged endog. parameters
for equation LBrent
=================================================================
                coef    std err          z      P>|z|     [0.025     0.975]
-----------------------------------------------------------------
const         0.0054      0.037      0.148      0.882     -0.066      0.077
L1.LBrent     0.1513      0.166      0.911      0.362     -0.174      0.477
L1.LWTI       0.1751      0.174      1.005      0.315     -0.166      0.517
L2.LBrent    -0.0630      0.165     -0.382      0.702     -0.386      0.260
L2.LWTI      -0.0899      0.173     -0.521      0.603     -0.428      0.248
Det. terms outside the coint. relation & lagged endog. parameters
```

```
for equation LWTI
==============================================================================
                coef    std err          z      P>|z|      [0.025      0.975]
------------------------------------------------------------------------------
const         0.0577      0.035      1.649      0.099      -0.011       0.126
L1.LBrent     0.1568      0.159      0.988      0.323      -0.154       0.468
L1.LWTI       0.1611      0.167      0.967      0.333      -0.165       0.488
L2.LBrent     0.0610      0.157      0.388      0.698      -0.248       0.370
L2.LWTI      -0.1713      0.165     -1.039      0.299      -0.494       0.152
        Loading coefficients (alpha) for equation LBrent
==============================================================================
              coef    std err          z      P>|z|      [0.025      0.975]
------------------------------------------------------------------------------
ec1         0.0094      0.088      0.107      0.915      -0.163       0.182
        Loading coefficients (alpha) for equation LWTI
==============================================================================
             coef    std err          z      P>|z|      [0.025      0.975]
------------------------------------------------------------------------------
ec1        0.1372      0.084      1.630      0.103      -0.028       0.302
  Cointegration relations for loading-coefficients-column 1
==============================================================================
              coef    std err          z      P>|z|      [0.025      0.975]
------------------------------------------------------------------------------
beta.1      1.0000          0          0      0.000       1.000       1.000
beta.2     -1.1103      0.017    -65.910      0.000      -1.143      -1.077
==============================================================================
>>> vecmresid=vecmfit.resid
>>> multi_ACFfig(vecmresid, nlags=18)
>>> plt.show()
>>> q,p=MultiQpvalue_plot(vecmresid,p=3,q=0,noestimatedcoef=10,
                          nolags=24)
>>> plt.show()
#noestimatedcoef=K**2*(p-1)+K*r=2**2*2+2*1=10
>>> vecmfit.sigma_u
array([[0.0091748 , 0.00836409],
       [0.00836409, 0.00837841]])
#Estimates of white noise process covariance matrix
```

For more information on cointegration and VECM, the reader can refer to Pfaff (2008), Lütkepohl (2005), Johansen (1996), Phillips and Ouliaris (1990), Engle and Granger (1987), and others.

### 9.4.5 Real Case of Spurious Regression and Noncointegration

We often simulate such two time series that one regressing on another is spurious. In this subsection, however, we are to analyze a real case of spurious regression and noncointegration.

*Example 9.15 (Real Case of Spurious Regression and Noncointegration)* The dataset "USmacronInRate" in the folder Ptsadata is from the Federal Reserve

Bank of St. Louis. It includes two US quarterly macroeconomic variables and two interest rates as follows:

- `rgnp`: The US real gross national product (GNP) measured in billions of chained 2005 dollars and seasonally adjusted
- `tb3m`: Rate of US 3-month treasury bills obtained by simple average of the monthly rates within a quarter
- `gs10`: Rate of US 10-year constant maturity interest rate obtained by the same method as that of `tb3m`
- `m1sk`: The US M1 money stock obtained from the average of monthly data, which are measured in billions of dollars and seasonally adjusted

The sampling period is from the first quarter of 1959 to the second quarter of 2012. The four variables constitute a tiny monetary system. Tsay (2014, Chapter 5) makes a cointegration analysis of them. Here we do not repeat the same analysis but will make a little different and interesting analysis. In order to make magnitude of these data consistent and reduce volatility, we take logarithm on `rgnp` and `m1sk` and obtain datasets `lrgnp` and `lm1sk`. Figure 9.37 exhibits the time series plots of the four series `lrgnp, tb3m, gs10, and lm1sk`. We see that `lrgnp` and `lm1sk` have very similar upward-trend patterns and `tb3m` and `gs10` are the other two series that have similar changing trends. It is easy to understand that `lrgnp` represents economic growth and `lm1sk` reflects state of savings. Their time series plots seem to suggest that `lm1sk` increases as `lrgnp` grows. And thus we could regress `lm1sk` on `lrgnp`. The regression results are shown in the following first table `OLS Regression Results`, and the regression model appears adequate. But through the ADF and KPSS tests, it is found that the residual series has a unit root and is nonstationary as the $p$-values for the ADF and KPSS tests are, respectively, 0.5212 and 0.0138, in addition, seeing the time plot of the residual series shown in Fig. 9.38. Therefore, the regression results are unacceptable. Now we are to see if the two series are cointegrated or not. Using the cointegration test function `coint`, we obtain the $p$-value for the cointegration test 0.3676 and note that the null hypothesis is no cointegration. It turns out that the two variables are noncointegrated. This illustrates that the estimated regression model is a spurious one. Furthermore, if we first-difference the two series and the differenced series are, respectively, denoted as `dlrgnp` and `dlm1sk`, then by the ADF and KPSS tests, the differenced series are both stationary, that is, `lrgnp` and `lm1sk` are both $I(1)$. Note that the results of regressing `dlm1sk` on `dlrgnp` unveils the truth to us. See the following second table `OLS Regression Results`, in which the $p$-value for the estimate of the coefficient of `dlrgnp` is 0.197. This means the estimate of the coefficient is actually zero. This result clearly suggests that economic growth has very little impact on savings in the American economic system although their time series plots are so similar.

```
>>> import numpy as np
>>> import pandas as pd
>>> import matplotlib.pyplot as plt
>>> import statsmodels.formula.api as smf
```

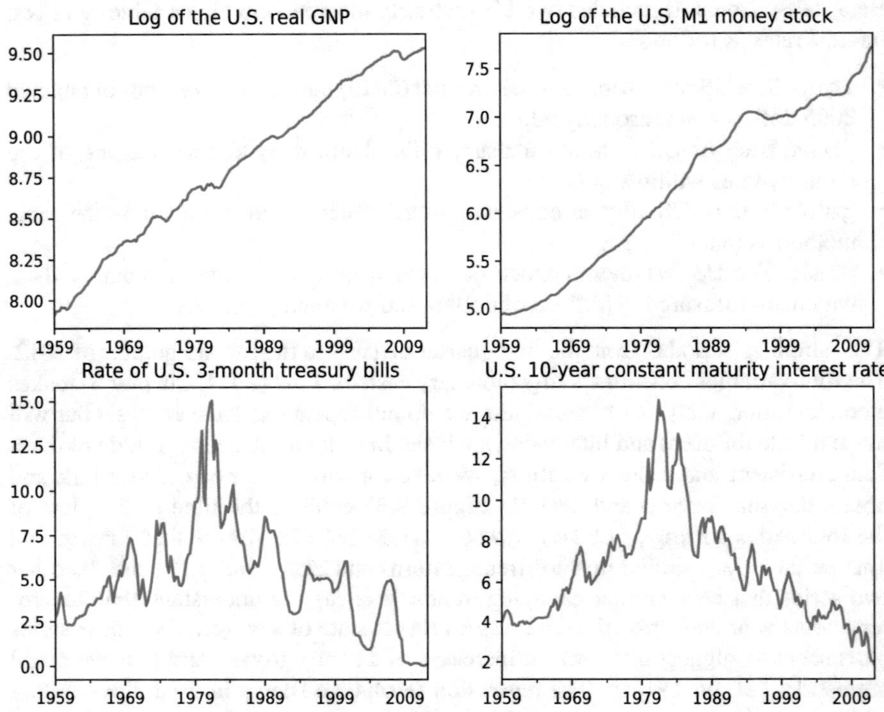


**Fig. 9.37** Time series plots of four US quarterly macroeconomic and interest rate variables

pyTSA_MacroUSCointegration

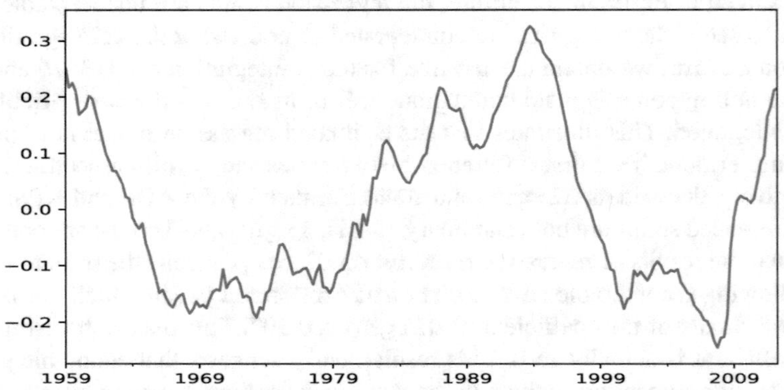


**Fig. 9.38** Time series plot of the model residuals of two US quarterly macroeconomic variables

pyTSA_MacroUSCointegration

```
>>> from statsmodels.tsa.stattools import coint, adfuller, kpss
>>> import statsmodels.api as sm
>>> from PythonTsa.datadir import getdtapath
>>> dtapath=getdtapath()
>>> usm=pd.read_table(dtapath + 'USmacronInRate.txt',
                      sep='\s+', header=0)
>>> usm=usm[['rgnp', 'tb3m', 'gs10', 'm1sk']]
>>> timeindex=pd.date_range('1959-01',periods=len(usm),freq='QS')
>>> usm.index = timeindex
>>> usm.rgnp=np.log(usm.rgnp)
>>> usm.m1sk=np.log(usm.m1sk)
>>> usm.columns=['lrgnp', 'tb3m', 'gs10', 'lm1sk']
>>> fig=plt.figure()
>>> usm.lrgnp.plot(ax= fig.add_subplot(221))
>>> plt.title('Log of the U.S. real GNP')
>>> usm.lm1sk.plot(ax= fig.add_subplot(222))
>>> plt.title('Log of the U.S. M1 money stock')
>>> usm.tb3m.plot(ax= fig.add_subplot(223))
>>> plt.title('Rate of U.S. 3-month treasury bills')
>>> usm.gs10.plot(ax= fig.add_subplot(224))
>>> plt.title('U.S. 10-year constant maturity interest rate')
>>> plt.show()
>>> olsres = smf.ols('usm.lm1sk ~ usm.lrgnp', data=usm).fit()
>>> print(olsres.summary())
                            OLS Regression Results
==============================================================================
Dep. Variable:              usm.lm1sk   R-squared:                       0.969
Model:                            OLS   Adj. R-squared:                  0.968
Method:                 Least Squares   F-statistic:                     6521.
Date:                Sun, 16 Aug 2020   Prob (F-statistic):         3.52e-161
Time:                        00:32:25   Log-Likelihood:                 102.52
No. Observations:                 214   AIC:                            -201.0
Df Residuals:                     212   BIC:                            -194.3
Df Model:                           1
Covariance Type:            nonrobust
==============================================================================
                 coef    std err          t      P>|t|      [0.025      0.975]
------------------------------------------------------------------------------
Intercept     -9.0752      0.190    -47.673      0.000      -9.450      -8.700
usm.lrgnp      1.7396      0.022     80.753      0.000       1.697       1.782
==============================================================================
Omnibus:                       67.306   Durbin-Watson:                   0.019
Prob(Omnibus):                  0.000   Jarque-Bera (JB):               16.637
Skew:                           0.400   Prob(JB):                     0.000244
Kurtosis:                       1.893   Cond. No.                         166.
==============================================================================
Notes:
[1] Standard Errors assume that the covariance matrix of
    the errors is correctly specified.
>>> olsresid=olsres.resid
>>> olsresid.plot(); plt.show()
# the 3 results below have been edited.
>>> adfuller(olsresid, regression='ct')
(-2.1443, 0.5212, 2, 211,
```

```
{'1%': -4.0023, '5%': -3.4315, '10%': -3.1394}, -1023.6568)
>>> kpss(olsresid, regression='ct', nlags='auto')
(0.2059, 0.0138, 9,
{'10%': 0.119, '5%': 0.146, '2.5%': 0.176, '1%': 0.216})
>>> coint(usm.lrgnp, usm.lm1sk, trend='c')
(-2.3120, 0.3676, array([-3.9486, -3.3650, -3.0644]))
>>> dusm=usm.diff(1).dropna()
>>> dusm.columns=['dlrgnp', 'dtb3m', 'dgs10', 'dlm1sk']
# the 4 results below have been edited.
>>> kpss(dusm.dlrgnp, regression='c', nlags='auto')
(0.3868, 0.0828, 6,
{'10%': 0.347, '5%': 0.463, '2.5%': 0.574, '1%': 0.739})
>>> adfuller(dusm.dlrgnp, regression='c')
(-7.0630, 5.1660e-10, 1, 211,
{'1%': -3.4617,'5%': -2.8753,'10%': -2.5741}, -1333.0989)
>>> kpss(dusm.dlm1sk, regression='c', nlags='auto')
(0.1415, 0.1, 8,
{'10%': 0.347, '5%': 0.463, '2.5%': 0.574, '1%': 0.739})
>>> adfuller(dusm.dlm1sk, regression='c')
(-3.4354, 0.0098, 11, 201,
{'1%': -3.4633,'5%': -2.8760,'10%': -2.5745},-1253.0813)
>>> olsres2=smf.ols('dusm.dlm1sk ~ dusm.dlrgnp',data=dusm).fit()
>>> print(olsres2.summary())
                    OLS Regression Results
==============================================================================
Dep. Variable:            dusm.dlm1sk   R-squared:                       0.008
Model:                            OLS   Adj. R-squared:                  0.003
Method:                 Least Squares   F-statistic:                     1.678
Date:                Wed, 27 Oct 2021   Prob (F-statistic):              0.197
Time:                        07:03:42   Log-Likelihood:                 625.39
No. Observations:                 213   AIC:                            -1247.
Df Residuals:                     211   BIC:                            -1240.
Df Model:                           1
Covariance Type:            nonrobust
==============================================================================
                  coef    std err          t      P>|t|      [0.025      0.975]
------------------------------------------------------------------------------
Intercept       0.0141      0.001     12.042      0.000       0.012       0.016
dusm.dlrgnp    -0.1297      0.100     -1.295      0.197      -0.327       0.068
==============================================================================
Omnibus:                       41.089   Durbin-Watson:                   0.719
Prob(Omnibus):                  0.000   Jarque-Bera (JB):               93.611
Skew:                           0.885   Prob(JB):                     4.71e-21
Kurtosis:                       5.723   Cond. No.                         113.
==============================================================================
Notes:
[1] Standard Errors assume that the covariance matrix
    of the errors is correctly specified.
```

In this chapter, we have discussed some advanced topics in time series analysis, including the concepts of stochastic trends and seasonality, unit root tests, stationarity tests, cointegration notion, and Granger's representation theorem. We also have analyzed the spurious regression phenomenon in regression analysis. It is believed that this introduction has laid a good foundation for readers to further study related topics.

## Problems

**9.1** Explain that the time series $X_t = \sum_{i=1}^{t} \varepsilon_i$ where $\varepsilon_i \sim \text{WN}(0, \sigma^2)$ have a stochastic trend.

**9.2** Respectively, generate a sample of size 720 from the stationary seasonal AR process $X_t = 0.2X_{t-12} + \epsilon_t$ (viz., monthly series) and the seasonal random walk $X_t = X_{t-12} + \epsilon_t$ with Python. And then (1) draw the seasonal plots for the two simulated samples, and what conclusions can be obtained? (2) Examine, respectively, stationarity of the concatenate series of the seasonal subseries of the two simulated samples.

**9.3** Use the function `simulSBM()` but not run `pd.DataFrame()` to reproduce Fig. 9.20.

**9.4** Do stationary seasonal time series such as $X_t = 0.3X_{t-4} + \varepsilon_t,\ \varepsilon_t \sim \text{WN}(0, 1)$ have stochastic seasonality and why?

**9.5** Given the time series $Y_t = 0.1 - 0.2t + X_t$ where $X_t$ is stationary, then the first-differenced series $\nabla Y_t = -0.2 + \nabla X_t$ and $\nabla Y_t$ is stationary since $\nabla X_t$ is still stationary. Now answer the question: is $Y_t$ difference-stationary?

**9.6** There is a time series dataset "RwalkwDrift0.3" in the folder Ptsadata. If its time series plot is drawn, we can observe that there appears to be a linearly ascending trend in the time series. Now try to fit the trend using a linear model $X_t = \alpha + \beta t + \varepsilon$, then analyze the regression results, and check the residual series.

**9.7** Show that Eq. (9.6) holds good.

**9.8** Make Phillips-Perron (unit root) tests for the logged series of the data in Example 9.7, and compare your results with the ones of Example 9.7. (Hint: the Python function for Phillips-Perron test can be found in the package 'arch'.)

**9.9** Suppose that $Y_t$ is $I(0)$. Show that $X_t = a + bt + Y_t$ is $I(1)$. Is $Z_t = a + bt^2 + Y_t$ a $I(2)$ process? Why?

**9.10** Verify that $Y_t = \sum_{h=1}^{t} X_h$ is $I(1)$ where $X_h$ is the causal AR(1) process defined in Example 9.9.

**9.11** Find the common stochastic trend of the two-dimensional cointegrated process in Example 9.11.

**9.12** Considering the two-dimensional VAR(1) process in Example 9.13, prove that both Eq. (9.14) and $X_{t1} - 2X_{t2} = (1 - \alpha)(X_{t-1,1} - 2X_{t-1,2}) + \varepsilon_{t1} - 2\varepsilon_{t2}$ hold.

**9.13** Check if the series `lrgnp` in Example 9.15 has a deterministic trend and stochastic trend.

**9.14** In Example 9.15, we have come to know that the series `lrgnp` and `lm1sk` are noncointegrated although the patterns of their time plots are very similar. Now analyze the cointegrating problem of the series `tb3m` and `gs10`.

# Chapter 10
# Modern Machine Learning Methods for Time Series Analysis

Artificial intelligence (AI) has gained considerable achievements over the last decades, and AI methods have been proposed as alternatives to statistical ones for time series forecasting and classification. This chapter introduces some latest advancements in this respect, including artificial neural networks and deep learning, Google's TensorFlow, and more. Now the Python package Keras has become a module of TensorFlow as a frontend. We also discuss how to use TensorFlow and write Python code to implement time series forecasting.

## 10.1 Introduction

In this section, we concisely introduce what is artificial intelligence (AI) and its development history, and then we briefly look at emerging AI methods for time series analysis including forecasting and classification.

### *10.1.1 Brief History of Artificial Intelligence*

It is generally acknowledged that the term *artificial intelligence* (AI) was coined by J. McCarthy (1927–2011) in 1955. Later, in the summer of 1956, J. McCarthy, M. Minsky, N. Rochester, and C. Shannon organized a 2-month workshop on AI at Dartmouth College in Hanover, New Hampshire, USA, which marked that artificial intelligence had become an independent discipline. J. McCarthy was the recipient of the 1971 Turing Award for his contributions to AI. And then, what is artificial intelligence? McCarthy (2007) states:



C. Huang, A. Petukhina, *Applied Time Series Analysis and Forecasting with Python*, Statistics and Computing,
https://doi.org/10.1007/978-3-031-13584-2_10

> It is the science and engineering of making intelligent machines, especially intelligent computer programs. It is related to the similar task of using computers to understand human intelligence, but AI does not have to confine itself to methods that are biologically observable.

Although different people have different minds about AI due to AI in relation to so many fields, this definition gives a clear and understandable statement on AI. In addition, note that *machine learning* (ML) is a subfield of AI and *deep learning* (DL) is a branch of machine learning. For better understanding, below given are the definitions of machine learning and deep learning. Machine learning is the science and engineering of having machines (computers) to act and think without being explicitly programmed, that is, it allows machines to humanly and rationally handle new situations by analysis, self-training, observation, and experience. The name of machine learning was first introduced by A. Turing in his 1950 paper "computing machinery and intelligence." Deep learning is to study how to utilize a (deep or multilayer) *artificial neural network* to implement the process of machine learning. Technically, deep learning makes machines learn appropriate representations of data so that these can be used to draw conclusions. The concept of the artificial neural network as well as more details on deep learning will be addressed in Sects. 10.2 and 10.3. As to differences among AI, ML, and DL, see, for example (Mueller and Massaron 2019). Besides, we should know that *statistical learning* is a kind of statistics with machine learning in mind. It is a powerful and indispensable set of tools in the field of AI. Meanwhile, AI also resorts to many other disciplines such as cognitive science, computer science, neuroscience, optimization theory, and so forth. For more on statistical learning, the reader may refer to Hastie et al. (2017), James et al. (2017), and others.

For a pretty long time, a number of researchers believed that human-like AI could be accomplished simply by having programmers handcraft a sufficiently large set of explicit rules for handling knowledge. And this approach is referred to as *symbolic AI* and was the prevailing paradigm in AI field from the 1950s to the late 1980s. It reached its peak during the *expert systems* flourishing of the 1980s. However, it turned out to be tough to figure out explicit rules for solving more complex, fuzzy problems, such as speech recognition, image classification, or language translation. It was machine learning that arose to take its place as a new approach to AI. Moreover, although machine learning only started to thrive in the 1990s, it has quickly become the most successful branch of AI. At the same time, a great number of scientific methods are adopted in the field of AI, including a wide variety of techniques of mathematical statistics. This enormously promotes the development of AI as well as machine learning.

In this century, there are very large datasets available almost everywhere, and a great number of new techniques for AI have been emerging. Thanks to this, AI has advanced more rapidly over the last two decades. In particular, the deep learning approach has brought major breakthroughs in AI. The branches of AI have been getting more integrated. A lot of AI applications have been deeply embedded in the infrastructures of many industries, and AI itself has become an industry. AI has had more and more influence on everyday life of humankind.

For more details on definitions and history of AI as well as machine learning, readers may see Russell and Norvig (2016, 2021), Nilsson (2010), and McCarthy (2007), among others.

### *10.1.2 AI in Time Series Analysis*

The 1990s witnessed more and more research efforts devoted to the field of time series forecasting by AI methods. For example, Zhang et al. (1998) presents an overview of artificial neural network methods as of 1998 for time series forecasting. Hochreiter and Schmidhuber (1997) first introduces the long short-term memory (LSTM) algorithm and discusses LSTM's advantages and limitations. The LSTM is an efficient, gradient-based neural network algorithm and is still today in use for time series predicting. In recent two decades, with the advent of deep learning techniques, new approaches of unsupervised learning for time series analysis have greatly been developed. There have been appearing a great amount of literature on this area. The following is merely a few latest examples.

Gamboa (2017) summarizes applications of deep learning techniques on time series forecasting, classification, as well as anomaly detection and points out that deep learning methods have a large quantity to contribute to the field of time series analysis. Deep learning methods tend to be "black-box" models that do not shed light on how they use the full range of inputs present in practical scenarios. Lim et al. (2020) proposes a novel attention-based deep learning architecture called temporal fusion transformer (TFT). It combines high-performance multi-horizon forecasting with interpretable insights into temporal dynamics. To learn temporal relationships at different scales, TFT uses recurrent layers for local processing and interpretable self-attention layers for long-term dependencies. TFT also utilizes specialized components to select relevant features and a series of gating layers to suppress unnecessary components, enabling high performance in a wide range of scenarios. Using a variety of real-world datasets, Lim et al. (2020) shows that TFT achieves state-of-the-art forecasting performance. For more details on AI time series forecasting, see Lim et al. (2020), Rangapuram et al. (2018), and Gamboa (2017), among others.

The conventional statistical time series analysis usually pays little attention to classification problems. However, it is vital to tackle classification problems in the field of AI since many applications of AI are related to classification. Naturally, time series classification (TSC) is also very important in AI applications. But AI time series classification is a young and challenging field. Fawaz et al. (2019) gives a review of deep learning methods for TSC and points out only a few TSC algorithms have considered deep neural networks to perform this TSC task, which is surprising as deep learning has seen very successful applications in other fields such as computer vision, text document classification, and speech recognition. By training 8730 deep learning models on 97 time series datasets, Fawaz et al. (2019) conducts the most exhaustive empirical study of deep neural networks for

TSC to date. The results show that deep learning can achieve the current state-of-the-art performance for TSC. In addition, Fons et al. (2020) indicates that data augmentation methods have been used extensively in computer vision on classification tasks and have achieved great success; however, their use in TSC is still at an early stage. Then Fons et al. (2020) evaluates several augmentation methods applied to stocks datasets using two state-of-the-art deep learning models. The results show that several data augmentation methods significantly improve financial performance when used in combination with a trading strategy. For more details on AI time series classification, the interested reader can refer to Fawaz et al. (2019); Fons et al. (2020) as well as the references therein, and Guennec et al. (2016), and others.

In short, time series analysis in the AI field includes forecasting, classification, anomaly detection, clustering (see Maharaj et al. (2019) and the references therein), and so forth. Time series analysis by AI methods is challenging yet promising, enticing, and worthy of consideration. Meanwhile, another important approach worthy of attention is to combine statistical methods and AI techniques together to analyze time series for better performance.

## 10.2 Artificial Neural Networks

In this section, first we briefly present the origination and history of artificial neural networks, and then we describe artificial neural network models in detail.

### *10.2.1 Artificial Neural Network Developments*

The concept of artificial neural networks originates from McCulloch and Pitts (1943), which is now generally recognized as the first work on AI. The seminal article describes a simplified neural network architecture for intelligence and quickly becomes a foundational work in the study of artificial neural networks although the neurons described in it are greatly simplified compared to real biological neurons. Now there have been various definitions of artificial neural networks, and they tend to be simply called neural networks or neural network models. Other names for artificial neural networks include connectionism models, neural computing (models), and parallel distributed processing (models). The main point is that an artificial neural network is a (mathematical) model of simulating the interaction of the human brain or other living creatures' brain to real-world objects. Technologically, the artificial neural network is a massively parallel interconnected network composed of adaptive simplified processing units. Its organization can simulate the interaction of biological nervous system to real-world objects. These (processing) units are also called *(artificial) neurons* and sometimes *nodes* especially in mathematics and statistics. Since 1943, researchers in AI and statistics have been interested in the properties of neural networks, such as their ability to carry out distributed computation, to tolerate noisy inputs, to learn, and so on.

From the 1940s to 1960s, a series of works on neural networks were made in addition to McCulloch and Pitts (1943), and the following are several important examples:

- In 1949, D. Hebb showed a simple updating rule for modifying the connection strengths between neurons. His rule, now called *Hebbian learning*, remains an influential model to date.
- In 1950, M. Minsky and D. Edmonds constructed the first neural network computer to simulate a network of 40 neurons.
- In 1957, F. Rosenblatt invented his perceptron, which had a simple learning algorithm for classifying images into categories. Although there were limitations in the perceptron, it has been recognized as a precursor of deep learning neural networks. Furthermore, F. Rosenblatt not just described the perceptron in detail but also put forward a theory of brain mechanisms in his 1962 book Principles of Neurodynamics: Perceptrons and the Theory of Brain Mechanisms.
- As early as the 1960s, the backpropagation learning algorithm for multilayer neural networks was invented. See, for example (Bryson and Ho 1969). The backpropagation approach implements a gradient descent (a mathematical optimization way) in parameter space for the minimization of the output error and is a critical algorithm in the field of deep learning. In 1970, a modern type of the backpropagation method was created by S. Linnainmaa, which is very efficient for discrete sparse neural networks. See, for example (Schmidhuber 2015). Interesting is that it was rediscovered by researchers in the mid-1980s.
- In 1965, A. Ivakhnenko and V. Lapa published "Cybernetic Predicting Devices," which introduced the first general, working learning algorithm for supervised deep feed-forward multilayer perceptrons. This has been acknowledged as the first work on deep learning neural networks.

From the 1970s to the mid-1980s, however, neural network research was at a low ebb, and research funding for it dwindled to almost zero partly because M. Minsky and S. Papert pointed out the computational limitations of a single artificial neuron in their 1969 book Perceptrons. In addition, to carry out neural network applications, we need good algorithms, large datasets, and faster computers. But, these conditions could not be met in that period. Moreover, the backpropagation algorithm seemed to be forgotten. Nevertheless, a few indomitable connectionists insisted on researching into neural network models and applications. Below are three examples to merely mention:

- Fukushima (1980) first gave the convolutional neural network (CNN) architecture (called neocognitron) inspired by neurophysiological insights, and now CNNs are widely used for computer vision.
- J. Hopfield introduced a type of neural networks now called the Hopfield net in his 1982 article "Neural Networks and Physical Systems with Emergent Collective Computational Abilities."

- Ackley et al. (1985) gave a counterexample to M. Minsky and S. Papert's widely accepted belief that no learning algorithm for neural networks was possible.

The year of 1986 saw the return of artificial neural networks. G. Hinton, D. Touretzky, and T. Sejnowski organized the first Connectionist Summer School at Carnegie Mellon in that year. Several different research groups reinvented the back-propagation learning algorithm. For example, in the journal Nature, D. Rumelhart, G. Hinton, and R. Williams published the paper "Learning Representations by Back-propagating Errors," in which the backpropagation algorithm was introduced in detail. See Rumelhart et al. (1986). In addition, the two volumes of "Parallel Distributed Processing" edited by D. Rumelhart and J. McClelland came out. All these rekindled researchers' enthusiasm for neural networks. From then on, many important advances have been made in developing new learning algorithms for neural networks as well as applications onto various fields. Here we can take only a few, for example:

- Hornik et al. (1989) as well as Hecht-Nielsen (1989) showed that a single-hidden-layer neural network with enough hidden nodes can approximate any multivariate continuous function with arbitrary accuracy.
- In 1990, Y. LeCun et al. published "handwritten digit recognition with a backpropagation network" in Advances in Neural Information Processing Systems. It is the first article about convolutional neural networks trained by the backpropagation algorithm in order to fulfil the task of classifying low-resolution images of handwritten digits.
- To overcome the vanishing gradient problem in the backpropagation algorithm, Schmidhuber (1992) used unsupervised pre-training for a stack of recurrent neural networks to facilitate supervised learning in very deep recurrent neural networks.
- In 1995, A. Bell and T. Sejnowski published "An Information Maximization Approach to Blind Separation and Blind Deconvolution," which describes an unsupervised learning algorithm for independent component analysis (ICA). And ICA provides a solution to the blind source (speech) separation problem and has been used in many applications.
- As mentioned in Sect. 10.1.2, Hochreiter and Schmidhuber (1997) first introduced LSTM recurrent neural networks, which are good at learning long-range dependent relations and have become one of critical algorithms for deep learning.
- Hinton et al. (2006) introduced a novel and fast algorithm for training deep neural networks by pre-training one hidden layer at a time and compared error rates of various learning algorithms. They made a deep neural network capable of recognizing handwritten digits with state-of-the-art precision (>98%). Later on, the term deep learning has become popular almost everywhere, and deep learning techniques have massively been applied on different domains.
- Glorot et al. (2011) showed that supervised training of very deep neural networks is much faster if the hidden layers are composed of the activation function ReLU (see Sect. 10.2.2).

- Go is a board game originated from ancient China and still popular today in many areas, especially East Asia. The board has $19 \times 19 = 361$ cross points, and the total number of the moves of Go is huge. In March 2016, Lee Sedol, the South Korean Go 18-time world champion, played and lost a five-game match against DeepMind's AlphaGo that used deep learning networks to evaluate board positions and possible moves. Thus AlphaGo has become the first AI program to defeat the human Go world champion. But AlphaGo is told the rules of the game beforehand. In 2019, DeepMind's MuZero replaced AlphaGo as well as AlphaZero. Moreover, an article published in the journal Nature on December 23, 2020, described MuZero, a significant step forward in the pursuit of general-purpose algorithms. MuZero can master Go, chess, shogi, and Atari without needing to be told the rules, due to its ability to plan winning strategies in unknown environments. For more details, the interested reader can refer to DeepMind's homepage: https://deepmind.com/.

### *10.2.2 Neural Network Models*

Since the 1940s, through investigation into the biological nervous system, researchers have proposed many types of neural network models. First of all, let us look at a foundational unit of artificial neural networks, that is, a neuron model or artificial neuron, which is often simply called neuron. Seeing Fig. 10.1, we can find that a neuron model consists of the following key components:

- An input vector $X = (x_1, x_2, \cdots, x_n)'$ that represents external information or signals received by the neuron.
- A corresponding weight (strength or parameter) vector $W = (w_1, w_2, \cdots, w_n)'$ with a constant $b$ that represents a bias.

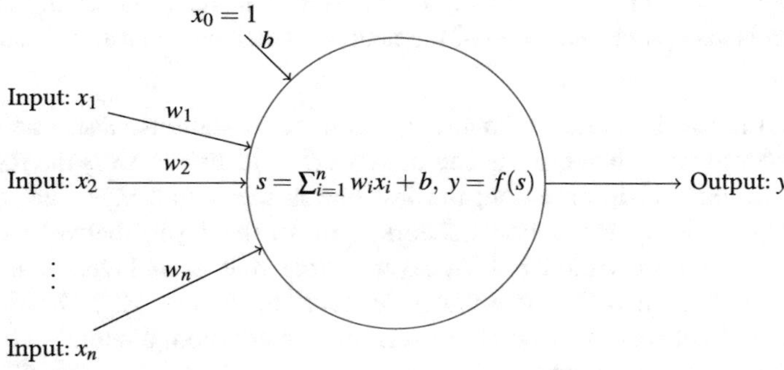


**Fig. 10.1** Schematic for a neuron model

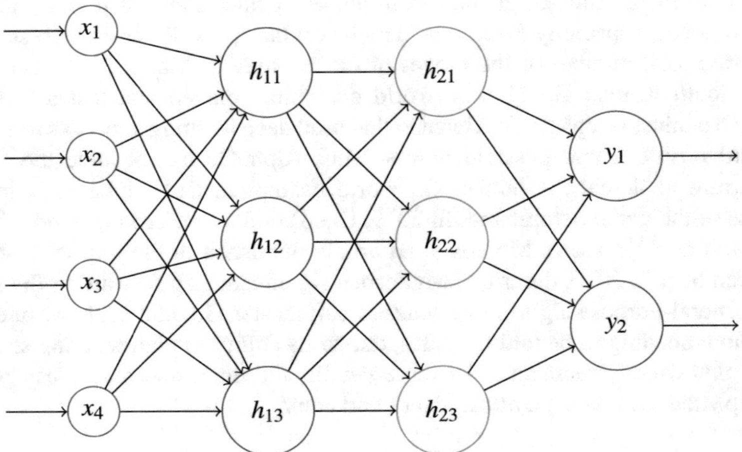


**Fig. 10.2** Schematic for multilayer feed-forward neural network models

- The weighted sum of the input signals $s = \sum_{i=1}^{n} w_i x_i + b$, which means that the input information is pre-processed by the neuron. And this sum is usually called *net input*.
- A function $y = f(s)$ that is known as the *activation function*. It can be seen as a neuron's decision-making mechanism. And its value is the result of decision-making and can be simply called *activation value*. There have now been several widely used activation functions such as the sigmoid function $\text{sigmoid}(s|a) = 1/(1 + e^{-as})$ where $a$ is its slope parameter and we usually let $a = 1$, the hyperbolic tangent function $\tanh(s) = (1 - e^{-2s})/(1 + e^{-2s})$, the ReLU (rectified linear unit) function $\text{ReLU}(s) = \max(s, 0)$, the Swish function $\text{swish}(s|a) = s/(1 + e^{-as})$, and so on. Note that the value of the activation function is also an output value of the neuron, which can be transmitted to other neurons.

A neural network consists of many interconnected neurons. These neurons are fixed in order in every layer of the neural network. The first layer is the input layer that takes in external information; the last one is the output layer that transmits the decision made by the neural network, and all the layers between the input and output layers are called hidden layers. Note that each layer of the neural network may have a different number of neurons. Figure 10.2 shows a feed-forward neural network with two hidden layers. In addition, a neural network can be viewed as a vector-valued function $(y_1, \cdots, y_m) = (g_1(X), \cdots, g_m(X))$ where $X$ is the input vector. The functionality of a neural network is determined by its network architecture or structure as well as neurons' mechanisms, and a network's architecture is formed mainly by the number of its neurons, number of its layers, and types of interconnections between neurons. There have been several types of neural networks such as feed-forward neural networks (FNNs), recurrent neural networks (RNNs) that have at least one feedback loop, and convolutional neural networks (CNNs). Here we will describe FNNs in detail, and for the other types of neural networks, readers can refer to Chollet (2021), Aggarwal (2018), and Goodfellow et al. (2016), among others.

A feed-forward neural network is a simple but useful architecture of neural networks. Its information (or signal) flow is unidirectional and has no feedback as Fig. 10.2 shows. Now we have a FNN with $K-1$ hidden layers where the input and output layers are denoted as layer 0 and $K$, respectively, and suppose that the input vector is $X=(x_1, x_2, \cdots, x_n)'$; the output vector is $Y=(y_1, y_2, \cdots, y_m)'$, and the weight vector for neuron $j$ in layer $i (1 \le i \le K)$ is $W_j^{(i)} = (w_{j1}^{(i)}, w_{j2}^{(i)}, \cdots, w_{jn_i}^{(i)})$ where $n_i$ is the number of the neurons in layer $i-1$ and $n_{K+1}=m$. Noting that $W_j^{(i)}$ is a row vector, then we can write the network as

$$\begin{aligned} H_i &= (h_{i1}, h_{i2}, \cdots, h_{i,n_{i+1}})' \\ &= \Big(f_i(W_1^{(i)} H_{i-1} + b_1^{(i)}), f_i(W_2^{(i)} H_{i-1} + b_2^{(i)}), \cdots, f_i(W_{n_{i+1}}^{(i)} H_{i-1} + b_{n_{i+1}}^{(i)})\Big)' \end{aligned}$$

where $H_0 = X$; $f_i(\cdot)$ is an activation function in layer $i$ and $B^{(i)} = (b_1^{(i)}, \cdots, b_{n_{i+1}}^{(i)})$ is the bias vector in layer $i$, $1 \le i \le K$. If the task is forecasting, then the estimated output vector is

$$\hat{Y} = (\hat{y}_1, \hat{y}_2, \cdots, \hat{y}_m)' = H_K = (h_{K1}, h_{K2}, \cdots, h_{K,n_{K+1}})'.$$

On the other hand, if the task is classification, then the output vector is

$$Y = (y_1, y_2, \cdots, y_m)' = \left(\frac{\exp(h_{K1})}{\sum_{t=1}^m \exp(h_{Kt})}, \frac{\exp(h_{K2})}{\sum_{t=1}^m \exp(h_{Kt})}, \cdots, \frac{\exp(h_{Km})}{\sum_{t=1}^m \exp(h_{Kt})}\right)'.$$

At this point, for each $1 \le t \le m$, $0 < y_t < 1$ and $\sum_{t=1}^m y_t = 1$. So they can be viewed as the probabilities of classified objects. Besides, the vector-valued function

$$f(s_1, s_2, \cdots, s_m) = \left(\frac{\exp(s_1)}{\sum_{t=1}^m \exp(s_t)}, \frac{\exp(s_2)}{\sum_{t=1}^m \exp(s_t)}, \cdots, \frac{\exp(s_m)}{\sum_{t=1}^m \exp(s_t)}\right)$$

is known as the *softmax function*. In the special case that $m=1$ and $Y$ is Boolean, the feed-forward neural network is simply reduced to the *perceptron* invented by F. Rosenblatt, and the output $Y = g(h_{K1})$ where g(s) is the *heaviside function*

$$g(s) = \begin{cases} 1 & \text{if } s > 0, \\ 0 & \text{if } s \le 0. \end{cases}$$

## 10.3 Deep Learning and Backpropagation Algorithms

In this section, we outline what deep learning means and then address a vital algorithm for deep learning: the backpropagation method.

### *10.3.1 What Is Deep Learning?*

Only from 2006 on has the term *deep learning* been a catchphrase although the first deep learning system was invented as early as 1965 (Goodfellow et al. 2016; Schmidhuber 2015). Here "learning" is about estimating or finding weights (parameters) that make the neural network exhibit the desired behavior, and the process of estimating is to train the neural network; "deep" usually means that the number of hidden layers is large in a trained neural network. Unfortunately, there has not been a clear rule for determining how many layers or neurons the neural network should possess in order to achieve a desired target. There have been many successful instances in which a neural network outperforms other neural networks simply by adjusting the numbers of layers and/or neurons. Therefore, in order to implement an adequate neural network, one probably has to spend a lot of time experimenting with different combinations of layers and neurons. What is more, as the numbers of layers and neurons increase, the scale of weights (parameters) will quickly become huge in a neural network. This challenges the computing power of a computer or computing system. At present, parallel computing (including processing simultaneously on a CPU and GPU) is widely used to enhance the computing power. For further information about deep learning, the reader may refer to Chollet (2021), Aggarwal (2018), Goodfellow et al. (2016), Schmidhuber (2015), and so forth.

### *10.3.2 Gradient Descent and Backpropagation Algorithms*

It has been long since the gradient descent method was found, and it is generally recognized that A. L. Cauchy firstly invented the method in 1847. The gradient descent algorithm (method) is used to minimize the loss function or the train error in machine (deep) learning modeling. This target of minimizing the error is realized by the backpropagation algorithm. The backpropagation works as follows. First, the inputs are feed-forwarded from the input layer through the hidden layer to the output layer, where the error is computed by comparing the network output with the actual values. Then, the error will be backpropagated to the hidden layer and the input layer, during which the connection weights and biases are adjusted to reduce the error. The process is finished by tuning toward the direction with the gradient, and such a process will be repeated in many rounds until the error is minimized or the train process is ceased to avoid over-fitting.

There has now been a great deal of literature on deep learning as well as neural networks. For more information on the past, present, and future of deep learning and neural networks, see, for example, LeCun et al. (2015), and Schmidhuber (2015) as well as literature therein, both of which give an excellent review of deep learning. In addition, Fan et al. (2019) elaborates on the theory of deep learning from the statistical perspective. Sejnowski (2018) gives a macro presentation on deep learning development over the past decades and describes its scientific and human everyday life impact.

## 10.4 Time Series Forecasting and TensorFlow

### *10.4.1 Time Series Forecasting*

As we see in Sect. 10.1.2, researchers have been investigating the topic of time series analysis by machine learning methods since the 1990s. RNNs were created in the 1980s. They are useful with time series data since each neuron or unit can use its internal memory to maintain information about the previous input. But a simple RNN suffers from a fundamental problem of not being able to capture long-term dependencies in a time series. This is the main reason RNNs faded out from practice until some good results were achieved with LSTM neural networks. Another special RNN is the GRU (gated recurrent unit) algorithm (network) introduced by K. Cho et al. in 2014. GRU is very similar to LSTM but has simpler architecture. The key difference is that GRU has fewer parameters than LSTM as it lacks an output gate. In the case of smaller datasets, GRUs tend to show better performance.

Besides, in order to speed up learning and lead to faster convergence as well as higher accuracy, it is important to transform or scale features (data) before training a neural network. Normalization and standardization are two common ways of doing so. An example is below in Sect. 10.5.2. For more critical advancements in time series forecasting by machine learning methods, the reader can refer to Lazzeri (2021); Lim and Zohren (2021); Torres et al. (2021), and others.

### *10.4.2 TensorFlow and Keras*

Recent years have seen many deep learning frameworks constructed by AI companies or researchers such as Caffe(2), Chainer, CNTK, Keras, MxNet, PaddlePaddle, (Py)Torch, TensorFlow, and Theano, among others. These frameworks are all open source and have greatly promoted applications of deep learning technology on various domains. The interested reader can visit their websites for more details. We here pay attention to TensorFlow and Keras.

TensorFlow is developed by Google. It is a free and end-to-end open-source platform for machine learning. It can be used to implement a range of tasks by dataflow programming but has a particular focus on training and inference of deep neural networks. The primary purpose of TensorFlow is to enable engineers and researchers to manipulate mathematical expressions over numerical tensors. It can run not only on CPUs but also on GPUs and TPUs so to accelerate processes of modeling and forecasting. Its homepage is https://www.tensorflow.org/.

The Python package Keras is created by F. Chollet. It is an API (application programming interface) designed for human beings, not machines. It provides a convenient way to define and train any kind of deep learning models and clear and actionable error messages. Readers can refer to the Keras documentation at https://keras.io/ and Chollet (2021) for more details. Note that Keras now defaults to a frontend of TensorFlow.

We will use TensorFlow of version 2.3.1 to analyze and predict time series below. If you have installed the solo Keras, we suggest that you should uninstall it since it may conflict with Tensorflow.keras (i.e., Keras in TensorFlow).

## 10.5 Implementation and Example

### *10.5.1 Implementation Steps*

There are several major differences between the classic statistical methodology and machine learning methodology for time series modeling and forecasting. In what follows, we present the basic steps of the machine learning methodology in deploying time series modeling and predicting. We assume that a cleaned time series dataset is given:

- *Step 1: Dividing the sample dataset into subsets*. The sample dataset is generally split into three subsets, train set, validation set, and test set, which are not overlapped. The train set is used to estimate a machine learning model. This process of estimating is just to train the machine learning model to learn from historical data in order to forecast future data points. The validation set is used to give an unbiased evaluation of the estimated model and fine-tune the estimated model hyper-parameters so to improve the estimated model. The test set is used to determine whether the estimated model is under-fitting or over-fitting to the train set by comparing the predictions (fitted values) with the actual values of the train and test sets and further evaluate the predicting power of the estimated machine learning model.
- *Step 2: Transforming or scaling data series*. As in the case of classic statistical modeling, to use machine learning approaches, we need to transform the values of datasets into a certain range such as the unit interval [0,1]. This process is often called normalization or standardization. There are several ways to do so. For more information, the reader may refer to the user guide on the website of

the Python package `scikit-learn` and see Pedregosa et al. (2011). Note that the transformed data series is not necessarily stationary. See Example 10.1.

- *Step 3: Being compatible with the machine learning algorithm*. Different deep learning frameworks have different requirements of data representations. Therefore we need to prepare the proper representation of the input dataset so that it can be used in the selected deep learning framework. Note that Steps 1–3 may be viewed as part of data pre-processing.
- *Step 4: Specifying and training*. According to the requirements of a deep learning framework that you choose, specify the related parameters and thus obtain the executable Python code. And then train and validate the machine learning model by running the code, and arrive at the trained (estimated) model.
- *Step 5: Testing and predicting*. Using the test dataset can make predictions and evaluate the trained model. There are several evaluating methods. For example, (1) draw the time plots of the actual values and fitted (predicted) values so to compare them, (2) calculate the evaluating index `mape`, which is often used in the machine learning field (see the user guide of the package `scikit-learn` for more details), and (3) analyze the residual series.

### 10.5.2 An Example

*Example 10.1 (Time Series Predicting by Machine Learning Methods)* The dataset "elec-temp" in the folder Ptsadata is an hourly series of electricity load and temperature from January 1, 2012 00:00:00 to December 31, 2014 23:00:00. It is a forecasting topic of the Global Energy Forecasting Competition 2014 (GEFCom2014) (Hong et al. 2016). We here consider only the electric load series. The load series is a "big" dataset and its length (viz., sample size) is 26,304. Moreover, there are at least three classes of seasonal periods in it. The first is the hourly period of 24 h (see Figs. 10.3 and 10.4), the second is the monthly period of 12 months (see Fig. 10.5), and the third is the quarterly period of 4 quarters (see Fig. 10.6). Thus, it is not easy to fit a classic statistical model to the load series data. We are now to train an RNN by deep learning methods and then predict data points through the trained model. We select the TensorFlow of version 2.3.1 as our deep learning platform.[1] First of all, split the electric load dataset `loadts` into the train, validation, and test sets (see Fig. 10.7). Then, normalize the train, validation, and test sets. From the ACF plot of the normalized train set shown in Fig. 10.8, it turns out that the normalized subseries is still nonstationary. Third, prepare the data representations to fulfill the requirements of the TensorFlow framework. Actually, we are to train an RNN within gated recurrent units (GRUs). We specify the number of units in the RNN layer `latent_dim`, the number of sample points per mini-batch

[1] The following Python code is also validated with Python of V. 3.9.7, TensorFlow of V. 2.7.0, and statsmodels of V. 0.13.1.

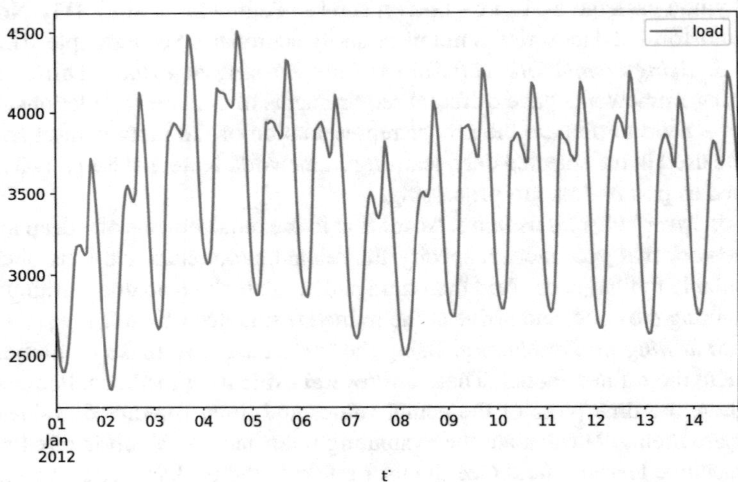


**Fig. 10.3** Time series plot of the electric load series for the first 2 weeks in Example 10.1

pyTSA_ML

batch_size, the maximum number of times the training algorithm will iterate epochs, and so on. And then fit the model sequential. Lastly, we evaluate the trained model. Figure 10.9 shows the comparison of the fitted (predicted) values with actual values, and the evaluating index mape is 0.0154. Hence, the trained model appears to possess a good forecasting power. Unfortunately, the ACF plot of the residual series shown in Fig. 10.10 illustrates that the residuals still have significant seasonal correlation. So, further improvements are worthy of paying attention.

```
>>> import numpy as np
>>> import pandas as pd
>>> import matplotlib.pyplot as plt
>>> import statsmodels.api as sm
>>> from sklearn.preprocessing import MinMaxScaler
>>> from tensorflow.keras.models import Model, Sequential
>>> from tensorflow.keras.layers import GRU, Dense
>>> from tensorflow.keras.callbacks import EarlyStopping
>>> from sklearn.metrics import mean_absolute_percentage_error
>>> from PythonTsa.plot_acf_pacf import acf_pacf_fig
>>> from PythonTsa.datadir import getdtapath
>>> dtapath=getdtapath()
>>> tsdta=pd.read_csv(dtapath + 'elec-temp.csv')
>>> tsdta['time']=pd.to_datetime(tsdta['time'])
>>> tsdta.index=tsdta['time']
>>> loadts=tsdta.drop(columns=['time', 'temp'])
>>> len(loadts)
    26304
>>> loadts1=loadts[loadts.index < '2012-01-14 23:00:00']
```

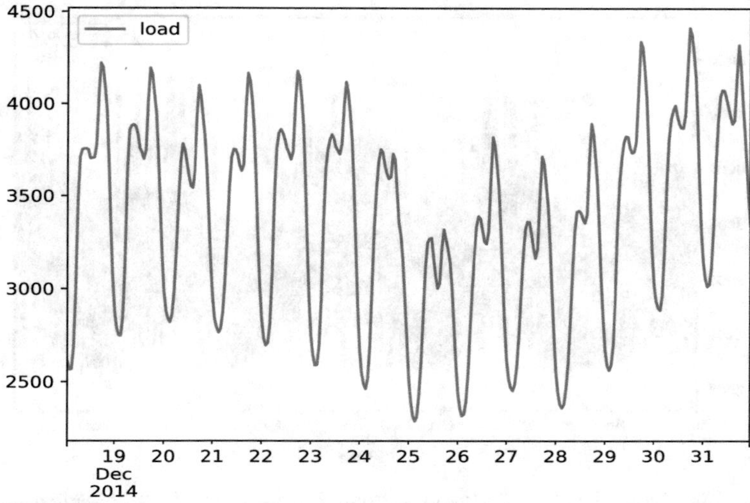


**Fig. 10.4** Time series plot of the electric load series for the last 2 weeks in Example 10.1

pyTSA_ML

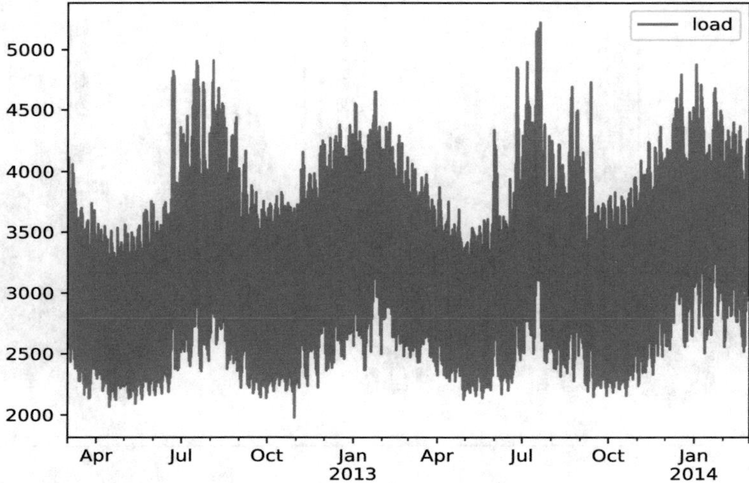


**Fig. 10.5** Time series plot of the electric loads from March 1, 2012 to February 28, 2014

pyTSA_ML

```
>>> loadts1.plot(); plt.show()
>>> loadts2=loadts[loadts.index >'2014-12-18 00:00:00']
>>> loadts2.plot(); plt.show()
>>> loadts3=loadts[(loadts.index > '2012-03-01 00:00:00')
                   &(loadts.index < '2014-03-01 00:00:00')]
>>> loadts3.plot(); plt.show()
```

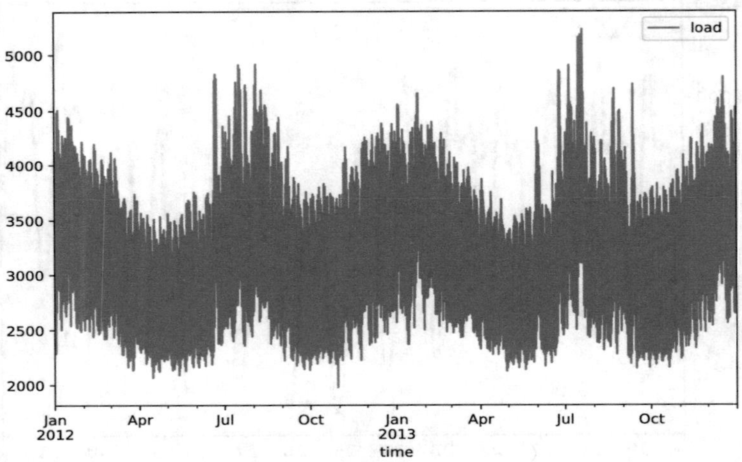


**Fig. 10.6** Time series plot of the electric loads from 2012 first quarter to 2013 fourth quarter

pyTSA_ML

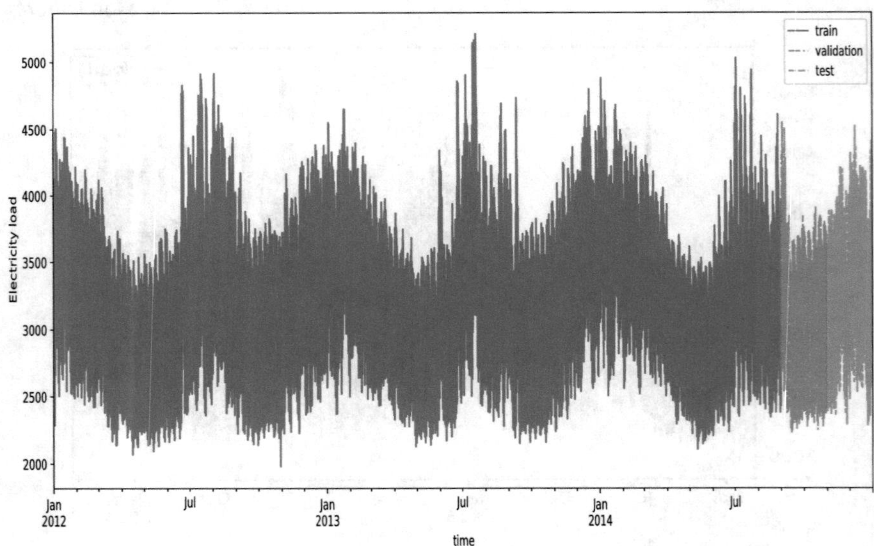


**Fig. 10.7** Time series plots of the train, validation, and test subseries in Example 10.1

pyTSA_ML

```
>>> loadts4=loadts[loadts.index <= '2013-12-31 23:00:00']
>>> loadts4.plot(); plt.show()
>>> validtime = '2014-09-01 00:00:00'
>>> testtime = '2014-11-01 00:00:00'
>>> loadts[loadts.index  <validtime].rename(columns
```

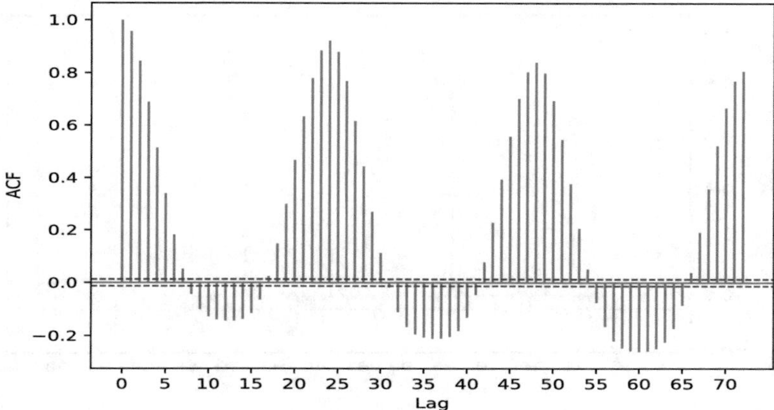


**Fig. 10.8** ACF plot of the normalized train subseries of the electric loads in Example 10.1

pyTSA_ML

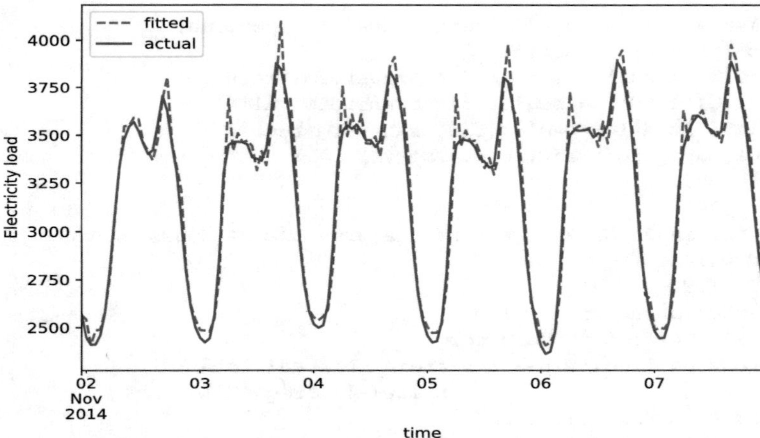


**Fig. 10.9** Comparison of the fitted (predicted) values with actual values in Example 10.1

pyTSA_ML

```
    ={'load':'train'}).join(loadts[(loadts.index >=
    validtime)&(loadts.index < testtime)].rename(columns
    ={'load':'validation'}), how='outer').join(loadts[testtime:]
    .rename(columns={'load':'test'}), how='outer').plot(y=['train',
     'validation', 'test'], style=['-','-.','--'])
>>> plt.ylabel('Electricity load')
>>> plt.show()
>>> train = loadts.copy()[loadts.index < validtime]
>>> valid = loadts.copy()[(loadts.index >= validtime)
            &(loadts.index < testtime)]
```

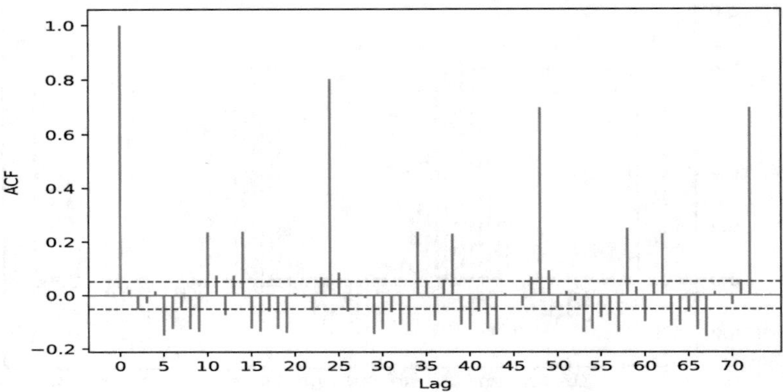


**Fig. 10.10** ACF plot of the residual series for the machine learning model in Example 10.1

pyTSA_ML

```
>>> test = loadts.copy()[loadts.index >= testtime]
>>> scaler = MinMaxScaler()
>>> train['load'] = scaler.fit_transform(train)
>>> valid['load'] = scaler.fit_transform(valid)
>>> test['load'] = scaler.fit_transform(test)
>>> acf_pacf_fig(train.load, lag=72)
>>> plt.show()
>>> T = 24
# Let the input be a vector of the previous 24 hours of the
# load values.
>>> HORIZON = 1
# one-step-ahead prediction
>>> train_shifted = train.copy()
>>> train_shifted['y_t+1'] = train_shifted['load']
                             .shift(-1, freq='H')
>>> for t in range(1, T+1):
        train_shifted[str(T-t)] = train_shifted['load']
        .shift(T-t, freq='H')
>>> y_col = 'y_t+1'
>>> X_cols = ['load_t-23','load_t-22','load_t-21','load_t-20',
             'load_t-19','load_t-18','load_t-17','load_t-16',
             'load_t-15','load_t-14','load_t-13','load_t-12',
             'load_t-11','load_t-10','load_t-9', 'load_t-8',
             'load_t-7','load_t-6', 'load_t-5', 'load_t-4',
             'load_t-3', 'load_t-2', 'load_t-1', 'load_t']
>>> train_shifted.columns = ['load_original']+[y_col]+X_cols
>>> train_shifted = train_shifted.dropna(how='any')
>>> y_train = np.array(train_shifted[y_col])
# equal to y_train = train_shifted[y_col].to_numpy()
>>> X_train = np.array(train_shifted[X_cols])
>>> X_train = X_train.reshape(X_train.shape[0], T, 1)
>>> valid_shifted = valid.copy()
>>> test_shifted = test.copy()
```

```
>>> valid_shifted['y_t+1'] = valid_shifted['load']
                              .shift(-1, freq='H')
>>> test_shifted['y_t+1'] = test_shifted['load']
                             .shift(-1, freq='H')
>>> for t in range(1, T+1):
        valid_shifted[str(T-t)] = valid_shifted['load']
        .shift(T-t, freq='H')
>>> for t in range(1, T+1):
        test_shifted[str(T-t)] = test_shifted['load']
        .shift(T-t, freq='H')
>>> valid_shifted.columns = ['load_original']+[y_col]+X_cols
>>> test_shifted.columns = ['load_original']+[y_col]+X_cols
>>> valid_shifted = valid_shifted.dropna(how='any')
>>> test_shifted = test_shifted.dropna(how='any')
>>> y_valid = np.array(valid_shifted[y_col])
>>> y_test = np.array(test_shifted[y_col])
>>> X_valid = np.array(valid_shifted[X_cols])
>>> X_test = np.array(test_shifted[X_cols])
>>> X_valid = X_valid.reshape(X_valid.shape[0], T, 1)
>>> X_test = X_test.reshape(X_test.shape[0], T, 1)
>>> latent_dim = 6
>>> batch_size = 32
>>> epochs = 15
>>> model = Sequential()
>>> model.add(GRU(latent_dim, input_shape=(T, 1)))
>>> model.add(Dense(HORIZON))
>>> model.compile(optimizer='RMSprop', loss='mse')
>>> model.summary()
                    Model: "sequential"
_________________________________________________________________
Layer (type)                 Output Shape              Param #
=================================================================
gru (GRU)                    (None, 6)                 162
_________________________________________________________________
dense (Dense)                (None, 1)                 7
=================================================================
Total params: 169
Trainable params: 169
Non-trainable params: 0
_________________________________________________________________

>>> GRU_earlystop = EarlyStopping(monitor='val_loss',
                    min_delta=0, patience=5)
>>> model_fit=model.fit(X_train,y_train,batch_size=batch_size,
                        epochs=epochs,
                        validation_data=(X_valid, y_valid),
                        callbacks=[GRU_earlystop], verbose=1)
>>> preds = model.predict(X_test)
>>> evdta = pd.DataFrame(preds, columns=['t+'+str(t)
            for t in range(1, HORIZON+1)])
>>> evdta['time'] = test_shifted.index
>>> evdta = pd.melt(evdta, id_vars='time',
            value_name='fitted', var_name='h')
```

```
>>> evdta['actual'] = np.transpose(y_test).ravel()
>>> evdta[['fitted', 'actual']] = scaler.inverse_transform
    (evdta[['fitted', 'actual']])
>>> evdta.head()
                 time    h       fitted  actual
0 2014-11-01 23:00:00  t+1  2532.896325  2566.0
1 2014-11-02 00:00:00  t+1  2533.742236  2449.0
2 2014-11-02 01:00:00  t+1  2350.824881  2411.0
3 2014-11-02 02:00:00  t+1  2470.362840  2412.0
4 2014-11-02 03:00:00  t+1  2482.103156  2460.0
>>> mean_absolute_percentage_error(evdta['actual'],
                                   evdta['fitted'])
0.015369812857679263
>>> evdta[evdta.time<'2014-11-08'].plot(x='time',
    y=['fitted', 'actual'],style=['--r', '-b'])
>>> plt.ylabel('Electricity load')
>>> plt.show()
>>> resid=evdta['actual']-evdta['fitted']
>>> acf_pacf_fig(resid, lag=72)
>>> plt.show()
```

## 10.6 Concluding Remarks

Time series analysis (including forecasting and classification) by machine learning methods has attracted more and more attention. Although deep learning has seen very successful applications in other fields such as computer vision, text document classification, and speech recognition, there is still a long way to go for time series analysis by deep learning. This is challenging but promising.

## Problems

**10.1** First plot the sigmoid (with $a = 1, 2, 3$), tanh, and ReLU functions using Python, respectively. Then show that $\tanh(s) = 2\text{sigmoid}(s|2) - 1$.

**10.2** Do your best to build a classic statistical model for the electric load series in Example 10.1, and then conduct a model residual analysis. If the model cannot be built, then give your reasons.

**10.3** It is common sense that the electric load is related to temperature. The dataset `tsdta` in Example 10.1 gives both a temperature series `temp` and a corresponding electric load series `load`. Now try to train a machine learning model with `tsdta` and make predictions using the trained model.

**10.4** In recent years, a number of papers on time series analysis (including forecasting and classification) by machine learning methods have been published. Interested readers can select and study them for understanding state-of-the-art developments in this domain.

# References

Ackley, D., Hinton, G., Sejnowski, T.: A learning algorithm for Boltzmann machines. Cogn. Sci. **9**, 147–169 (1985)

Aggarwal, C.C.: Neural Networks and Deep Learning. Springer, Switzerland (2018)

Akaike, H.: Fitting autoregressive models for prediction. Ann. Inst. Stat. Math. **21**, 243–247 (1969)

Akaike, H.: Information theory and an extension of the maximum likelihood principle. In: Proceedings of the 2nd International Symposium on Information Theory, Budapest (1973)

Akaike, H.: On entropy maximization principle. In: Krishnaiah, P. R. (ed.) Applications of statistics, pp. 27–41. North-Holland Publishing Company, Amsterdam (1977)

Anderson, T.W., Walker, A.M.: On the asymptotic distribution of the autocorrelations of a sample from a linear stochastic process. Ann. Math. Stat. **35**, 1296–1303 (1964)

Applebaum, D.: Lévy Processes and Stochastic Calculus, 2nd edn. Cambridge University Press, New York (2009)

Billingsley, P.: Convergence of Probability Measures, 2nd edn. Wiley, New York (1999)

Box, G.E.P., Jenkins, G.M., Reinsel, G.C., Ljung, G.M.: Time Series Analysis: Forecasting and Control, 5th edn. Wiley, Hoboken, NJ (2016)

Box, G.E.P., Pierce, D.: Distribution of residual autocorrelations in autoregressive-integrated moving average time series models. J. Am. Stat. Assoc. **65**, 1509–1526 (1970)

Brockwell, P.J., Davis, R.A.: Time Series: Theory and Methods, 2nd edn. Springer, New York (1991)

Brockwell, P.J., Davis, R.A.: Introduction to Time Series and Forecasting, 3rd edn. Springer, Switzerland (2016)

Brüggemann, R., Lütkepohl, H., Saikkonen, P.: Residual autocorrelation testing for vector error correction models. J. Econ. **134**, 579–604 (2006)

Bryson, A., Ho, Y.: Applied Optimal Control: Optimization, Estimation, and Control. Blaisdell Publishing Company, Waltham, MA (1969)

Campbell, J.Y., Lo, A.W., MacKinlay, A.C.: The Econometrics of Financial Markets, 2nd edn. Princeton University Press, Princeton, NJ (2012)

Casals, J., Garcia-Hiernaux, A., Jerez, M., Sotoca, S., Trindade, A.A.: State-Space Methods for Time Series Analysis: Theory, Applications and Software. CRC Press, London (2016)

Casella, G., Berger, R.L.: Statistical Inference, 2nd edn. Duxbury Press, Belmont (2002)

Chacón, J.E., Duong, T.: Multivariate Kernel Smoothing and Its Applications. CRC Press, London (2018)

Chan, N.H.: Time Series: Applications to Finance with R and S-plus, 2nd ed. John Wiley, New Jersey (2010)



C. Huang, A. Petukhina, *Applied Time Series Analysis and Forecasting with Python*, Statistics and Computing,
https://doi.org/10.1007/978-3-031-13584-2

Chollet, F.: Deep Learning with Python, 2nd ed. Manning Publications, New York (2021). https://www.manning.com/
Cowpertwait, P., Metcalfe, A.: Introductory Time Series with R. Springer, New York (2009)
Dagum, E.B., Bianconcini, S.: Seasonal Adjustment Methods and Real Time Trend-Cycle Estimation. Springer, Switzerland (2016)
Dickey, D.A., Fuller, W.A.: Distribution of the estimates for autoregressive time series with a unit root. J. Am. Stat Assoc. **74**, 427–431 (1979)
Douc, R., Moulines, E., Stoffer, D.S.: Nonlinear Time Series: Theory, Methods, and Applications with R Examples. CRC Press, Boca Raton (2014)
Durbin, J.: The fitting of time series models. Review of the International Institute of Statistics **28**, 233–244 (1960)
Durbin, J., Koopman, S.J.: Time Series Analysis by State Space Methods, 2nd edn. Oxford University Press, Oxford, UK (2012)
Engle, R.F.: Autoregressive conditional heteroscedasticity with estimates of the variance of United Kingdom inflation. Econometrica **50**, 987–1007 (1982)
Engle, R.F., Granger, C.W.J.: Co-integration and error correction: representation, estimation, and testing. Econometrica **55**, 251–276 (1987)
Fan, J., Yao, Q.: Nonlinear Time Series: Nonparametric and Parametric Methods. Springer, New York (2003)
Fan, J., Ma, C., Zhong, Y.: A Selective Overview of Deep Learning (2019). https://arxiv.org/abs/1904.05526v2. Cited 6 Feb 2021
Fawaz, H.I. et al.: Deep learning for time series classification: a review. Data Min. Knowl. Disc. **33**, 917–963 (2019)
Fons, E., et al.: Evaluating Data Augmentation for Financial Time Series Classification (2020). https://arxiv.org/abs/2010.15111v1. Cited 7 Nov 2020
Franke, J., Härdle, W.K., Hafner, C.M.: Statistics of Financial Markets: An Introduction, 5th edn. Springer, Switzerland (2019)
Frühwirth-Schnatter, S.: Finite Mixture and Markov Switching Models. Springer, New York (2006)
Fukushima, K.: Neocognitron: A self-organizing neural network model for a mechanism of pattern recognition unaffected by shift in position. Biol. Cybern. **36**, 193–202 (1980)
Gamboa, J.: Deep Learning for Time-Series Analysis (2017). https://arxiv.org/abs/1701.01887. Cited 7 Dec 2018
Ghosh, S.: Kernel Smoothing: Principles, Methods and Applications. Wiley, Hoboken, NJ (2018)
Ghysels, E., Osborn, D.R.: The Econometric Analysis of Seasonal Time Series. Cambridge University Press, Cambridge (2001)
Glorot, X., Bordes, A., Bengio, Y.: Deep sparse rectifier neural networks. In: Proceedings of the 14th International Conference on Artificial Intelligence and Statistics, pp. 315–323 (2011)
Glosten, L.R., Jagannathan, R., Runkle, D.E.: On the relation between the expected value and the volatility of nominal excess return on stocks. J. Financ. **48**, 1779–1801 (1993)
Goodfellow, I., Bengio, Y., Courville, A.: Deep Learning. MIT Press, Cambridge, MA (2016)
Gómez, V.: Multivariate Time Series With Linear State Space Structure. Springer, Switzerland (2016)
Granger, C.W.J.: Investigating causal relations by econometric models and cross-spectral methods. Econometrica **37**, 424–438 (1969)
Granger, C.W.J., Newbold, P.: Spurious regressions in econometrics. J. Econ. **2**, 111–120 (1974)
Guennec, A.L., et al.: Data augmentation for time series classsification using convolutional neural networks. In: ECML/PKDD Workshop on Advanced Analytics and Learning on Temporal Data, Sep 2016, Riva Del Garda, Italy (2016)
Hamilton, J.D.: A new approach to the economic analysis of nonstationary time series and the business cycle. Econometrica **57**, 357–384 (1989)
Hamilton, J.D.: Time Series Analysis. Princeton University Press, Princeton, NJ (1994)
Hannan, E.J., Quinn B.G.: The Determination of the order of an autoregression. J. R. Stat. Soc. Ser. B **41**, 190–195 (1979)

Hansen, J.E., Lebedeff, S.: Global trends of measured surface air temperature. J. Geophys. Res. **92**, 13345–13372 (1987)
Harvey, A.C.: Forecasting, Structural Time Series Models and the Kalman Filter. Cambridge University Press, Cambridge (1989)
Hastie, T., Tibshirani, R., Friedman, J.: The Elements of Statistical Learning: Data Mining, Inference, and Prediction, 2nd edn. and 12th printing. Springer, New York (2017)
Hecht-Nielsen, R.: Theory of the backpropagation neural network. In: International Joint Conference on Neural Networks, pp 593–605. IEEE, New York (1989)
Hilpisch, Y.: Derivatives analytics with Python: data analysis, models, simulation, calibration and hedging. Wiley, West Sussex, UK (2015)
Hinton, G.E., Osindero, S., Teh, Y.: A fast learning algorithm for deep belief nets. Neural Comput. **18**, 1527–1554 (2006)
Hobijn, B., Frances, B.H., Ooms, M.: Generalizations of the KPSS-test for stationarity. Statistica Neerlandica **52**, 483–502 (2004)
Hochreiter, S., Schmidhuber, J.: Long short-term memory. Neural Comput. **9**(8), 1735–1780 (1997)
Holt, C.E.: Forecasting trends and seasonals by exponentially weighted averages. O.W.R. Memorandum no.52, Carregie Institute of Technology, Pittsburgh (1957)
Hong, T., et al.: Probabilistic energy forecasting: Global Energy Forecasting Competition 2014 and beyond. Int. J. Forecast. **32**(3), 896–913 (2016)
Hornik, K., Stinchcombe, M., White, H.: Multilayer feedforward networks are universal approximators. Neural Netw. **2**(5), 359–366 (1989)
Hosking, J.R.M.: The multivariate portmanteau statistic. J. Am. Stat. Assoc. **75**, 602–608 (1980)
Hunter, J.D.: Matplotlib: a 2D graphics environment. Comput. Sci. Eng. **9**(3), 90–95 (2007)
Hurvich, C.M., Tsai, C.L.: Regression and time series model selection in small samples. Biometrika **76**, 297–307 (1989)
Hyndman, R.J., Athanasopoulos, G.: Forecasting: Principles and Practice, 2nd edn. OTexts, Melbourne (2018)
James, G. et al.: An Introduction to Statistical Learning: with Applications in R, 1st edn. and 7th printing. Springer, New York (2017)
Johansen, S.: Likelihood-Based Inference in Cointegrated Vector Autoregressive Models. Oxford University Press, New York (1995)
Johansen, S.: Likelihood-based inference for cointegration of some nonstationary time series. In Cox, D.R., Hinkley, D.V., Barndorff-Nielsen, O.E. (eds.) Time Series Models In Econometrics, Finance and Other Fields. Chapman and Hall, New York (1996)
Kalman, R.E.: A new approach to linear filtering and prediction problems. Trans. ASME J. Basic Eng. (Series D) **82**, 35–45 (1960)
Kalman, R.E., Bucy, R.S.: New results in linear filtering and prediction theory. Trans. ASME J. Basic Eng. (Series D) **83**, 95–108 (1961)
Karatzas, I., Shreve, S.E.: Brownian Motion and Stochastic Calculus, 2nd edn. Springer, New York (1998)
Kim, C.J., Nelson, C.R.: State-Space Models with Regime Switching: Classical and Gibbs-Sampling Approaches with Applications. The MIT Press, Cambridge (1999)
Klebaner, F.C.: Introduction to Stochastic Calculus with Applications, 2nd edn. Imperial College Press, London (2005)
Kolmogorov, A.N.: Foundations of the Theory of Probability, 2nd English edn. Chelsea Publishing Company, New York (1956). (The original German monograph is Grundbegriffe der Wahrscheinlichkeitsrechnung published in 1933)
Konishi, S., Kitagawa, G.: Information Criteria and Statistical Modeling. Springer, New York (2008)
Kullback, S., Leibler, R.A.: On information and sufficiency. Ann. Math. Statist. **22**, 79–86 (1951)
Kwiatkowski, D., Phillips, P.C.B., Schmidt, P., Shin, Y.: Testing the null hypothesis of stationarity against the alternative of a unit root: How sure are we that economic time series have a unit root?. J. Econ. **54**, 159–178 (1992)

Lazzeri, F.: Machine Learning for Time Series Forecasting with Python. Wiley, Indianapolis, Indiana (2021)

LeCun, Y., Bengio, Y., Hinton, G.E.: Deep learning. Nature **521**, 436–444 (2015)

Levinson, N.: The Weiner RMS error criterion in filter design and prediction. J. Math. Phys. **25**, 261–278 (1947)

Li, W.K., McLeod, A.I.: Distribution of the residual autocorrelations in multivariate ARMA time series models. J. R. Stat. Soc. **B43**, 231–239 (1981)

Li, W.K.: Diagnostic Checks in Time Series. Chapman and Hall/CRC, Boca Raton, FL (2004)

Lim, B., et al.: Temporal Fusion Transformers for Interpretable Multi-horizon Time Series Forecasting (2020). https://arxiv.org/abs/1912.09363v3. Cited 4 Nov 2020

Lim, B., Zohren, S.: Time series forecasting with deep learning: a survey. Phil. Trans. R. Soc. A **379**(2194) (2021)

Ljung, G., Box, G.E.P.: On a measure of lack of fit in time series models. Biometrika **66**, 67–72 (1978)

Lütkepohl, H.: New Introduction to Multiple Time Series Analysis. Springer, Berlin (2005)

Maharaj, E.A., D'Urso, P., Caiado, J.: Time Series Clustering and Classification. CRC Press, Boca Raton, FL (2019)

Mahdi, E., Mcleod, I.: Improved multivariate portmanteau test. J. Time Ser. Anal. **33**, 2 (2012)

McCarthy, J.: What is Artificial Intelligence? (revised November 12, 2007). http://jmc.stanford.edu/articles/whatisai/whatisai.pdf. Cited 6 Oct 2020

McCulloch, W.S., Pitts, W.: A logical calculus of the ideas immanent in nervous activity. Bull. Math. Biophys. **5**(4), 115–133 (1943)

McCulloch, R.E., Tsay, R.S.: Statistical analysis of economic time series via Markov switching models. J. Time Ser. Anal. **15**, 523–539 (1994)

Mueller, J.P., Massaron, L.: Deep Learning for Dummies. John Wiley, Hoboken, NJ (2019)

Needham, J.: Science and Civilisation in China, Vol. III. Cambridge University Press, Cambridge, UK (1959)

Nelson, D.: Conditional heteroskedasticity in asset returns: A new approach. Econometrica **59**, 347–370 (1991)

Nilsson, N.J.: The Quest for Artificial Intelligence: A History of Ideas and Achievements. Cambridge University Press, New York (2010)

Øksendal, B.: Stochastic Differential Equations: An Introduction with Applications, 6th edn. Springer, Berlin (2003)

Ord, K., Fildes, R., Kourentzes, N.: Principles of Business Forecasting, 2nd edn. Wessex Press, New York (2017)

Pearson, K.: The problem of the random walk. Nature **72**, 294 (1905)

Pedregosa, F., et al.: Scikit-learn: machine learning in Python. JMLR **12**, 2825–2830 (2011)

Pfaff, B.: Analysis of Integrated and Cointegrated Time Series with R, 2nd edn. Springer, New York (2008)

Phillips, P.C.B.: Understanding spurious regression in econometrics. J. Econ. **33**, 311–340 (1986)

Phillips, P.C.B., Ouliaris, S.: Asymptotic properties of residual based tests for cointegration. Econometrica **58**, 73–93 (1990)

Phillips, P.C.B., Perron, P.: Testing for a unit root in time series regression. Biometrika **75**, 335–346 (1988)

Quenouille, M.H.: Approximate tests of correlation in time-series. J. R. Stat. Soc. B **11**, 68–84 (1949)

Rangapuram, S.S. et al.: Deep state space models for time series forecasting. In: 32nd Conference on Neural Information Processing Systems (NIPS 2018), Montréal, Canada (2018)

Rumelhart, D., Hinton, G., Williams, R.: Learning representations by back-propagating errors. Nature **323**, 533–536 (1986)

Russell, S.J., Norvig, P.: Artificial Intelligence: A Modern Approach, 3rd edn. Pearson Education, Essex England (2016)

Russell, S.J., Norvig, P.: Artificial Intelligence: A Modern Approach, 4rd edn. Pearson Education, Hoboken, NJ (2021)

Schmidhuber, J.: Learning complex, extended sequences using the principle of history compression. Neural Comput. **4**(2), 234–242 (1992)
Schmidhuber, J.: Deep learning in neural networks: an overview. Neural Netw. **61**, 85–117 (2015)
Schwarz, G.E.: Estimating the dimension of a model. Ann. Stat. **6**, 461–464 (1978)
Seabold, S., Perktold, J.: Statsmodels: Econometric and statistical modeling with Python. In: Proceedings of the 9th Python in Science Conference, Austin (2010)
Sejnowski, T.J.: The Deep Learning Revolution. The MIT Press, Cambridge, MA (2018)
Sheppard, K.: Bashtage/arch: Release 4.15. Zenodo (2020). https://doi.org/10.5281/zenodo.593254. Cited 24 June 2020
Shreve, S.E.: Stochastic Calculus for Finance II: Continuous-Time Models. Springer, New York (2004)
Shumway, R.H., Stoffer, D.S.: Time Series Analysis and Its Applications With R Examples, 4th edn. Springer, New York (2017)
Slutsky, E.: The summation of random causes as the source of cyclic processes (in Russian). Problems of Economic Conditions **3**, (1927). English translation in Econometrca **5**, 105–146 (1937)
Stigler, S.M.: Gauss and the Invention of Least Squares. Ann. Stat. **9**(3), 465–474 (1981)
Thu, H.N.: Diagnostic checks in multiple time series modelling. In: Rojas, I., et al. (eds.) Advances in Time Series Analysis and Forecasting, Contributions to Statistics (2017)
Tong, H.: Non-linear Time Series: A Dynamical Systems Approach. Oxford University Press, Oxford (1990)
Torres, J.F. et al.: Deep learning for time series forecasting: a survey. Big Data **9**(1), 3–21 (2021)
Tsay, R.S.: Analysis of Financial Time Series, 3rd edn. Wiley, Hoboken, NJ (2010)
Tsay, R.S.: Multivariate Time Series Analysis with R and Financial Applications. Wiley, Hoboken, NJ (2014)
Tsay, R.S., Chen, R.: Nonlinear Time Series Analysis. Wiley, Hoboken, NJ (2019)
Turing, A.: Computing machinery and intelligence. Mind **59**, 433–460 (1950)
Waerden, B.L.v.d.: Algebra I, 7th edn. Springer, Berlin (2003)
Wei, W.W.S.: Multivariate Time Series Analysis and Applications. Wiley, Hoboken, NJ (2019)
Winters, P.R.: Forecasting sales by exponentially weighted moving averages. Manag. Sci. **6**, 324–342 (1960)
Yule, G.U.: On a method of investigating periodicities disturbed series, with special reference to Wolfer's sunspot numbers. Philos. Trans. R. Soc. Lond. A **226**, 267–298 (1927)
Zakoian, J.M.: Threshold heteroscedastic models. J. Econ. Dyn. Control **18**, 931–955 (1994)
Zhang, G., Patuwo, B.E., Hu, M.Y.: Forecasting with artificial neural networks: the state of the art. Int. J. Forecast. **14**(1), 35–62 (1998)

# Index


**A**
Accumulated (cumulative) responses, 237
ACF and PACF behavior
    AR model, 103
    ARMA model, 103
    MA model, 103
ACF and PACF properties
    pure SARMA model, 149
Activation function, 348
Activation value, 348
Akaike information criterion (AIC), 115
Almost surely (a.s.), 303
AlphaGo, 347
ARCH effect, 117
ARCH model, 183
ARIMA$(p, d, q)$ model, 108
ARIMA$(p, d, q)$ process, 108
ARMA$(p, q)$ process, 98
ARMAX models, 165
AR polynomial, 88
AR$(p)$ process, 88
AR$(\infty)$ representation, 86
Artificial intelligence (AI), 341
Artificial neural network, 341
(Artificial) neurons, 344
Augmented Dickey-Fuller (ADF) test, 309
Autocorrelation function (ACF), 16
Autocovariance function, 16
Autoregressive (AR) model, 88
Autoregressive conditional heteroscedasticity (ARCH), 183
Autoregressive moving average (ARMA) model, 98

**B**
Backpropagation algorithm, 350
Backshift operator, 72
Bayesian information criterion (BIC), 115
Best linear predictor, 118
Box-Cox transformation, 108
Box-Jenkins method, 144
Box-Pierce test, 43

**C**
Capital asset pricing model (CAPM), 259
Causality, 94
Causality of AR model, 94
Causality theorem, 94
Centered moving average (filtering), 58
Coefficients of impulse response, 237
Cointegrating matrix, 326
Cointegrating series, 326
Cointegrating vector, 326
Cointegration, 322
Cointegration rank, 326
Common stochastic trend, 326
Convolutional neural network (CNN), 348
Correlation matrix (function), 216
Correlogram, 18
Covariance matrix (function), 216
Cumulative distribution function (cdf), 303
Curse of dimensionality, 232
Cutting off, 84




C. Huang, A. Petukhina, *Applied Time Series Analysis and Forecasting with Python*, Statistics and Computing,
https://doi.org/10.1007/978-3-031-13584-2

**D**
Data pre-processing, 352
Data visualization, 29
Deep learning (DL), 341
Deterministic seasonality (component), 53
Deterministic trend (component), 53
Diagnosis of models, 116
Difference stationarity, 306
Differencing, 72
Donsker's theorem, 305
Durbin-Levinson recursion algorithm, 91

**E**
EGARCH model, 204
Endogenous variable, 258
Ergodicity, 17
Exogenous (control) variable, 258
Exploratory time series data analysis, 37
Extended trace of correlation matrix, 220

**F**
Feed-forward neural network (FNN), 348
Field of events, 302
First level Brownian bridge, 317
Fitted values, 118
Forecasting, 118
Forecast origin, 118
Fourier series, 166

**G**
GARCH model, 185
GJRGARCH model, 205
Gradient descent, 350
Granger causality, 235
Granger's representation theorem, 325

**H**
Hannan-Quinn information criterion (HQIC), 115
Harmonic seasonal regression model, 166
Heaviside function, 349
History of time series analysis, 5
Holt-Winters smoothing, 58
h-step-ahead predictor(forecast), 118

**I**
Identifiability problem, 232
Impulse response, 236
Initialization, 258
Innovations algorithm, 109
Innovation (shock) term, 80
Input variable, 265
In-sample predictors, 118
Invertibility, 86
Invertibility of ARMA model, 100
Invertibility theorem, 86
$I(d)$ process, 318

**K**
Kalman filtering, 261
Kalman forecasting, 262
Kalman gain (matrix), 262
Kalman recursion, 261
Kalman smoothing, 262
Keras, 351
KPSS stationarity test, 73, 316

**L**
Lead time, 118
Least squares estimator (LSE), 114
Leverage effect coefficient, 205
Ljung-Box test, 43
Loading matrix, 326
Local-level model, 263
Log return, 28
Long-run effects, 237
Long short-term memory (LSTM) algorithm, 343

**M**
Machine learning (ML), 341
MA polynomial, 80
MA($q$) process, 80
MA($\infty$) representation, 94
Markov switching autoregressive model, 272
Markov switching model, 272
Markov switching (dynamic) regression model, 272
Maximum likelihood estimators (MLE), 114
Mean equation (model), 186
Measurable space, 302
Measurement equation, 258
Method of conditional least squares, 112
Method of least squares (LS), 112
Method of maximum likelihood (ML), 113
Method of moments, 110
Minimum mean squared error criterion, 118
Moment estimates (estimators), 110
Moment function, 16
Moving average (MA) model, 80

Multiplier analysis, 236
Multivariate causality theorem, 230
Multivariate invertibility theorem, 230
Multivariate Li-McLeod test, 221
Multivariate Ljung-Box test, 221
Multivariate portmanteau test, 221
Multivariate (vector) time series, 216
MuZero, 347

**N**
Neocognitron, 345
Net input, 347
Neural network model, 347
Node, 344
Normality test, 117
Normalization, 352

**O**
Objectives of time series analysis, 6
Observation equation, 258
One-step transition probability (matrix), 272
Order determination, 115
Orthogonal complement, 325
Out-of-sample forecasting, 118
Output gap (GDP gap), 276

**P**
Partial autocorrelation function (PACF), 37
Perceptron, 345
Phillips-Perron (PP) unit root test, 310
Portmanteau (Q) test, 43
Prewhitening, 110
Probability measure, 303
Probability space, 303
Properties of ARMA model, 100
Properties of autoregressive (AR) model, 92
Properties of GARCH model, 185
Properties of MA model, 84
Python, 7
Python extension packages, 12

**R**
Random event, 302
Random variable, 303
Random walk, 24
Rectified linear unit (ReLU) function, 348
Recurrent neural network (RNN), 348
REGARMA models, 165
Regime switching model, 272
Residuals, 117

**S**
Sample autocorrelation function, 18
Sample autocovariance function, 18
Sample correlation matrix (function), 219
Sample covariance matrix (function), 219
Sample PACF, 39
SARIMA$(p, d, q)(P, D, Q)_s$ model, 147
SARIMA models building, 155
SARIMA$(p, d, q)(P, D, Q)_s$ process, 147
SARIMAX model, 265
Seasonal AR polynomial, 147
Seasonal correlation, 147
Seasonal differencing of order $D$, 144
Seasonal MA polynomial, 147
Seasonal period, 147
Second level Brownian bridge, 317
$\sigma$-algebra, 302
Simple time series composition, 47
Softmax function, 349
Spurious regression, 319
Standard Brownian motion, 304
Standardization, 352
Standardized residuals, 117
State equation, 258
State space model (method), 258
State space representation (form), 259
State variable, 258
Stationarity, 17
Stationarity of ARMA model, 100
Stationarity of AR model, 94
Stationarity of vector time series, 217
Stationary model, 80
Statistical learning, 342
Stochastic process, 303
Stochastic seasonality, 297
Stochastic trend, 289
Stylized facts (features) of financial time series
  ARCH effect, 182
  asymmetry, 183
  fat (heavy) tails, 182
  volatility clustering, 182

**T**
Tailing off, 84
TensorFlow, 351
TGARCH model, 205
Time homogeneous, 258
Time-invariant, 258
Time (series) plot, 2
Time series, 4
Time series classification (TSC), 343
Time series decomposition, 53
Time series smoothing, 53

Total multipliers, 237
Trace of a matrix, 220
Training a model, 353
Transition equation, 258
Trend stationarity, 306

**U**
Unit root test, 309

**V**
Variance (volatility) equation (model), 186
VARMA($p, q$) model, 227
VARMA($p, q$) process, 227
VAR matrix polynomial, 227
VAR($p$) model, 227
VAR model building, 233
VAR($\infty$) representation, 230
Vector error correction model (VECM), 326
Vector time series
  AIC, 234
  BIC, 234
  FPE, 234
  HQIC, 234
Vector white noise, 218
VMA matrix polynomial, 227
VMA($q$) model, 227
VMA($\infty$) representation, 230

**W**
White noise, 21
Wiener proces, 304

**Y**
Yule-Walker equations, 93
Yule-Walker estimates, 93
Yule-Walker estimator
  large-sample property, 111

**Z**
ZGARCH model, 205

Printed in the United States
by Baker & Taylor Publisher Services